Institute for Statistics and Mathematics
Vienna University of Economics and Business
Augasse 2-6, 1090 Vienna

DEPENDENCE MODELING
Vine Copula Handbook

DEPENDENCE MODELING
Vine Copula Handbook

editors

Dorota Kurowicka
Delft University of Technology, The Netherlands

Harry Joe
University of British Columbia, Canada

NEW JERSEY · LONDON · SINGAPORE · BEIJING · SHANGHAI · HONG KONG · TAIPEI · CHENNAI

Published by

World Scientific Publishing Co. Pte. Ltd.
5 Toh Tuck Link, Singapore 596224
USA office: 27 Warren Street, Suite 401-402, Hackensack, NJ 07601
UK office: 57 Shelton Street, Covent Garden, London WC2H 9HE

British Library Cataloguing-in-Publication Data
A catalogue record for this book is available from the British Library.

DEPENDENCE MODELING
Vine Copula Handbook

Copyright © 2011 by World Scientific Publishing Co. Pte. Ltd.

All rights reserved. This book, or parts thereof, may not be reproduced in any form or by any means, electronic or mechanical, including photocopying, recording or any information storage and retrieval system now known or to be invented, without written permission from the Publisher.

For photocopying of material in this volume, please pay a copying fee through the Copyright Clearance Center, Inc., 222 Rosewood Drive, Danvers, MA 01923, USA. In this case permission to photocopy is not required from the publisher.

ISBN-13 978-981-4299-87-9
ISBN-10 981-4299-87-1

Typeset by Stallion Press
Email: enquiries@stallionpress.com

Printed in Singapore by World Scientific Printers.

Preface

This book emerges from three workshops held over the last three years involving all the principal contributors to the vine-copula methodology. Vines are possibly the most important recent development in dependence modeling. Their flexibility in modeling various dependence structures as well as their potential to construct rich set of distributions promises wide application capabilities. As research and applications in vines have been growing rapidly, there is a need for an authoritative handbook collating the basic results, standardizing terminology and methods. Specifically, this handbook

(1) traces historical developments, standardizing notation and terminology,
(2) summarizes findings on bivariate and multivariate copulae,
(3) summarizes findings on regular vines, and
(4) gives an overview of applications.

Many of the results presented here are quite new and not readily available in journals. New research directions in relation to vines are also discussed.

For available vine-copula software, please visit http://risk.ewi.tudelft.nl.

D. Kurowicka

Contents

Preface		v
1.	Introduction: Dependence Modeling D. Kurowicka	1
2.	Multivariate Copulae M. Fischer	19
3.	Vines Arise R. M. Cooke, H. Joe and K. Aas	37
4.	Sampling Count Variables with Specified Pearson Correlation: A Comparison between a Naive and a C-Vine Sampling Approach V. Erhardt and C. Czado	73
5.	Micro Correlations and Tail Dependence R. M. Cooke, C. Kousky and H. Joe	89
6.	The Copula Information Criterion and Its Implications for the Maximum Pseudo-Likelihood Estimator S. Grønneberg	113
7.	Dependence Comparisons of Vine Copulae with Four or More Variables H. Joe	139

8. Tail Dependence in Vine Copulae   165
   H. Joe

9. Counting Vines   189
   O. Morales-Nápoles

10. Regular Vines: Generation Algorithm and Number of Equivalence Classes   219
    H. Joe, R. M. Cooke and D. Kurowicka

11. Optimal Truncation of Vines   233
    D. Kurowicka

12. Bayesian Inference for D-Vines: Estimation and Model Selection   249
    C. Czado and A. Min

13. Analysis of Australian Electricity Loads Using Joint Bayesian Inference of D-Vines with Autoregressive Margins   265
    C. Czado, F. Gärtner and A. Min

14. Non-Parametric Bayesian Belief Nets versus Vines   281
    A. Hanea

15. Modeling Dependence between Financial Returns Using Pair-Copula Constructions   305
    K. Aas and D. Berg

16. Dynamic D-Vine Model   329
    A. Heinen and A. Valdesogo

17. Summary and Future Directions   355
    D. Kurowicka

Index   359

CHAPTER 1

Introduction: Dependence Modeling

Dorota Kurowicka

Delft University of Technology
Mekelweg 4, 2628CD Delft, The Netherlands
d.kurowicka@tudelft.nl

1.1	Introduction		1
1.2	Investment Example		2
1.3	Vines		6
	1.3.1	Graphical representation	7
	1.3.2	Vine density	9
	1.3.3	Estimation	10
	1.3.4	Properties and applications	11
1.4	Outline		13
1.5	Glossary and Notation		15
References			16

1.1 Introduction

The vines described in this book are not *climbing* or *trailing plants*. Nor do they refer to the Australian rock band. Vines are graphical structures that represent joint probability distributions. They were named for their close visual resemblance to grapes (compare Figs. 1.1 and 1.5)

Vines first appeared in mathematical publications in the late 1990s. It took time before the community of researchers interested in this model grew sufficiently and before vines were recognized in applications. Vines are still young but now have sufficiently matured to deserve a comprehensive presentation. This book is a joint effort of the vine-community and contains established as well as the newest results concerning vines. Samuel Johnson once said: "What is written without effort is in general read without pleasure".

Figure 1.1. Grapes.

A lot of effort has gone into this book and we hope that it will be read with pleasure.

1.2 Investment Example

It is recognized that modeling dependence is of great importance for financial and engineering applications. We present the motivation for this book in a very simple financial example. When we invest $1000 for five years, the five-year return is:

$$5yR = 1000(1 + r_1)(1 + r_2)(1 + r_3)(1 + r_4)(1 + r_5)$$

where r_1, r_2, r_3, r_4, r_5 are the interest rates in those five years. Interest rates are not known with certainty. We can find their distribution in principle from data but here they were assumed to be uniformly distributed on $[0.05, 0.15]$.

To find the distribution of our fortune after five years, we require the joint distribution of interest rates. If we assume that interest rates are independent, then their joint distribution is a product of marginal distributions and the five-year return can be easily calculated. If we recognize some sort of dependence between interest rates, we must build a joint distribution with given margins and given dependencies. A popular model recently used for this purpose is a copula. A copula is a distribution on the unit hypercube with uniform margins. It is also called a dependence function as it allows the separation of information coming from margins and dependence in the joint distribution. Different bivariate copulae, with the ability to model various features of a joint distribution (correlation, tail dependence), are available (see e.g., Refs. 12 and 20). In Fig. 1.2, scatter plots of the normal and

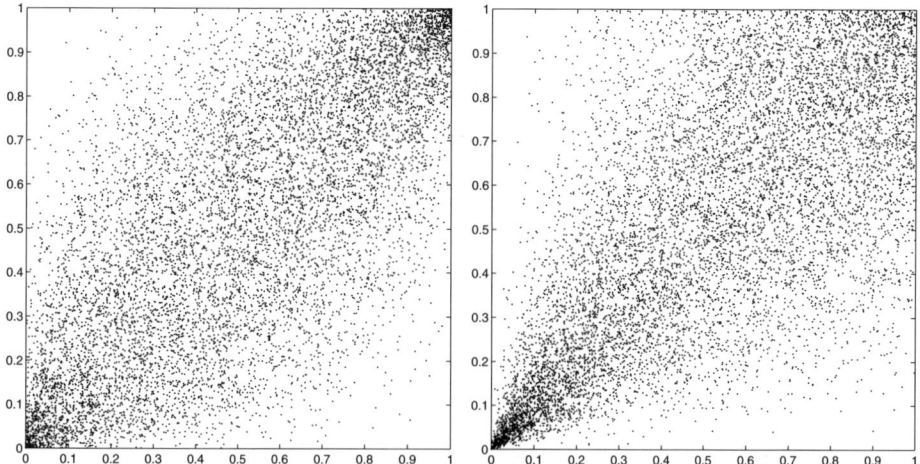

Figure 1.2. Scatter plots of normal (left) and Clayton (right) copulae with correlation 0.8.

Clayton copulae are shown. We can see how different these distributions are even for the same correlation value.

The choice of copula is an important question as this can affect the results significantly. In the bivariate case, this choice is based on statistical tests (see e.g., Ref. 9) when joint data are available. If only information about rank correlation is given, then the minimum information copula with given correlation is advocated.[16] Bivariate copulae are well studied, understood and applied (see e.g., Refs. 12 and 20).

Multivariate copulae are often limited in the range of dependence structures that they can handle. The most popular choice is the normal copula that can model the full range of correlation structures but does not allow for tail dependence. The student-t copula enjoys the flexibility of taking into account dependence in the tails of the distribution. However, it has only one parameter that controls tail dependence in all bivariate margins. Some models involving constructions with Archimedean copulae are also available (see e.g., Ref. 15 and Chapter 2). The choice of multivariate copula is usually based on its simplicity, popularity and possibility of modeling a given dependence structure.

Graphical models with bivariate copulae as building blocks of the joint distribution have recently become the tool of choice in dependence modeling. Coming back to our example, let us assume that successive yearly interest rates have a rank correlation of 0.7. Different dependence structures can be considered to satisfy the specified information. One possibility

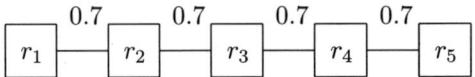

Figure 1.3. Tree for investment.

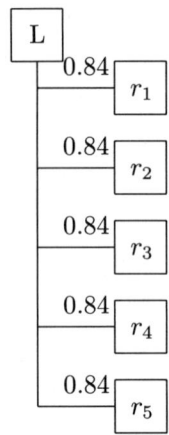

Figure 1.4. Tree for investment with latent variable.

is the dependence tree given in Fig. 1.3. With this structure, the correlation between r_1 and r_5 is much smaller than that between r_1 and r_2.

The second possibility is to correlate all yearly interests to a latent variable with a rank correlation of 0.84 (see Fig. 1.4). This dependence structure is symmetric and all interest rates are correlated at approximately 0.7.

Dependence structures that can be realized with trees are very limited. For joint distribution on n variables, we can specify only $n-1$ correlations and realize them with different bivariate copulae. A new graphical model introduced in 1997, called a regular vine, allows the specification of a joint distribution on n variables with given margins by specifying $\binom{n}{2}$ bivariate copulae and conditional copulae. Dependence trees are special cases of vines where conditional copulae are the independence copulae.

Returning to the investment example, we may reflect that if the interest is high in year 2, it is unlikely to be high in both years 1 and 3. We can capture this with a D-vine structure with a rank correlation of -0.7 between r_i and r_{i+2} conditional on r_{i+1}, $i = 1, 2, 3$, as shown in Fig. 1.5.

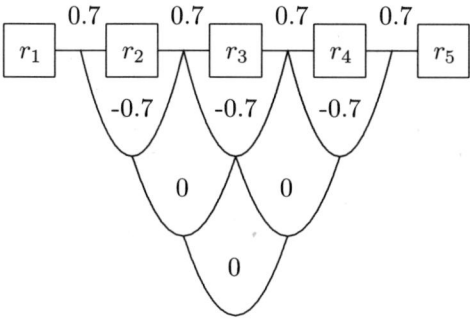

Figure 1.5. D-vine for investment.

Table 1.1. Quantiles, means and variances of distributions for five-year return in case of independence and for different dependence structures realized with the normal copula.

Model	5%-quant.	50%-quant.	95%-quant.	Mean	Variance
5yRind	1459.03	1609.42	1769.03	1611.13	9038.33
5yRtree	1367.43	1615.72	1889.93	1618.66	27384.70
5yRlatent	1348.15	1611.04	1913.49	1618.89	34352.21
5yRvine	1403.41	1607.61	1831.79	1610.58	16817.76

The results of our investment after five years in the case of independence between interest rates and for different dependence structures realized with the normal copula are shown in Table 1.1 and Fig. 1.6.

We see that the distributions of our fortune after five years in the case of different dependence structures are different even when realized by the same copula. Different choices of copula can be made and they will lead to different results (see results for Clayton copula in Table 1.2).

To decide if our investment is worth considering, imagine that there is another investment that in five years yields $1900 with probability 95%. From Table 1.1 we see that only if the latent model with normal copula was appropriate for the joint distribution of interest rates would we prefer our original investment. However, Table 1.2 shows that with none of the dependence structures realized with the Clayton copula can we reach $1900 with probability 95%. In this case, the competing investment would be preferred. From this simple example we can appreciate the importance of getting the dependence right. Flexible models that allow representation of a variety of

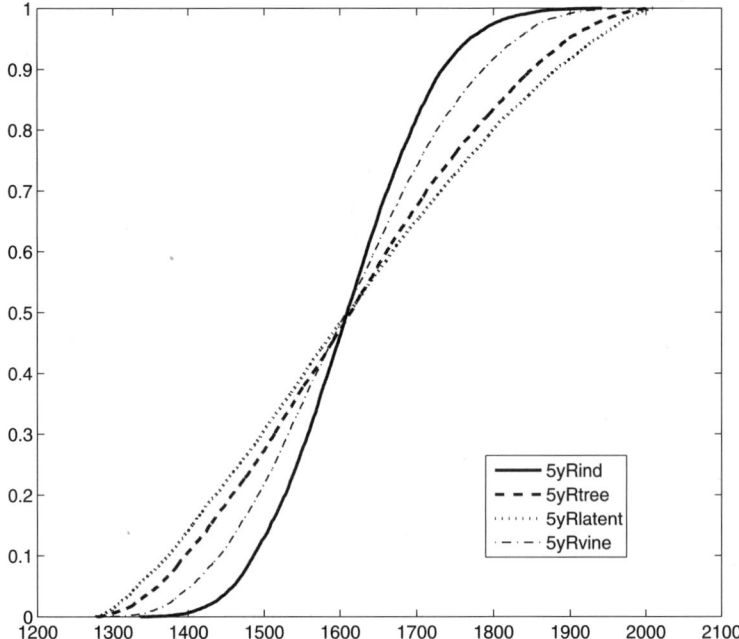

Figure 1.6. CDFs for five-year return in case of independence and for different dependence structures realized with the normal copula.

Table 1.2. Quantiles, means and variances of distributions for five-year return in case of independence and for different dependence structures realized with the Clayton copula.

Model	5%-quant.	50%-quant.	95%-quant.	Mean	Variance
5yRind	1459.33	1609.30	1768.85	1611.13	9039.08
5yRtree	1323.06	1638.60	1863.00	1618.64	27808.26
5yRlatent	1311.26	1639.51	1885.18	1619.20	33761.83
5yRvine	1337.70	1643.43	1792.84	1612.98	18665.36

dependence structures and different choices of copulae is essential for this kind of problem. Vines are very promising in this respect.

1.3 Vines

A vine on n elements $\mathcal{V} = (T_1, \ldots, T_{n-1})$ is a nested set of trees where the edges of the tree j are nodes of the tree $j+1$ and each tree has the maximum number of edges. A regular vine on n elements is one in which two edges in tree j are joined by an edge in tree $j+1$ only if these edges share a common node.

For each edge of a vine, we define *constraint, conditioned* and *conditioning* sets of this edge as follows: the nodes of the first tree reachable from a given edge via the membership relation are called the constraint set of that edge. When two edges are joined by an edge in the next tree, the intersection of the respective constraint sets form the conditioning set, and the symmetric difference of the constraint sets is the conditioned set of this edge. Formal definitions can be found in Refs. 4, 3 and 13 are summarized in Chapter 3. Copulae can be assigned to the edges of the vine such that the conditioned variables correspond to the conditioned set, and the conditioning variables to the conditioning set of an edge.

1.3.1 *Graphical representation*

In Fig. 1.5, a special type of vine on five elements, the D-vine, is shown. Copulae and conditional copulae that can be assigned to the edges of this vine are (from left to right), in T_1, $c_{12}, c_{23}, c_{34}, c_{45}$; in T_2, $c_{13|2}, c_{24|3}, c_{35|4}$; $c_{14|23}, c_{25|34}$ in the third tree; and only one copula $c_{15|234}$ in the fourth. For the D-vine the graphical representation in Fig. 1.5 is quite clear; however, for other regular vines, this type of graphical representation can be a bit messy (see the C-vine in Fig. 1.7).

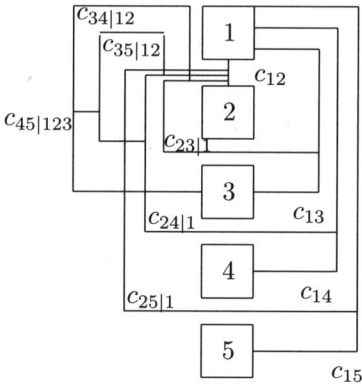

Figure 1.7. C-vine on five variables with copulae and conditional copulae assigned to the edges.

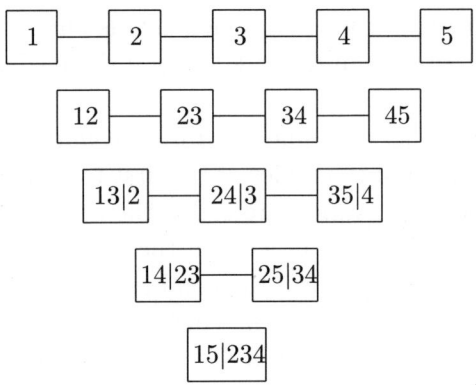

Figure 1.8. Trees for D-vine in Fig. 1.5 where nodes of each tree are enumerated by conditioned and conditioning sets of copula that can be assigned to it.

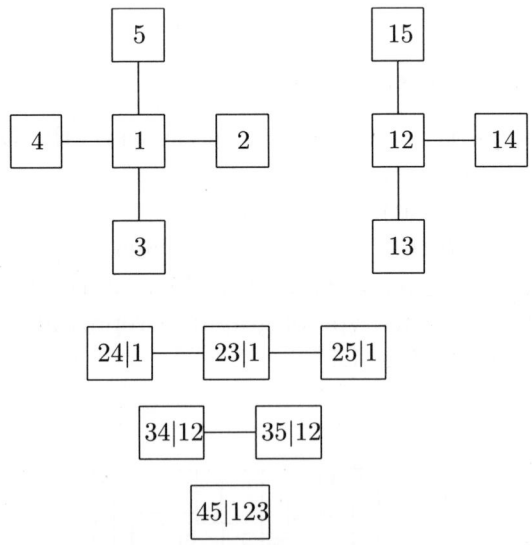

Figure 1.9. Trees for C-vine in Fig. 1.7 where nodes of each tree are enumerated by conditioned and conditioning sets of copula that can be assigned to it.

Often it is clearer to show all trees in a vine separately (see Figs. 1.8 and 1.9 where all trees are shown separately for the D-vine in Fig. 1.5 and the C-vine in Fig. 1.7 respectively).

A representation where all trees are kept separately is not very compact in terms of storing information necessary to represent a given vine. For storing information necessary for any regular vine, an n by n array can be used instead (for more information, see Ref. 18 or Chapters 9 and 10. We follow

Table 1.3. Array for the D-vine in Fig. 1.5 (left) and the C-vine in Fig. 1.7 (right).

3	3	3	2	4	**1**	1	1	1	1
	2	2	3	3		**2**	2	2	2
		4	4	2			**3**	3	3
			1	1				**4**	4
				5					**5**

here the notation used in Chapter 10). In Table 1.3, arrays containing all information for the D-vine in Fig. 1.5 and the C-vine in Fig. 1.7 are shown.

The diagonal elements in arrays printed in bold denote an ordering of variables in the vine, called the natural order. From the rightmost column of the left array in Table 1.3, we can read that variable 5 is in the conditioned set of the top node of the vine together with variable 1, and the conditioning set consists of variables $\{2, 3, 4\}$. In the third tree, the variable 5 forms with 2 the conditioned set of an edge, and the conditioning set of this edge is $\{3, 4\}$. In the second tree, 5 is paired with 3 and conditioned on 4, and in the first tree 5 is connected with 4. Notice the very simple structure of the array for the C-vine.

The bottom part of both arrays is empty and it can store, for instance, information about the parameters of copulae that are assigned to the edges of the vine.

1.3.2 Vine density

Bedford and Cooke[3] show that the joint density of a regular vine copula with margins $f_1, \ldots, f_n$ is a product of (conditional) copula densities assigned to the edges of the vine and a product of marginal densities. For the C-vine in Fig. 1.7, the density is of the form:

$$f_{1\ldots5} = f_1 \ldots f_5 \prod_{i=1}^{4} \prod_{j=i+1}^{5} c_{ij|i+1\ldots j-1}(F_{i|i+1\ldots j-1}, F_{j|i+1\ldots j-1}). \quad (1.1)$$

On the other hand, given a positive joint density $f_{1\ldots5}$ with standard factorization $\prod_{i=1}^{5} f_{i|1\ldots i-1}$, we can see that the ith term of this factorization $(i > 1)$ can be expressed as:

$$f_{i|1\ldots i-1} = c_{i,i-1|1\ldots i-2}(F_{i|1\ldots i-2}, F_{i-1|1\ldots i-2})f_{i-1|1\ldots i-2}.$$

This recursive representation leads to the conclusion that every positive multivariate density function can be expressed as a product of bivariate copula acting on several different conditional probability distributions. Hence, every positive joint density can be represented as a density of any regular vine copula.

For most densities, the conditional copulae will depend on conditioning variables as shown for the trivariate Frank's copula density in Example 1.1. Distributions belonging to the elliptical family factorize on a vine such that conditional copulae do not depend on conditioning variables.

Example 1.1. Consider Frank's copula with density

$$c(u,v;\theta) = \theta\eta \frac{e^{-\theta(u+v)}}{(\eta - (1-e^{-\theta u})(1-e^{-\theta v}))^2} \quad (1.2)$$

where $\theta > 0$, $\eta = 1 - e^{-\theta}$. It is shown in Ref. 12 that this copula can be extended to the multivariate case. The trivariate copula density belonging to Frank's family is:

$$c_{123}(u_1, u_2, u_3; \theta) = \theta^2 \eta^2 e^{-\theta(u_1+u_2+u_3)}$$
$$\times \frac{\eta^2 + (1-e^{-\theta u_1})(1-e^{-\theta u_2})(1-e^{-\theta u_3})}{(\eta^2 - (1-e^{-\theta u_1})(1-e^{-\theta u_2})(1-e^{-\theta u_3}))^3}. \quad (1.3)$$

All three bivariate margins of (1.3) are Frank's with the same parameter θ. If we consider a D-vine with c_{12} and c_{13} of the form (1.2), then to get a vine copula representation of the density (1.3) we must only find the conditional copula $c_{13|2}$, which is:

$$c_{13|2}(u, v; \theta, u_2) = \frac{2uv + (u - 3uv + v)e^{-\theta u_2} + (1-u)(1-v)e^{-2\theta u_2}}{(1 + (1-u)(1-v)(e^{-\theta u_2} - 1))^3}.$$

The conditional copula does not belong to Frank's family and it depends on the conditioning variable.

1.3.3 *Estimation*

When estimating a vine copula from data it is usually assumed that conditional copulae do not depend on conditioning variables.[a] Moreover, the type of a vine structure is fixed and only few families of bivariate copulae are taken into consideration.

The estimation of copula parameters using the maximum likelihood principle, for the vine copula or pair-copula construction (PCC),[2] is performed

---

[a]To our knowledge, it is not known how severe this assumption is. For more information, see Ref. 10.

sequentially starting from the first tree. This landmark advance in associating bivariate copulae to a vine and estimating copula parameters from data demonstrated the superiority of vines and opened up large areas of application in mathematical finance, risk analysis and uncertainty modeling in engineering[1] (for more information about PCC, see Chapter 3, and Chapters 13, 15 and 16 for examples of applications).

The assumption of constant conditional copulae and consideration of only a few types of bivariate families in fitting a vine to data cause a phenomenon where some types of vines fit the data better than the others. To find the best vine structure, we would in principle have to estimate all possible vines. In dimensions higher than seven or eight, this is infeasible as the number of vines grows rapidly with dimension (see Ref. 18 and Chapter 9). Moreover, because of sequential estimation in PCC, estimates for parameters of conditional copulae in higher-order trees are less reliable. For higher-dimensional cases, some simplifying assumptions for fitting vines to data will have to be made. Some ideas on this subject are based on optimal truncations of a vine (see Chapter 11). Searching for the best vine model can also be approached from a Bayesian perspective (see Chapter 12).

If joint data are not available, there exist protocols to elicit copula parameters from experts.[19]

1.3.4 *Properties and applications*

Some properties of vines are already well established (see Refs. 3, 4 and 13 and Chapter 3). In this volume, new properties of vine distributions are presented. In Chapter 8, the similarities and differences in dependence for different regular vines on n variables are studied. It is shown that for $n \leq 4$, only two types of vines are available, namely C-vines and D-vines. In higher dimensions, other vine structures appear. Distributions corresponding to different vines are compared from the perspective of marginal dependence. It is shown that under some conditions the bivariate marginal dependence from the C-vine is the highest of all vines on five nodes.

Tail dependence of vine copula is of great interest. It is shown in Chapter 9 that vine copulae have flexible asymmetry in the joint upper and lower tails. This flexibility can be achieved by appropriate choice of bivariate copulae. Figure 1.10 shows contours of bivariate densities for (X_1, X_3) with standard normal margins and a copula which is a bivariate margin of a D-vine with copulae c_{12}, c_{23} and $c_{13|2}$. Rank correlations on the D-vine are $r_{12} = 0.5$, $r_{23} = 0.6$ and $r_{13|2} = 0.7$ for all cases. They are, however, realized

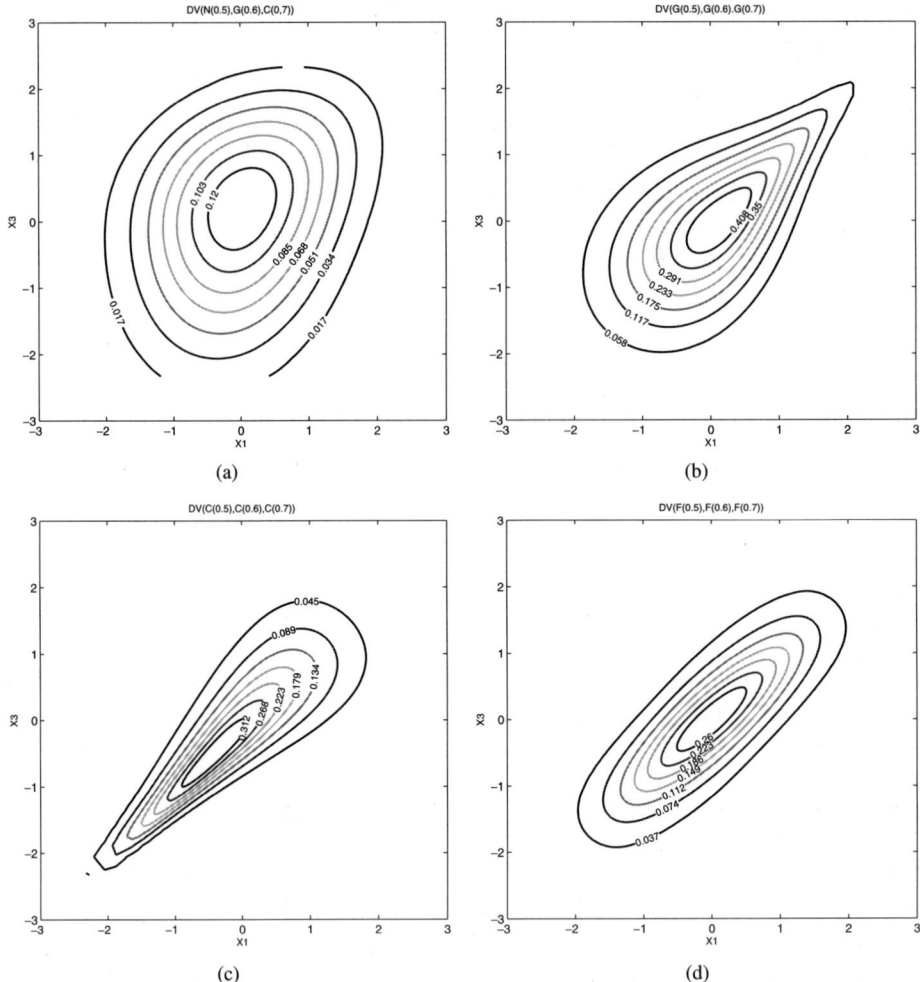

Figure 1.10. Contours of densities for (X_1, X_3) with standard normal margins and copula obtained from the D-vine DV(1,2,3) with Spearman correlations $r_{12} = 0.5$, $r_{23} = 0.6$ and $r_{13|2} = 0.7$ realized by copulae (a) Normal, Gumbel, Clayton (b) Gumbel, Gumbel, Gumbel (c) Clayton, Clayton, Clayton and (d) Frank, Frank, Frank.

by different bivariate copulae. A variety of distributions can be obtained with different choices of bivariate copulae.

Copulae are naturally used for continuous variables. They can also be useful in simulating models for high-dimensional count variables. The normal copulae were utilized for the construction of multivariate discrete distributions with specified correlation structure. However, due to the positive definiteness constraint, it is challenging to determine the appropriate normal copula parameters for a specified target correlation. Parameters of the normal copula can, however, be reparameterized into an algebraically independent set of partial correlations assigned to the edges of a vine. This idea allows efficient sampling of high-dimensional count variables and is presented in Chapter 4.

Vines are undirected graphical structures representing joint distributions. In Chapter 14, a directed graphical model, Non-Parametric Bayesian Belief Net (NPBBN), is presented and compared with vines. The most important difference between NPBBNs and regular vines appears to be in the conditional independencies that they can represent. The choice between representing a multivariate distribution using a regular vine or a NPBBN depends on many factors, some of which are discussed in Chapter 14.

1.4 Outline

The focus of this book is on vine copulae or PCCs. However, in Chapter 2, different multivariate copula classes and construction schemes of multivariate models are also reviewed. Popular multivariate copulae belonging to the elliptical family as well as Archimedean copulae and their generalizations are presented.

In Chapter 3, an introduction to the main idea of vines as graphical models is presented. This chapter traces the early history of vines and presents the motivation for their construction. Important properties and applications of vines are included.

In Chapter 4, vine copulae are used to sample multivariate count variables with target correlation structure.

Chapter 6 surveys the asymptotic theory of estimation of a copula from a frequentistic perspective and presents the problems involved in frequentistic model selection among several candidate copulae using the Maximum Pseudo Likelihood Estimator (MPLE). Frequentistic copula model selection has recently been addressed through the development of the Copula Information Criterion (CIC) — a model selection formula that extends the Akaike Information Criterion (AIC), based on maximum likelihood, to the MPLE.

Chapters 5 and 8 study tail dependence of various multivariate distributions. In Chapter 8, multivariate tail dependence functions are introduced and applied to vine copulae.

Chapter 9 explores how many different vines and regular vines are available. It is shown that the number of possible vines grows rapidly with dimension. A few algorithms for generating regular vines are presented in this chapter. It also contains a catalogue of regular vines up to dimension 9.

Chapter 10 shows an algorithm to generate all regular vines. This algorithm is used to obtain some general results about the number of equivalence classes for regular vines.

In Chapter 7, different types of five-dimensional vines are studied. Six equivalence classes for five-dimensional vines have been found. They are compared from the perspective of bivariate marginal dependence obtained when the bivariate copulae at each level were assumed to be the same. An interesting pattern of dependence emerges from this study, which may be helpful in the use of vine copulae for modeling multivariate data.

For high-dimensional problems, simplifying assumptions have to be made to reduce the complexity and computational burden involved in fitting vine copulae. Chapter 11 explores truncations of vines and proposes a heuristic search algorithm for the 'best vine' for the target correlation structure.

Chapter 12 reviews available MCMC estimation and model selection algorithms as well as their possible extensions for D-vine pair-copula constructions based on bivariate t-copulae. Theory presented in this chapter is then applied in Chapter 13 to Australian electricity load data.

Chapter 14 compares regular vines with the directed graphical models that represent joint distribution, called Non-Parametric Bayesian Belief Nets.

Chapter 15 compares three constructions for modeling higher-dimensional dependence: the Student copula, the partially nested Archimedean construction and the pair-copula construction. It is shown through two applications that the PCC provides a better fit to financial data than the two other structures.

In Chapter 16, the dependence structure of multivariate financial returns is modeled with a time-varying D-vine copula. Two different data sets, one with six exchange rates and another with five Asian equity indices were used for the analysis. The D-vine structure allows us to model the symmetric dependence of exchange rates and the asymmetric one of Asian equities. For both cases, the dependence structure was found to vary in time.

This book concludes with a short summary and a few future research directions.

All chapters are self-contained and provided with their own set of references. However, the notation in all chapters has been unified as much as possible. The general notation for this book is presented in the next section.

1.5 Glossary and Notation

Random vectors, distributions, densities, copulae

$\mathbf{X} = (X_1, \ldots, X_n)$ n-dimensional random vector;

$F_{\mathbf{X}} = F_{1,2,\ldots,n}$ cumulative distribution function (cdf) of $\mathbf{X}$, $F_{\mathbf{X}}(\mathbf{x}) = F_{1,2,\ldots,n}(x_1, \ldots, x_n)$ value at $\mathbf{x} = (x_1, \ldots, x_n)$;

$f_{\mathbf{X}} = f_{1,2,\ldots,n}$ probability density function (pdf) of $\mathbf{X}$;

$\bar{F} = \bar{F}_{\mathbf{X}}$ the survival function of $\mathbf{X}$;

$F_{X_i} = F_i$ and $f_{X_i} = f_i$ marginal cdfs and pdfs, respectively;

F_S marginal distribution, where $S \subset \{1, \ldots, n\}$,

$F_{X_i|X_1,\ldots,X_k} = F_{i|1\ldots k}$ conditional distribution of X_i given $X_1, \ldots, X_k$; its value $F_{i|1\ldots k}(x_i|x_1, \ldots, x_k)$;

$C_{\mathbf{X}} = C_{1,2,\ldots,n}$ copula for $\mathbf{X}$ with density $c_{\mathbf{X}}$ where $F_{\mathbf{X}} = C_{\mathbf{X}}(F_1, \ldots, F_n)$ and $f_{\mathbf{X}} = f_1 \cdots f_n \cdot c_{\mathbf{X}}(F_1, \ldots, F_n)$;

$C_{i_1,i_2|S}(\cdot|x_k : k \in S)$ bivariate conditional copula of $F_{i_1|S}(\cdot|x_k : k \in S)$ and $F_{i_2|S}(\cdot|x_k : k \in S)$ where $F_{i_1|S}$ and $F_{i_2|S}$ univariate conditional cdfs and $i_j \notin S$ for $j = 1, 2$.

Correlations

$\rho(X_i, X_j) = \rho_{ij}$ product moment (pearson) correlation of X_i, X_j

$r_{X_i,X_j} = r_{ij} = \rho(F_i(X_i), F_j(X_j))$ Spearman rank correlation of X_i, X_j.

Graphs

$G = (N, E)$ graph, N vertex set, E edge set;
$T = (N, E)$ tree;
$\mathcal{V} = \mathcal{V}(n) = (T_1, \ldots, T_{n-1})$ vine on n elements.

Information

$I(f, g)$ or $KL(f, g)$ information (Kullback–Leibler) divergence between f and g;
$MI(f)$ mutual information of f.

Time series

$AR(m)$ autoregressive process of order m.

References

1. Aas K. and Berg D. (2009). Models for construction of multivariate dependence — A comparison study. *The European Journal of Finance*, 15:639–659.
2. Aas K., Czado C., Frigessi A. and Bakken H. (2009). Pair-copula constructions of multiple dependence. *Insurance: Mathematics and Economics*, 44(2):182–198.
3. Bedford T.J. and Cooke R.M. (2001). Probability density decomposition for conditionally dependent random variables modeled by vines. *Annals of Mathematics and Artificial Intelligence*, 32:245–268.
4. Bedford T.J. and Cooke R.M. (2002). Vines — A new graphical model for dependent random variables. *Annals of Statistics*, 30(4):1031–1068.
5. Chollete L., Heinen A. and Valdesogo A. (2009). Modeling international financial returns with a multivariate regime switching copula. *Journal of Financial Econometrics*, 7(4):437–480.
6. Cooke R.M. (1997). Markov and entropy properties of tree- and vine-dependent variables. In *Proceedings of the ASA Section on Bayesian Statistical Science*. Washington: American Statistical Association.
7. Czado C., Min A., Baumann T. and Dakovic R. (2009). Pair-copula constructions for modeling exchange rate dependence. Technical report, Technische Universität München.
8. Fischer M., Köck C., Schlüter S. and Weigert F. (2009). Multivariate copula models at work. *Quantitative Finance*, 9(7):839–854.
9. Genest C., Rémillard B. and Beaudoin D. (2009). Omnibus goodness-of-fit tests for copulas: A review and a power study. *Insurance: Mathematics and Economics*, 44: 199–213.
10. Hobæk Haff I., Aas K. and Frigessi A. (2009). On the simplified pair-copula construction — Simply useful or too simplistic? *Journal of Multivariate Analysis*, 101:1296–1310.
11. Joe H. (1996). Families of m-variate distributions with given margins and $m(m-1)/2$ bivariate dependence parameters. In L. Rüschendorf, B. Schweizer and M. D. Taylor (eds.), *Distributions with Fixed Marginals and Related Topics*, 28:120–141.
12. Joe H. (1997). *Multivariate Models and Dependence Concepts*. Chapman & Hall, London.
13. Kurowicka D. and Cooke R.M. (2006). *Uncertainty Analysis with High Dimensional Dependence Modelling*. Wiley, New York.
14. Kurowicka D. and Cooke R.M. (2007). Sampling algorithms for generating joint uniform distributions using the vine-copula method. *Computational Statistics and Data Analysis*, 51:2889–2906.
15. McNeil A.J., Frey R. and Embrechts P. (2006). *Quantitative Risk Management: Concepts, Techniques and Tools*. Princeton University Press, Princeton.
16. Meeuwissen A. and Bedford T.J. (1997). Minimally informative distributions with given rank correlation for use in uncertainty analysis. *Journal of Statistical Computation and Simulation*, 57:143–175.
17. Min A. and Czado C. (2010). Bayesian inference for multivariate copulas using pair copula constructions. *Journal of Financial Econometrics*. In press.
18. Morales-Nápoles O., Cooke R.M. and Kurowicka D. (2008). The number of vines and regular vines on n nodes. Submitted to *Discrete Applied Mathematics*.
19. Morales-Napoles O., Kurowicka D. and Roelen A. (2007). Elicitation procedures for conditional and unconditional rank correlations. *Reliability Engineering and Systems Safety*, 95(5):699–710.

20. Nelsen R.B. (2006). *An Introduction to Copulas*, 2nd ed., Springer Series in Statistics. Springer, New York.
21. Schirmacher D. and Schirmacher E. (2008). Multivariate dependence modeling using pair-copulas. Technical report, presented at the 2008 ERM Symposium, Chicago.
22. Yule G.U. and Kendall M.G. (1965). *An Introduction to the Theory of Statistics*, 14th ed. Charles Griffin & Co., Belmont, California.

CHAPTER 2

Multivariate Copulae

Matthias Fischer

Department of Statistics & Econometrics
Lange Gasse 20. 90403 Nürnberg, Germany
matthias.fischer@wiso.uni-erlangen.de

Though dating back to 1959 when the term "copulae" was coined, copula models only started their triumphal procession in the mid-1990s. Application of copulae was primarily restricted to the world of finance and insurance but now the copula concept has found its way into nearly all relevant statistical and mathematical literature where multivariate dependence structures are involved. Whereas the bivariate case was central in most of the publications and seems to be well-explored at present, there is still an ongoing and active debate on the construction of multivariate copula models. Apart from pair-copula constructions, which are the focus of this book and intensively discussed in the following chapters, this chapter briefly reviews both different copula classes and construction schemes of multivariate models.

2.1	Copulae	20
2.2	Elliptical Copulae and Generalizations	22
	2.2.1 Elliptical copulae	22
	2.2.2 Generalized t-copulae	25
2.3	Archimedean Copulae and Generalizations	27
	2.3.1 Classical Archimedean copulae	27
	2.3.2 Non-exchangeable Archimedean copulae	27
	2.3.3 Generalized multiplicative Archimedean copulae	29
	2.3.4 Koehler–Symanowski copulae	31
2.4	Combinations of Arbitrary Copulae into a New One	32
2.5	Summary	34
References		34

2.1 Copulae

Loosely speaking, a *copula* incorporates the information on the dependence structure of $n > 1$ random variables $X_1, \ldots, X_n$. For reasons of simplicity, let us assume that the corresponding distribution functions $F_1, \ldots, F_n$ are *continuous* with the inverse functions $F_1^{-1}, \ldots, F_n^{-1}$ (details on discrete margins can be found, for instance, in Genest and Neslehova[20]). It follows from the probability integral transform that $U_i \equiv F_i(X_i)$ is uniformly distributed on $(0,1)$ for $i = 1, \ldots, n$. Conversely, $X_i = F_i^{-1}(U_i)$ for $i = 1, \ldots, n$. With this in mind,

$$P(X_1 \leq F_1^{-1}(x_1), \ldots, X_n \leq F_n^{-1}(x_n)) = P(U_1 \leq x_1, \ldots, U_n \leq x_n)$$
$$\equiv C(x_1, \ldots, x_n).$$

Obviously, the function C is a distribution function with support on $[0,1]^n$ with uniform margins, a so-called *copula*.[a] Conversely, we obtain the following decomposition:

$$P(X_1 \leq x_1, \ldots, X_n \leq x_n) = P(F_1(X_1) \leq F_1(x_1), \ldots, F_n(X_n) \leq F_n(x_n))$$
$$= C(F_1(x_1), \ldots, F_n(x_n)).$$

Under the above assumptions, there is a one-to-one correspondence between the copula C and the distribution of $\boldsymbol{X} = (X_1, \ldots, X_n)'$, as stated in the fundamental theorem of Sklar.

Theorem 2.1 (Sklar[43]). *Given random variables $X_1, \ldots, X_n$ with continuous distribution functions $F_1, \ldots, F_n$ and joint distribution function F, there exists a unique copula C such that for all $\boldsymbol{x} = (x_1, \ldots, x_n)' \in \mathbb{R}^n$:*

$$F(x_1, \ldots, x_n) = C(F_1(x_1), \ldots, F_n(x_n)). \tag{2.1}$$

Conversely, given any distribution functions $F_1, \ldots, F_n$ and copula C, F defined through Eq. (2.1) is an n-variate distribution function with marginals $F_1, \ldots, F_n$.

According to (2.1), the copula "couples" the marginal distributions to the joint distribution function F. Hence, Eq. (2.1) enables us to construct the joint distribution function F as follows: At the first stage, the marginal distribution $F_1, \ldots, F_n$ have to be specified, whereas, at the second stage,

---

[a]A formal definition of multivariate copulae is provided by Nelsen.[36]

the underlying copula model has to be selected. On the other hand, Eq. (2.1) can be re-written as follows:

$$F(F_1^{-1}(x_1), \ldots, F_n^{-1}(x_n)) = C(u_1, \ldots, u_n). \tag{2.2}$$

Equation (2.2) reveals how to extract the copula of a (given) multivariate distribution. Take, for instance, elliptical copulae which are discussed in the next subsection. We conclude this section with an example that contains simple but prominent copulae.

Example 2.1.

- *Independence copula*: Assume that the random variables $X_1, \ldots, X_n$ are independent. According to (2.1), the underlying ("independence") copula is given by

$$C^{\perp}(\mathbf{u}) \equiv C(u_1, \ldots, u_n) = u_1 \cdots u_n.$$

- *Copula bounds*: Every multivariate copula is bounded from above and below by the so-called Fréchet–Hoeffding bounds, i.e.,

$$\max\{u_1 + \cdots + u_n - (n-1), 0\} \leq C(u_1, \ldots, u_n) \leq \min\{u_1, \ldots, u_n\}.$$

Note that only the upper bound is a valid copula for $n > 2$.

By the end of this chapter, we will have looked at much more flexible, parametric copula classes and construction schemes for multivariate copulae, without claiming to be fully comprehensive. For a long time, both practitioners and theorists have relied solely on the multivariate Gaussian distribution and Gaussian copula, respectively, where the dependence structure is completely determined by pairwise correlations. More generally, elliptical copulae (see Section 2.2) still maintain many of their attractive properties. But while elliptical distributions are able to model moderate and/or heavy tails, they fail to capture asymmetric dependence structures. Among the classes of non-elliptical copulae, Archimedean copulae and its generalizations enjoy great popularity and are the subject of Section 2.3. Within this chapter, the focus is primarily on these two copula classes and on selected construction schemes of multivariate copulae published recently (e.g., Refs. 17, 30, 35). Beyond that, there exist a bundle of multivariate copulae which are excluded from this overview. To name only a few, we refer to multivariate extreme-value copulae (see, for instance, McNeil et al.[32] or Joe[24]), multivariate Farlie–Gumbel–Morgenstern copulae (see, for instance, Drouet and Kotz[12]) or multivariate Marshall–Olkin copulae (see, for instance, Joe[24]).

For a detailed introduction to copulae we refer the reader to the textbooks.[12,24,36] Application of copulae to finance can be found in

Refs. 7, 9, 32. Furthermore, overviews of copulae and some background theory are provided in Genest and Favre,[21] Embrechts et al.[14,15] or, from a more critical point of view, in Mikosch.[34]

2.2 Elliptical Copulae and Generalizations

2.2.1 *Elliptical copulae*

The class of *elliptical copulae* (EC) constitutes the prime example of implicit copulae stated in (2.2). EC are copulae associated with elliptical distributions[b] and are widely used in statistics and econometrics, especially in finance. Note that EC are not elliptical distributions themselves. EC have the virtue that they extend easily to arbitrary dimensions n and are rich in parameters, at least $n(n-1)/2$. However, radial asymmetries and asymmetric tail behavior cannot be captured within this class. Due to their implicit definition, explicit expressions for the copula are not available. Evaluating an elliptical copula requires the calculation of multiple integrals without closed-form solutions, which must be done numerically. Applications and limitations of EC are discussed in more detail by Frahm et al.,[19] whereas Hult and Lindskog[23] and Abdous et al.[1] deal with extremal dependence and tail dependence of elliptically contoured distributions. Within the elliptical class, both Gaussian and t-copulae play a predominant role.

Example 2.2 (Gaussian copula). Let $\Phi_{\boldsymbol{R}}^n$ denote the standardized n-variate normal distribution with correlation matrix $\boldsymbol{R}$. Applying (2.2), the *Gaussian copula* is defined as follows:

$$C(\mathbf{u}; \boldsymbol{R}) = \Phi_{\boldsymbol{R}}^n(\Phi^{-1}(u_1), \ldots, \Phi^{-1}(u_n)),$$

where Φ^{-1} denotes the quantile function of a univariate standard normal distribution. Per construction, the Gaussian copula generates the standard Gaussian joint distribution if and only if the margins follow a standard normal distribution. The corresponding copula density is given by

$$c(\mathbf{u}; \boldsymbol{R}) = \frac{\frac{1}{(2\pi)^{n/2}\sqrt{|\boldsymbol{R}|}} \exp(-0.5\boldsymbol{\zeta}'\boldsymbol{R}^{-1}\boldsymbol{\zeta})}{\prod_{j=1}^n \frac{1}{\sqrt{2\pi}}\exp(-0.5\zeta_j^2)} = \frac{\exp(-0.5\boldsymbol{\zeta}'(\boldsymbol{R}^{-1} - \mathbf{I}_n)\boldsymbol{\zeta})}{\sqrt{|\boldsymbol{R}|}},$$

with $\boldsymbol{\zeta} \equiv (\zeta_1, \ldots, \zeta_n)'$ and $\zeta_i = \Phi^{-1}(u_i)$ for $i = 1, \ldots, n$. Restricting to the bivariate case, a bivariate Gaussian variable admits no tail dependence

---

[b]A detailed treatment of elliptically contoured distribution is provided by Fang et al.[16]

(see, e.g., Ref. 15). Extensions to the Gaussian copula can be found in Andersen and Sidenius.[2]

Example 2.3 (Student-t copula). Let $\mathbf{Z} \sim \mathcal{N}_n(\mathbf{0}, \Sigma)$ and $R = \sqrt{\nu}/\sqrt{S}$ with $S \sim \chi^2(\nu)$, i.e. a chi-squared variable with ν degrees of freedom. Then the $\mathbb{R}^n$-valued random vector

$$\mathbf{Y} \equiv R\mathbf{Z} = (RZ_1, \ldots, RZ_n)$$

has a t-distribution with ν degrees of freedom. If $\nu > 2$, $\mathrm{Cov}(\mathbf{Y}) = \frac{\nu}{\nu-2}\Sigma$. Again, applying Sklar's theorem and defining $\boldsymbol{\rho} \equiv (\rho_{ij})_{1 \leq i,j \leq n}$ with $\rho_{ij} \equiv \Sigma_{ij}/\sqrt{\Sigma_{ii}\Sigma_{jj}}$, the implicit copula expression is given by

$$C_t(\mathbf{u}; \nu, \boldsymbol{\rho}) = \mathbf{t}^n_{\nu, \boldsymbol{\rho}}(t_\nu^{-1}(u), t_\nu^{-1}(v)),$$

where t_ν^{-1} denotes the inverse function of the classical univariate t-distribution. The associated density function of the t-copula is given by

$$c_t(\mathbf{u}; \boldsymbol{\rho}, \nu) = \frac{1}{\sqrt{|\boldsymbol{\rho}|}} \frac{\Gamma(\frac{\nu+n}{2})}{\Gamma(\frac{\nu}{2})} \left(\frac{\Gamma(\frac{\nu}{2})}{\Gamma(\frac{\nu+1}{2})} \right)^n \frac{\prod_{j=1}^n \left(1 + \frac{t_\nu^{-1}(u_j)^2}{\nu}\right)^{\frac{\nu+1}{2}}}{\left(1 + \frac{\zeta'\boldsymbol{\rho}^{-1}\zeta}{\nu}\right)^{\frac{\nu+n}{2}}}. \quad (2.3)$$

Restricting again to the bivariate case, the t-copula has tail dependence coefficient

$$\lambda = \lambda_U = \lambda_L = 2t_{\nu+1}\left(-\frac{\sqrt{\nu+1}\sqrt{1-\rho}}{\sqrt{1-\rho}}\right) > 0,$$

provided that $\rho \geq -1$. Venter[44,45] deals with the estimation, application and limitations of the Student t-copula, whereas Kole et al.[28] perform stress testing under Student's t-dependence.

Still within the elliptical class, Mendes and Arslan[33] favor a generalized t-copula which allows for different degrees and types of linear and non-linear dependence. In particular, they derive expressions for its coefficients of upper and lower tail dependence and suggest applications in finance, including portfolio optimization and computation of measures of contagion.

Example 2.4 (GT-copula). Arslan[3] introduces a new family of multivariate generalized distributions as a scale mixture of a multivariate power exponential distribution (see Gómez et al.[22]) and an inverse generalized gamma distribution with a scale parameter, and shows that this family of distributions belongs to the family of elliptically contoured distributions that includes the multivariate normal distribution and the multivariate t-distribution as special or limiting cases. The corresponding copula

("GT-copula") is intensively discussed by Mendes and Arslan[33] who show that the bivariate copula density is given by

$$c(u_1, u_2; \rho, \nu, \beta) = \frac{K}{\sqrt{1-\rho^2}} \cdot \frac{\left[\frac{\nu}{2} + \left(\frac{\zeta_1^2+\zeta_2^2-2\rho\zeta_1\zeta_2}{1-\rho^2}\right)^\beta\right]^{-\frac{\nu}{2}-\frac{1}{\beta}}}{f(\zeta_1; \beta, \nu/2) f(\zeta_2; \beta, \nu/2)}, \qquad (2.4)$$

with $K \equiv \frac{\beta \Gamma(n/2) q^q}{\pi^{n/2} B(q, n/2\beta)}$, $\zeta_i \equiv F_{GT}^{-1}(u_i)$ for $i = 1, 2$ and where the marginal density and distribution function of MGT marginals, respectively, are

$$f(x; \beta, q) = K \int_{x^2}^{\infty} \frac{(y-x^2)^{-1/2}}{(q+y^\beta)^{q+1/\beta}} dy \qquad (2.5)$$

and

$$F(x; \beta, q) = \frac{1}{2} + \int_{x^2}^{\infty} \frac{\arcsin(x/\sqrt{y})}{(q+y^\beta)^{q+1/\beta}} dy. \qquad (2.6)$$

Unfortunately, explicit formulae for the integrals in (2.5) and (2.6) are not available and numerical procedures are required in order to evaluate both copula and copula density.

Example 2.5 (Elliptical generalized hyperbolic (GH) copulae). Dating back to Barndorff-Nielsen,[4,5] both univariate and multivariate GH distributions have become very popular in the last decade, especially in finance (see, for instance, Prause[38]). This distribution family exhibits heavier tails than the Gaussian distribution but lighter ones than the t-distribution, both of which appear as limit cases. All moments of the GH distribution exist and the moment-generating function is available in closed form. Though multivariate GH distributions share the desirable characteristics of the univariate one (i.e., flexibility, semi-heavy tails), this distribution family possesses no parameter configuration for which the case of marginal independence can be modeled. Above that, the bivariate GH distribution is tail-independent (see, e.g., Schmidt[41]). In general, the multivariate version arises as a normal mean-variance mixture, i.e., as a multivariate normal distribution with (random) mean vector $\boldsymbol{\mu} + \boldsymbol{\beta}\tau\Delta$ and (random) covariance matrix $\tau\Delta$, where τ itself follows a univariate generalized inverse Gaussian distribution (see, e.g., Jørgensen[25]). The corresponding GH density is given by

$$f_n(\boldsymbol{x}; \boldsymbol{\Theta}) = \frac{\left(\frac{\psi}{\psi+\boldsymbol{\beta}\Delta\boldsymbol{\beta}'}\right)^{\lambda/2} \left(\frac{(\psi+\boldsymbol{\beta}\Delta\boldsymbol{\beta}')}{\chi}\right)^{n/4}}{(2\pi)^{n/2} K_\lambda(\sqrt{\psi\chi})} \cdot \frac{K_{\lambda-n/2}(\sqrt{(\psi+\boldsymbol{\beta}\Delta\boldsymbol{\beta}')(\chi+z)})}{(1+z/\chi)^{n/4-\lambda/2} e^{-\boldsymbol{\beta}'(\boldsymbol{x}-\boldsymbol{\mu})}}$$

with $z \equiv (\boldsymbol{x}-\boldsymbol{\mu})'\Delta^{-1}(\boldsymbol{x}-\boldsymbol{\mu})$, Δ being a positive definite matrix with determinant 1, parameter vector $\boldsymbol{\Theta} \equiv (\boldsymbol{\mu}, \chi, \boldsymbol{\beta}, \psi, \lambda, \Delta)'$ and where $K_\lambda(x)$ denotes the modified Bessel function of the third kind. Ellipticity is achieved only if the asymmetry parameter vector $\boldsymbol{\beta}$ is set to zero. Despite the popularity of the GH distribution, the literature on the corresponding GH copula itself is relatively sparse (e.g., Schmidt[41,42] and Lentzas[29]). Under a slightly different parametrization (see McNeil et al.[32]) that has the property that mixing parameters remain invariant under linear affine transformations, Lentzas[29] derives the copula density of a GH distribution as follows:

$$c_{GH}(\boldsymbol{u}) = \frac{k \cdot K_{\lambda-\frac{n}{2}}\left(\sqrt{(\chi + (\boldsymbol{\zeta}-\boldsymbol{\mu})'\boldsymbol{\Sigma}^{-1}(\boldsymbol{\zeta}-\boldsymbol{\mu}))(\psi + \boldsymbol{\gamma}'\boldsymbol{\Sigma}^{-1}\boldsymbol{\gamma})}\right)}{e^{-(\boldsymbol{\zeta}-\boldsymbol{\mu})'\boldsymbol{\Sigma}^{-1}\boldsymbol{\gamma}}((\chi + (\boldsymbol{\zeta}-\boldsymbol{\mu})'\boldsymbol{\Sigma}^{-1}(\boldsymbol{\zeta}-\boldsymbol{\mu}))(\psi + \boldsymbol{\gamma}'\boldsymbol{\Sigma}^{-1}\boldsymbol{\gamma}))^{\frac{n}{2}-\lambda}}$$

$$\times \left(\prod_{i=1}^n \frac{k_i \cdot K_{\lambda-\frac{1}{2}}\left\{\sqrt{\left(\chi + \frac{(\zeta_i-\mu_i)^2}{\Sigma_{ii}}\right)\left(\psi + \frac{\gamma_i^2}{\Sigma_{ii}}\right)}\right\}}{e^{-\frac{\gamma_i(\zeta_i-\mu_i)}{\Sigma_{ii}}}\sqrt{\left(\chi + \frac{(\zeta_i-\mu_i)^2}{\Sigma_{ii}}\right)\left(\psi + \frac{\gamma_i^2}{\Sigma_{ii}}\right)}^{\frac{1}{2}-\lambda}} \right)^{-1}$$

with $\boldsymbol{\zeta} \equiv (\zeta_1, \ldots, \zeta_n)$, $\zeta_i \equiv F_{GH}^{-1}(u_i)$ and constants given by

$$k \equiv \frac{(\sqrt{\psi\chi})^{-\lambda}\psi^\lambda(\psi + \boldsymbol{\gamma}'\boldsymbol{\Sigma}^{-1}\boldsymbol{\gamma})^{\frac{n}{2}-\lambda}}{(2\pi)^{\frac{n}{2}}|\boldsymbol{\Sigma}|^{\frac{1}{2}}K_\lambda(\sqrt{\psi\chi})}, \quad k_i \equiv \frac{(\sqrt{\psi\chi})^{-\lambda}\psi^\lambda\left(\psi + \frac{\gamma_i^2}{\Sigma_{ii}}\right)^{\frac{1}{2}-\lambda}}{\sqrt{2\pi\Sigma_{ii}}K_\lambda(\sqrt{\psi\chi})}.$$

Note that the hyperbolic quantile function has to be approximated numerically which complicates the evaluation of the GH copula. Lentzas[29] also deals with different estimation methods (ML estimation, rank correlation ML, Monte Carlo rank correlation ML, simulated GMM and based on the EM algorithm) for the unknown parameters of a GH copula.

2.2.2 Generalized t-copulae

The t-copula is often chosen when a multivariate model with extreme dependence is needed. However, the use of the standard t-copula is often criticized due to its restriction of having only a single parameter for the degrees of freedom that may limit its capability to model the tail dependence structure in a multivariate case. This motivates the next two examples, the grouped t-copula and the IT-copula.

Example 2.6 (Grouped t-copula). In order to increase the flexibility of the popular t-copula, Daul et al.[10] and Demarta and McNeil[11] introduce the grouped t-copula. Their aim is to describe the dependence among risk factors of different classes. For a given partition of $\{1,\ldots,n\}$ into m subsets of sizes $s_1,\ldots,s_m$ with $s_1+\cdots+s_m=n$,

$$\mathbf{Y} \equiv (R_1 Z_1, \ldots, R_1 Z_{s_1}, R_2 Z_{s_1+1}, \ldots, R_2 Z_{s_1+s_2}, \ldots, R_m Z_n)'.$$

The random vector $(Y_1,\ldots,Y_{s_1})'$ has s_1-dimensional t-distribution with ν_1 degrees of freedom and, for $k = 1,\ldots,m-1$, $(Y_{s_1+\cdots+s_k+1},\ldots,Y_{s_1+\cdots+s_{k+1}})'$ has s_{k+1}-dimensional t-distribution with ν_{k+1} degrees of freedom. Finally, the grouped t-copula is the distribution function of the random vector

$$\mathbf{U} = (t_{\nu_1}(Y_1),\ldots,t_{\nu_1}(Y_{s_1}),t_{\nu_2}(Y_{s_1}+1),\ldots,t_{\nu_2}(Y_{s_1+s_2}),\ldots,t_{\nu_m}(Y_n))'$$

where again t_{ν_i} denotes the distribution function of a classical Student's t-distribution with ν_i degrees of freedom. Daul et al.[10] also show how to estimate the unknown parameters and give some application to credit risk modeling.

Example 2.7 (The IT-copula). Instead of grouping variables a priori in such a way that each group has a standard t-copula with its specific degrees of freedom parameter, both Luo and Shevchenko[31] and Barnett et al.[6] propose the so-called "individual" t-copula, or IT-copula, where each group boils down to one variable or risk factor only. Starting from the stochastic random vector $\mathbf{X} \equiv (R_1 Z_1,\ldots,R_n Z_n)'$ with R_i, Z_i as in Example 2.3, the IT-copula is defined as the cumulative distribution function of the random vector $\mathbf{U} \equiv (t_{\nu_1}(X_1),\ldots,t_{\nu_n}(X_n))'$. Clearly, both t-copula and grouped t-copula are special cases of this construction. Luo and Shevchenko[31] derive the corresponding explicit integral representation with $\overline{\nu} \equiv (\nu_1,\ldots,\nu_n)$

$$C(\mathbf{u}; \overline{\nu}, \mathbf{\Sigma}) = \int_0^1 \Phi_n(z_1(u_1,s),\ldots,z_n(u_n,s))\,ds \qquad (2.7)$$

with $z_i(u_i,s) \equiv t_{\nu_i}(u_i)/G_{\nu_i}^{-1}(s)$, where $G_\nu^{-1}(x)$ corresponds to the distribution function of $\sqrt{\nu/S}$ for a χ_ν^2-variable S and Φ_n denotes the classical multivariate Gaussian distribution function. From (2.7), the density derives as

$$c(\mathbf{u}; \overline{\nu}, \mathbf{\Sigma}) = \frac{\int_0^1 \phi_n(z_1(u_1,s),\ldots,z_n(u_n,s)) \prod_{i=1}^n (G_{\nu_i}^{-1}(s))^{-1}\,ds}{\prod_{i=1}^n f_{\nu_i}(t_{\nu_i}^{-1}(u_i))}. \qquad (2.8)$$

Obviously, the multivariate copula density involves an additional one-dimensional integration which makes fitting this copula more computationally demanding than fitting a standard t-copula. For details on model calibration and application to risk quantification, we refer to Luo and Shevchenko[31] and Barnett et al.[6]

2.3 Archimedean Copulae and Generalizations

2.3.1 *Classical Archimedean copulae*

Let $\varphi\colon [0,1] \to [0,\infty]$ be a continuous, strictly decreasing and convex function with $\varphi(1) = 0$, $\varphi(0) \leq \infty$ and let $\varphi^{[-1]}$ be the so-called pseudo-inverse of φ defined by

$$\varphi^{[-1]}(t) \equiv \begin{cases} \varphi^{-1}(t) & 0 \leq t \leq \varphi(0), \\ 0 & \varphi(0) \leq t \leq \infty. \end{cases}$$

It can be shown (see, e.g., Nelsen[36]) that

$$C(u_1, u_2) = \varphi^{[-1]}(\varphi(u_1) + \varphi(u_2))$$

defines a class of bivariate copulae, the so-called Archimedean copulae. The function φ is called the (additive) generator of the copula. Furthermore, if $\varphi(0) = \infty$, the pseudo-inverse describes an ordinary inverse function (i.e., $\varphi^{[-1]} = \varphi^{-1}$) and in this case φ is known as a strict generator.

Given a strict generator $\varphi\colon [0,1] \to [0,\infty]$, bivariate Archimedean copulae can be extended to the n-dimensional case. For every $n \geq 2$, the function $C\colon [0,1]^n \to [0,1]$ defined as

$$C(\mathbf{u}) = \varphi^{-1}(\varphi(u_1) + \varphi(u_2) + \cdots + \varphi(u_n)) \qquad (2.9)$$

is an n-dimensional Archimedean copula if and only if φ^{-1} is completely monotonic on $\mathbb{R}_+$, i.e., if $\varphi^{-1} \in \mathcal{L}_\infty$ with

$$\mathcal{L}_m \equiv \{\phi\colon \mathbb{R}_+ \to [0,1] | \phi(0) = 1,\ \phi(\infty) = 0,\ (-1)^k \phi^{(k)}(t) \geq 0\ ,\ k \leq m\}.$$

The Gumbel copula is derived from the generator $\varphi(t) = (-\ln t)^\theta, \theta \geq 1$ and the Clayton copula is generated by

$$\varphi(t) = \frac{1}{\theta}(t^{-\theta} - 1), \quad \theta > 0. \qquad (2.10)$$

For an overview of further Archimedean copulae and the properties of the aforementioned ones, we refer the reader to the monographs of Nelsen[36] and Joe.[24]

2.3.2 *Non-exchangeable Archimedean copulae*

In order to increase flexibility and to allow for non-exchangeable dependence structures, several generalizations have emerged in the recent literature: A simple one — the so-called fully nested Archimedean (FNA) copulae — can be found in Joe[24] (p. 89), Whelan[46] and Savu and Trede,[39] and requires

$n-1$ generator functions $\varphi_1, \ldots, \varphi_{n-1}$ with $\varphi_1^{-1}, \ldots, \varphi_{n-1}^{-1} \in \mathcal{L}_\infty$ and $\varphi_{i+1} \circ \varphi_i^{-1}(t) = \varphi_{i+1}(\varphi_i^{-1}(t)) \in \mathcal{L}_\infty^*$ for

$$\mathcal{L}_n^* = \{\phi \colon \mathbb{R}_+ \to \mathbb{R}_+ | \phi(0) = 0, \phi(\infty) = \infty, (-1)^{k-1}\phi^{(k)}(t) \geq 0, k \leq n\}.$$

The structure of FNA n-copulae is rather simple: One first couples u_1 and u_2, then the copula of u_1 and u_2 with u_3 to form a new copula, which is coupled afterwards with u_4 and so on. Hence, the FNA four-copula is of the form

$$C(\mathbf{u}) = \varphi_3^{-1}[\varphi_3(\varphi_2^{-1}[\varphi_2(\varphi_1^{-1}[\varphi_1(u_1) + \varphi_1(u_2)]) + \varphi_2(u_3)]) + \varphi_3(u_4)]. \tag{2.11}$$

Figure 2.1 illustrates one possible FNA copula for dimension $n = 5$.

Alternatively, mixing ordinary Archimedean and FNA copulae, partially nested Archimedean (PNA) copulae may be used. Again, for ease of notation, we focus on the four-variate case:

$$\begin{aligned} C(\mathbf{u}) &= \varphi^{-1}[\varphi(\varphi_{12}^{-1}[\varphi_{12}(u_1) + \varphi_{12}(u_2)]) \\ &\quad + \varphi(\varphi_{34}^{-1}[\varphi_{34}(u_3) + \varphi_{34}(u_4)])]. \end{aligned} \tag{2.12}$$

Note that $\varphi, \varphi_{12}, \varphi_{34}$ are generators with $\varphi^{-1}, \varphi_{12}^{-1}, \varphi_{34}^{-1} \in \mathcal{L}_\infty$ and $\varphi \circ \varphi_{12}^{-1}, \varphi \circ \varphi_{34}^{-1} \in \mathcal{L}_\infty^*$. Obviously, one first couples the pairs u_1, u_2 and u_3, u_4 with distinct generators. The resulting copula pair is then coupled using a third generator φ (which in turn might be coupled with an additional variable u_5 using a fourth generator ψ for an extension to the five-dimensional case). Another possible structure of a PNA copula is illustrated in Fig. 2.2.

Third, copula C from (2.12) is also an example of a so-called hierarchical Archimedean (HA) copula. Borrowing the notation of Savu and Trede,[39] the basic idea of this approach is to build a hierarchy of Archimedean copulae with L different levels, indexed by $l = 1, \ldots, L$. At each level l, there are n_l distinct objects, indexed by $j = 1, \ldots, n_l$. In a first step (i.e., in level 1), the

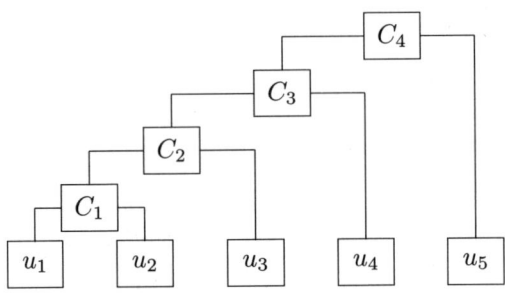

Figure 2.1. FNA copula for $n = 5$.

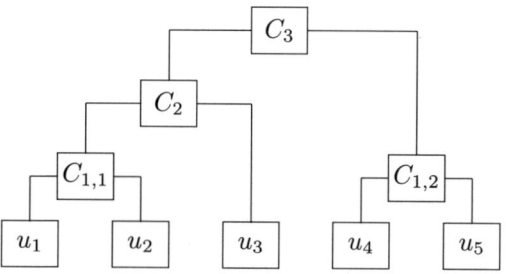

Figure 2.2. PNA copula for $n = 5$.

variables $u_1, \ldots, u_n$ are grouped into n_1 ordinary multivariate Archimedean copulae:

$$C_{1,j}(\mathbf{u}_{1,j}) = \varphi_{1,j}^{-1}\left(\sum_{\mathbf{u}_{1,j}} \varphi_{1,j}(\mathbf{u}_{1,j})\right), \quad j = 1, \ldots, n_1$$

with (possibly different) generators $\varphi_{1,j}$ and where $\mathbf{u}_{1,j}$ denotes the set of elements of $u_1, \ldots, u_n$ belonging to $C_{1,j}$. All copulae of the first level are again grouped into copulae at level $l = 2$. These copulae $C_{2,j}$ with generator function $\varphi_{2,j}$, $j = 1, \ldots, n_2$ are generalized Archimedean copulae, whose dependence structure is only of partial exchangeability and consists of copulae from the previous level (as elements), denoted by

$$C_{2,j}(\mathbf{C}_{2,j}) = \varphi_{2,j}^{-1}\left(\sum_{\mathbf{C}_{2,j}} \varphi_{2,j}(\mathbf{C}_{2,j})\right),$$

where $\mathbf{C}_{2,j}$ represents the set of all copulae from level $l = 1$ entering copula $C_{2,j}$. This procedure continues until only a single hierarchical Archimedean copula $C_{L,1}$ is achieved at level L. In order to ensure that $C_{L,1}$ is a proper copula, we have to proclaim that $\varphi_{l,j}^{-1} \in \mathcal{L}_\infty$ for $l = 1, \ldots, L$ and $j = 1, \ldots, n_l$, and that $\varphi_{l+1,i} \circ \varphi_{l,j}^{-1} \in \mathcal{L}_\infty^*$ for all $l = 1, \ldots, L$ and $j = 1, \ldots, n_l$, $i = 1, \ldots, n_{l+1}$ such that $C_{l,j} \in \mathbf{C}_{l+1,i}$. Moreover, a hierarchy is established if the number of copulae decreases at each level, if the top level contains only a single object and if at each level the dimensions of the copulae add up to n. Figure 2.3 displays the possible construction of a five-dimensional HA copula.

2.3.3 *Generalized multiplicative Archimedean copulae*

In this section, we focus on methods recently proposed by Morillas[35] and Liebscher.[30] Both approaches are based on a second functional representation of Archimedean copulae via so-called multiplicative generators (see

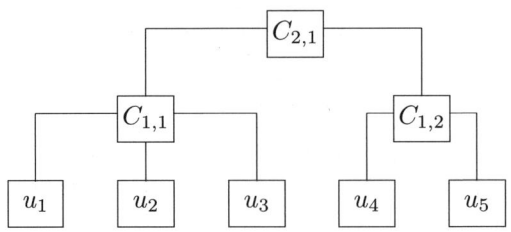

Figure 2.3. HA copula for $n = 5$.

Nelsen[36]). Setting $\vartheta(t) \equiv \exp(-\varphi(t))$ and $\vartheta^{[-1]}(t) \equiv \varphi^{[-1]}(-\ln t)$, Eq. (2.9) can be rewritten as

$$C(u_1, \ldots, u_n) = \vartheta^{[-1]}(\vartheta(u_1) \cdot \vartheta(u_2) \cdot \cdots \cdot \vartheta(u_n)). \qquad (2.13)$$

The function ϑ is called a multiplicative generator of C. Due to the relationship between φ and ϑ, the function $\vartheta: [0, 1] \to [0, 1]$ is continuous, strictly increasing and concave with $\vartheta(1) = 1$ and $\vartheta^{[-1]}(t) = 0$ if $0 \le t \le \vartheta(0)$ and $\vartheta^{[-1]}(t) = \vartheta^{-1}(t)$ if $\vartheta(0) \le t \le 1$.

Equation (2.13) can also be expressed using the independence copula $C^\perp(\mathbf{u}) = \prod_{i=1}^{n} u_i$:

$$C(u_1, \ldots, u_n) = \vartheta^{[-1]}(C^\perp(\vartheta(u_1), \ldots, \vartheta(u_n))).$$

Morillas[35] substitutes $C^\perp$ by an arbitrary n-copula C in order to obtain

$$C_\vartheta(u_1, \ldots, u_n) = \vartheta^{[-1]}(C(\vartheta(u_1), \vartheta(u_2), \ldots, \vartheta(u_n))) \qquad (2.14)$$

and proves that C_ϑ is an n-copula if $\vartheta^{[-1]}$ is *absolutely monotonic of order n* on $[0, 1]$, i.e. if $\vartheta^{[-1]}(t)$ satisfies $(\vartheta^{[-1]})^{(k)}(t) = \frac{d^k \vartheta^{[-1]}(t)}{dt^k} \ge 0$ for $k = 1, 2, \ldots, n$ and $t \in (0, 1)$.

Examples of generator functions are stated in Morillas.[35] Notice that not every generator given there is absolutely monotonic for arbitrary $n > 1$. As one can easily verify, the generator $\vartheta(t) = t^r/(2 - t^r), r \in (0, 1/3]$ (see Table 1, no. 9 in Morillas[35]) has no absolutely monotonic pseudo-inverse of order $n \ge 3$ because the third derivative of $\vartheta^{[-1]}$ becomes negative. Hence, this generator is suitable only for a construction of generalized bivariate copulae. For the basic properties of such Morillas copulae we refer to Morillas.[35]

Another way of generalizing Archimedean copulae is the method proposed by Liebscher.[30] He introduces the following copula representation:

$$C(u_1, \ldots, u_n) = \Psi\left(\frac{1}{m}\sum_{j=1}^{m} \psi_{j1}(u_1) \cdot \psi_{j2}(u_2) \cdot \cdots \cdot \psi_{jn}(u_n)\right), \qquad (2.15)$$

where Ψ and ψ_{jk}: $[0,1] \to [0,1]$ are functions satisfying the following conditions: First, it is assumed that $\Psi^{(n)}$ exists with $\Psi^{(k)}(u) \geq 0$ for $k = 1, 2, \ldots, n$ and $u \in [0, 1]$, and that $\Psi(0) = 0$. Second, ψ_{jk} is assumed to be differentiable and monotone increasing with $\psi_{jk}(0) = 0$ and $\psi_{jk}(1) = 1$ for all k, j. Third, Liebscher's construction requires that

$$\Psi\left(\frac{1}{m} \sum_{j=1}^{m} \psi_{jk}(v)\right) = v \quad \text{for } k = 1, 2, \ldots, n \text{ and } v \in [0, 1].$$

The three conditions guarantee that C as defined in (2.15) is actually a copula.

It is easily seen that the approaches of Morillas and Liebscher coincide for $m = 1$, $\vartheta^{[-1]} = \Psi$ in (2.15) and $C_\vartheta = C^\perp$ in (2.14).

Liebscher[30] also states a general method for deriving appropriate functions ψ_{jk}. Let h_{jk}: $[0,1] \to [0,1], j = 1, \ldots, m, k = 1, \ldots, n$ be a differentiable and bijective function such that $h'_{jk}(u) > 0$ for $u \in (0, 1)$, $h_{jk}(0) = 0$, $h_{jk}(1) = 1$ and $m \cdot u = \sum_{j=1}^{m} h_{jk}(u)$, $u \in [0, 1]$, $k = 1, \ldots, n$. Let $\psi = \Psi^{-1}$ be the differentiable inverse function of Ψ. An appropriate choice is setting $\psi_{jk}(u) = h_{jk}(\psi(u))$, since $\psi'_{jk}(u) = h'_{jk}(\psi(u)) \cdot \psi'(u) > 0$ for $j = 1, \ldots, m$ and $u \in [0, 1]$.

2.3.4 Koehler–Symanowski copulae

Just like Archimedean copulae, Koehler–Symanowski (KS) copulae admit closed-form representations. Although KS copulae are not Archimedean in general, the (Archimedean) Clayton copula with generator function given in (2.10) is included as a KS copula under certain parameter restrictions. More generally, Koehler and Symanowski[27] introduce a multivariate distribution as follows: For the index set $V = \{1, 2, \ldots, n\}$, let $\mathcal{V}$ denote the power set of V and $\mathcal{I} \equiv \{I \in \mathcal{V} \text{ with } |I| \geq 2\}$. Let further $\mathbf{X}$ denote an n-dimensional random vector with univariate marginal distributions $F_i(x_i), i \in V$. For all subsets $I \in \mathcal{I}$, let $\alpha_I \in \mathbb{R}_0^+$ and $\alpha_i \in \mathbb{R}_0^+$ for all $i \in V$ such that $\alpha_{i+} = \alpha_i + \sum_{I \in \mathcal{I}} \alpha_I > 0$ for $i \in I$. Then the common distribution function F is defined by

$$F(\boldsymbol{x}) = \frac{\prod_{i \in V} F_i(x_i)}{\prod_{I \in \mathcal{I}} \left[\sum_{i \in I} \prod_{j \in I, j \neq i} F_j(x_j)^{\alpha_{j+}} - (|I| - 1) \prod_{i \in I} F_i(x_i)^{\alpha_{i+}}\right]^{\alpha_I}}.$$

The terms $K_I = \sum_{i \in I} \prod_{j \in I, j \neq i} F_j(x_j)^{\alpha_{j+}} - (|I|-1) \prod_{i \in I} F_i(x_i)^{\alpha_{i+}}$ are called association terms. Moreover, Koehler and Symanowski[27] showed that the joint density function exists if the marginal density functions f_i exist for all $i \in V$. Due to the design of the Koehler–Symanowski (KS) distribution, the corresponding copula has a similar functional form: setting $u_i = F_i(x_i)$ for all $i \in V$, the KS copula is

$$C(u_1, \ldots, u_d) = \frac{\prod_{i \in V} u_i}{\prod_{I \in \mathcal{I}} \left[\sum_{i \in I} \prod_{j \in I, j \neq i} u_j^{\alpha_{j+}} - (|I|-1) \prod_{i \in I} u_i^{\alpha_{i+}} \right]^{\alpha_I}}.$$

In contrast to the cumulative distribution function, the functional representation of the density is quite complicated due to complex factors with additive components. Koehler and Symanowski[27] gave an explicit formula for the special case of a so-called KS(2) distribution (see also Caputo[8]), where all parameters α_I are set equal to zero for $|I| > 2$. The corresponding copula is termed a KS(2) copula. Assuming that $\alpha_{ij} \equiv \alpha_{ji} \geq 0$ for all $(i,j) \in V \times V$ and $\alpha_{i+} = \alpha_{i1} + \alpha_{i2} + \cdots + \alpha_{in} > 0$ for all $i \in V$, the KS(2) copula simplifies to

$$C(u_1, u_2, \ldots, u_n) = \prod_{i=1}^n u_i \prod_{i<j} K_{ij}^{-\alpha_{ij}} \qquad (2.16)$$

with $K_{ij} \equiv u_i^{1/\alpha_{i+}} + u_j^{1/\alpha_{j+}} - u_i^{1/\alpha_{i+}} u_j^{1/\alpha_{j+}} = K_{ji}$.

Palmitesta and Provasi[37] apply the KS(2) copula to financial return data. They also argue that this copula has the ability to model complex dependence structures among subsets of marginal distribution but they do not present any goodness-of-fit measure or comparison with other copulae. In contrast, Fischer et al.[18] show that the goodness-of-fit can be improved considerably if a four-dimensional association term is included as well.

2.4 Combinations of Arbitrary Copulae into a New One

Morillas' construction scheme in (2.14) can be seen as a distortion of a single but arbitrary copula. Similarly, one might be interested in constructing a new copula C from d given copulae $C_1, \ldots, C_d$ in order to increase flexibility and/or introduce asymmetry. A simple way is to consider linear combinations, where the weights sum up to one, i.e.,

$$C(\mathbf{u}) \equiv \alpha_1 C_1(\mathbf{u}) + \cdots + \alpha_d C_d(\mathbf{u}) \quad \text{with} \quad \alpha_1 + \cdots + \alpha_d = 1. \qquad (2.17)$$

Putting things differently, the copula in (2.17) results from a weighted arithmetic mean of $C_1, \ldots, C_d$. Klein et al.,[26] more generally, deal with conditions on the copulae such that the weighted Hölder mean of two copulae is again a copula. Recently, Liebscher[30] has discussed products of n-copulae of the form

$$C(u_1, \ldots, u_n) = \prod_{j=1}^{d} C_j(g_{j1}(u_1), \ldots, g_{jn}(u_n))$$

with a set of $d \cdot n$ admissible functions $g_{11}, \ldots, g_{1n}, \ldots, g_{d1}, \ldots, g_{dn}$, each of which, being bijective, monotonously increasing or identically equal to 1, satisfy

$$\prod_{j=1}^{d} g_{ji}(v) = v, \quad i = 1, \ldots, n. \tag{2.18}$$

Note that (2.18) reduces to $g_{1i}(v) = v$ for $d = 1$ and $i = 1, \ldots, n$, and C is recovered. In accordance with Liebscher,[30] the possible choices are

$$g_{ji}(v) \equiv v^{\theta_{ji}} \text{ with } \theta_{ji} > 0 \quad \text{and} \quad \sum_{j=1}^{d} \theta_{ji} = 1 \text{ for } i = 1, \ldots, n$$

or, for $\theta > 0$ and $\alpha \in (0, 1)$,

$$g_{1i}(v) \equiv f(v), \quad g_{2i}(v) \equiv \frac{v}{f(v)}, \quad f(v) = \left(\frac{1 - e^{-\theta_i v}}{1 - e^{-\theta_i}}\right)^{\alpha}.$$

Finally, Fischer and Köck[17] develop a construction scheme which includes both Morillas copulae in (2.14) and Liebscher copulae in (2.15) as special cases. The key idea of Morillas[35] is to replace the independence copula (which is implicitly assumed within the multiplicative Archimedean framework) with an arbitrary copula C and to prove that the new function is a copula, too. Taking a closer look at (2.15), one might be tempted to replace the product with an arbitrary n-copula in order to extend Liebscher's proposal. Assuming that Ψ is absolutely monotonic of order d and ψ_{ij} is differentiable and monotone increasing with $\psi_{ij}(0) = 0$, $\psi_{ij}(1) = 1$ and that $\Psi(\frac{1}{m} \sum_{j=1}^{m} \psi_{jk}(v)) = v$, and $C_1, \ldots, C_m$ are arbitrary copulae with existing copula densities, Fischer and Köck[17] showed that

$$C(u_1, \ldots, u_n) = \Psi\left(\frac{1}{m} \sum_{j=1}^{m} C_j(\psi_{j1}(u_1), \ldots, \psi_{jn}(u_n))\right), \quad m \geq 1, \; n \geq 2$$

is again a copula.

2.5 Summary

Whereas copulae seem to be well-explored in the bivariate case, there are several open issues in the multivariate setting. In particular, the construction of multivariate copula models which allow us to rebuild various types of dependencies and admit closed-form representations (at least for the copula density) in order to perform fast and easy parameter estimation is a challenging task. Within this chapter, we reviewed both popular copula classes and different construction schemes which emerged in the previous literature. Apart from pair-copula constructions, which are the focus of this book, special emphasis was put on elliptical copulae and selected generalizations as well as on generalized Archimedean copulae.

References

1. Abdous B., Fougéres A. and Ghoudi K. (2005). Extreme behaviour for bivariate elliptical distributions. *The Canadian Journal of Statistics*, 33(2):317–334.
2. Andersen L. and Sidenius J. (2004). Extensions to the Gaussian copula: Random recovery and random factor loadings. *The Journal of Credit Risk*, 1(1):29–70.
3. Arslan O. (2004). Family of multivariate generalized t distributions. *Journal of Multivariate Analysis*, 89(2):329–337.
4. Barndorff-Nielsen O.E. (1977). Exponentially decreasing distributions for the logarithm of particle size. *Proceedings of the Royal Society of London A*, 353:401–419.
5. Barndorff-Nielsen O.E. (1978). Hyperbolic distributions and distributions on hyperbolae. *Scandinavian Journal of Statistics*, 5:151–157.
6. Barnett J., Kreps R.E., Major J.A. and Venter G.G. (2009). Multivariate copulas for financial modeling. *Variance*, 1(1):103–119.
7. Breymann W., Dias A. and Embrechts P. (2003). Dependence structures for multivariate high-frequency data in finance. *Quantitative Finance*, 1:1–14.
8. Caputo A. (1998). Some properties of the family of Koehler Symanowski distributions. Working Paper No. 103, Ludwig Maximilian University Munich, available at http://epub.ub.uni-muenchen.de/1493/1/paper_103.pdf.
9. Cherubini U., Luciano E. and Vecchiato W. (2004). *Copula Methods in Finance*. Wiley Finance, Chichester.
10. Daul S., De Giorgi E., Lindskog F. and McNeil A. (2003). The grouped t-copula with an application to credit risk. *RISK*, 16(11):73–76.
11. Demarta S. and McNeil A. (2005). The t-copula and related copulas. *International Statistical Review*, 73(1):111–129.
12. Drouet D. and Kotz S. (2001). *Correlation and Dependence*. Imperial College Press, Singapore.
13. Embrechts P., McNeil A. and Straumann D. (1999). Correlation: Pitfalls and alternatives. *Risk*, 5:69–71.
14. Embrechts P., McNeil A. and Straumann D. (2002). Correlation and dependence in risk management: Properties and pitfalls. In *Risk Management: Value at Risk and Beyond*, M.A.H. Dempster (ed.), Cambridge University Press, Cambridge, pp. 176–223.

15. Embrechts P., Lindskog F. and McNeil A. (2003). Modelling dependence with copulas and applications to risk management. In *Handbook of Heavy Tailed Distributions in Finance*, S.T. Rachev (ed.), Elsevier, Amsterdam, pp. 329–384.
16. Fang K., Kotz S. and Ng K. (1990). *Symmetric Multivariate and Related Distributions*. Monographs on Statistics and Applied Probability, Vol. 36. Chapman & Hall, London.
17. Fischer M. and Köck C. (2009). Constructing and generalizing given multivariate copulas: A unifying approach. *Statistics*, in revision.
18. Fischer M., Köck C., Schlüter S. and Weigert F. (2009). Multivariate copula models at work. *Quantitative Finance*, 9(7):839–854.
19. Frahm G., Junker M. and Szimayer A. (2003). Elliptical copulas: Applicability and limitations. *Statistics & Probability Letters*, 63(3):275–286.
20. Genest C. and Neslehova J. (2007). A primer on copulas for count data. *Astin Bulletin*, 37(2):475–515.
21. Genest C. and Favre A. (2007). Everything you always wanted to know about copula modeling but were afraid to ask. *Journal of Hydrologic Engineering*, 12:347–368.
22. Gómez E., Gómez-Villegas M.A. and Martín M.A. (1998). A multivariate generalization of the power exponential family of distributions. *Communications in Statistics — Theory and Methods*, 27(3):589–600.
23. Hult H. and Lindskog P. (2002). Multivariate extremes, aggregation and dependence in elliptical distribution. *Advances in Applied Probability*, 34(3):587–609.
24. Joe H. (1997). *Multivariate Models and Dependence Concepts*, Monographs on Statistics and Applied Probability, Vol. 37. Chapman & Hall, London.
25. Jørgensen H. (1982). *Statistical Properties of the Generalized Inverse Gaussian Distribution*. Lecture Notes in Statistics, Vol. 9. Springer, New York.
26. Klein I., Fischer M. and Pleier T. (2009). Some results on weighted power mean copulas. Working Paper, University of Erlangen-Nuremberg.
27. Koehler K.J. and Symanowski J.T. (1995). Constructing multivariate distributions with specific marginal distributions. *Journal of Multivariate Analysis*, 55: 261–282.
28. Kole E., Koedijk C.G. and Verbeek M. (2003). Stress testing with Student's t dependence. ERIM Report Series 2003-056FA.
29. Lentzas G. (2008). Multivariate models with generalized hyperbolic margins and generalized hyperbolic copula. Working Paper, submitted.
30. Liebscher E. (2008). Construction of asymmetric multivariate copulas. *Journal of Multivariate Analysis*, 99:2234–2250.
31. Luo X. and Shevchenko P.V. (2010). The t copula with multiple parameters of degrees of freedom: Bivariate characteristics and application to risk management. *Quantitative Finance*. In press.
32. McNeil A.J., Frey R. and Embrechts P. (2005). *Quantitative Risk Management: Concepts, Techniques and Tools*, Princeton Series in Finance. Princeton University Press, Princeton, NJ.
33. Mendes B.V.M. and Arslan O. (2003). A new family of multivariate skew distributions based on the GT copula. In *Proceedings of the First Brazilian Conference in Statistical Modeling in Finance and Insurance*, Ubatuba.
34. Mikosch T. (2006). Copulas: Tales and facts. *Extremes*, 9(1):3–20.
35. Morillas P.M. (2005). A method to obtain new copulas from a given one. *Metrika*, 61:169–184.
36. Nelsen R.B. (2006). *An Introduction to Copulas*. Springer Series in Statistics. Springer, Berlin.

37. Palmitesta P. and Provasi C. (2005). Aggregation of dependent risks using the Koehler–Symanowski copula function. *Computational Economics*, 25:189–205.
38. Prause K. (1999). The generalized hyperbolic model: Estimation, financial derivatives and risk measures. PhD thesis, University of Freiburg, Freiburg.
39. Savu C. and Trede M. (2006). Hierarchical Archimedean copulas. In *International Conference on High Frequency Finance*, Konstanz, Germany, May.
40. Schmidt R. (2003). Credit risk modelling and estimation via elliptical copulae. In: *Credit Risk: Measurement, Evaluation and Management (Contributions to Economics)*, G. Bol, G. Nakhaeizadeh and S.T. Rachev (eds.), Physica-Verlag, Heidelberg, pp. 267–289.
41. Schmidt R. (2003). Dependencies of extreme events in finance: Modelling, statistics and data analysis PhD thesis, University of Ulm, Ulm.
42. Schmidt R., Hrycej T. and Stützle E. (2006). Multidimensional data modelling with generalized hyperbolic distributions. *Computational Statistics and Data Analysis*, 50:2065–2096.
43. Sklar A. (1959). Fonctions de répartition á n dimensions et leurs marges. *Publications de l'Institut de Statistique de l'Université de Paris*, 8:229–231.
44. Venter G.G. (2002). Tails of copulas. *Proceedings of the Casualty Actuarial Society*, 89:68–113.
45. Venter G.G. (2003). Fit to a t-estimation, application and limitations of the t-copula. ASTIN Colloquium, Berlin, Germany, available at http://www.actuaries.org/ASTIN/Colloquia/Berlin/venter1.pdf.
46. Whelan N. (2004). Sampling from Archimedean copulas. *Quantitative Finance*, 4: 339–352.

CHAPTER 3

Vines Arise

Roger M. Cooke,* Harry Joe† and Kjersti Aas‡
*Resources for the Future, and Department of Mathematics
Delft University of Technology
Cooke@Rff.org
†Department of Statistics, University of British Columbia
Harry.Joe@ubc.ca
‡Norwegian Computing Centre
Kjersti.Aas@nr.no

An introduction to the main idea of vines as graphical models is presented, along with various notation and graphs for representing vines. The early history of vines is summarized, together with the motivation for their construction. The relation to compatibility of subsets of marginal distributions is given to provide some intuition. Important properties and applications of vines are included.

3.1	Introduction	38
3.2	Regular Vines	39
3.3	Vine Types	43
	3.3.1 Vine copula or pair-copula construction	43
	3.3.2 Partial correlation vine	46
3.4	Historical Origins	50
3.5	Compatibility of Marginal Distributions	52
3.6	Sampling	55
	3.6.1 Sampling a D-vine	55
	3.6.2 Sampling an arbitrary regular vine	56
	3.6.3 Density approach sampling	57
3.7	Parametric Inference for a Specific Pair-Copula Construction	58
	3.7.1 Inference for a C-vine	59
	3.7.2 Inference for a D-vine	61

3.8 Model Inference 62
    3.8.1 Sequential selection 63
    3.8.2 Information-based model inference 64
3.9 Applications 67
    3.9.1 Multivariate data analysis 67
    3.9.2 Non-parametric Bayesian belief nets 68
References 69

3.1 Introduction

A vine is a graphical tool for labeling constraints in high-dimensional distributions. A regular vine is a special case in which all constraints are two-dimensional or conditional two-dimensional. Regular vines generalize trees, and are themselves specializations of something called Cantor trees.[5] Combined with copulae, regular vines have proven to be a flexible tool in high-dimensional dependence modeling. Copulae[27,45] are multivariate distributions with uniform univariate margins. Representing a joint distribution as univariate margins plus copulae allows us to separate the problems of estimating univariate distributions from those of estimating dependence. This is handy inasmuch as univariate distributions in many cases can be adequately estimated from data, whereas dependence information is rough-hewn, involving summary indicators and judgments.[3,33] Whereas the number of parametric multivariate copula families with flexible dependence is limited, there are many parametric families of bivariate copulae. Regular vines owe their increasing popularity to the fact that they leverage from bivariate copulae and enable extensions to arbitrary dimensions. Sampling theory and estimation theory for regular vines are well-developed,[2,39] and model inferential methods are being developed.[2,36,37] Regular vines have proven useful in other problems such as (constrained) sampling of correlation matrices,[28,40,41] building non-parametric continuous Bayesian belief nets,[17,18] and characterizing the set of rank correlation matrices.[29]

This chapter traces the historical development of vines and summarizes their most important properties. We focus on formulating the main results and indicating their role in the development; for proofs the reader is referred to the original articles. Section 3.2 gives precise definitions while Section 3.3 describes different types of vines. Section 3.4 on historical origins gives an informal rendering of the main ideas and Section 3.5 makes the links to the compatibility of marginal distributions. Sections 3.6–3.9 treat sampling, model inference and applications, respectively.

3.2 Regular Vines

Graphical models called *vines* were introduced in Cooke,[9] Bedford and Cooke[5] and Kurowicka and Cooke.[34] A vine $\mathcal{V}$ on n variables is a nested set of connected trees $\mathcal{V} = \{T_1, \ldots, T_{n-1}\}$ where the edges of tree j are the nodes of tree $j+1$, $j = 1, \ldots, n-2$. A *regular vine* on n variables is a vine in which two edges in tree j are joined by an edge in tree $j+1$ only if these edges share a common node, $j = 1, \ldots, n-2$. The formal definitions follow (based on Section 4.4.1 of Kurowicka and Cooke[36]).

Definition 3.1 (Regular vine). $\mathcal{V}$ is a regular vine on n elements with $E(\mathcal{V}) = E_1 \cup \cdots \cup E_{n-1}$ denoting the set of edges of $\mathcal{V}$ if

1. $\mathcal{V} = \{T_1, \ldots, T_{n-1}\}$,
2. T_1 is a connected tree with nodes $N_1 = \{1, \ldots, n\}$, and edges E_1; for $i = 2, \ldots, n-1$, T_i is a tree with nodes $N_i = E_{i-1}$,
3. **(proximity)** for $i = 2, \ldots, n-1$, $\{a, b\} \in E_i, \#(a \triangle b) = 2$ where $\triangle$ denotes the symmetric difference operator and $\#$ denotes the cardinality of a set.

An edge in tree T_j is an unordered pair of nodes of T_j or, equivalently, an unordered pair of edges of T_{j-1}. By definition, the *order* of an edge in tree T_j is $j-1$, $j = 1, \ldots, n-1$. The degree of a node is the number of edges attached to that node. A regular vine is called a *canonical* or *C-vine* if each tree T_i has a unique node of degree $n-i$, and hence has the maximum degree. A regular vine is called a *D-vine* if all nodes in T_1 have degrees no higher than 2 (see Figs. 3.2 and 3.3).[a]

The constraint, conditioning and the conditioned set of an edge are defined as follows:

Definition 3.2.

1. For $e \in E_i, i \leq n-1$, the **constraint set** associated with e is the **complete union** U_e^* of e, that is, the subset of $\{1, \ldots, n\}$ reachable from e by the membership relation.

---

[a]The term *canonical vine* first appeared in Bedford and Cooke,[4] with the abbreviation of C-vine appearing in Kurowicka and Cooke;[36] the term D-vine first appeared in Kurowicka and Cooke.[36,38] The designation "D" has nothing to recommend it, beyond being the letter following "C" but it is linked to "drawable" on p. 93 of Kurowicka and Cooke[36] (the suggestion that D-vine is an irreverent pun is unfounded).

2. For $i = 1, \ldots, n-1, e \in E_i$, if $e = \{j, k\}$ then the **conditioning set** associated with e is

$$D_e = U_j^* \cap U_k^*$$

and the **conditioned set** associated with e is

$$\{C_{e,j}, C_{e,k}\} = \{U_j^* \setminus D_e, U_k^* \setminus D_e\}.$$

Note that for $e \in E_1$, the conditioning set is empty. One can see that the order of an edge is the cardinality of its conditioning set. For $e \in E_i, i \leq n-1, e = \{j, k\}$ we have $U_e^* = U_j^* \cup U_k^*$.

Figure 3.1 shows a regular vine (left) and a non-regular vine (right). Figure 3.2 shows a D-vine on five variables with the constraint sets added. Conditioning variables are shown to the right of "|", conditioned variables to the left. The trees at each echelon are drawn in a different style. Figure 3.3 shows similar information for the C-vine. Although the D-vine looks simpler, in many ways the C-vine is simpler mathematically. Compare Algorithms 3.1 and 3.2 in Section 3.7 for maximum likelihood estimation.

The following propositions of regular vines are proved in Refs. 5, 34 and 38:

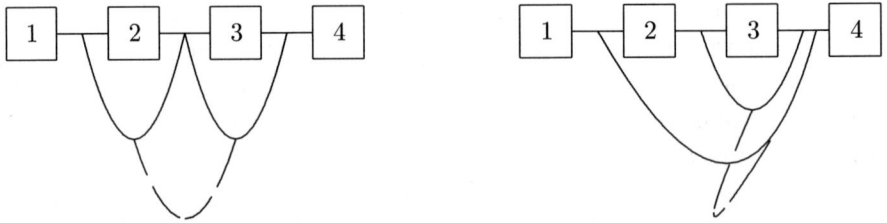

Figure 3.1. A regular (left) and a non-regular (right) vine on four variables.

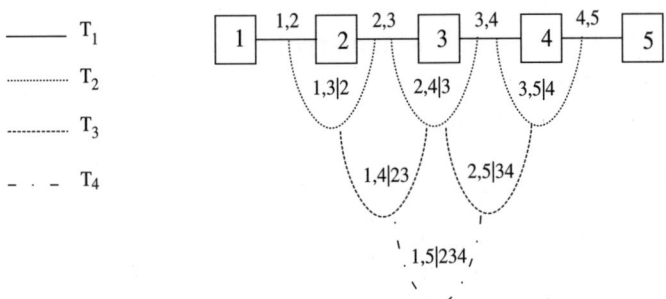

Figure 3.2. A D-vine on five variables with constraint sets.

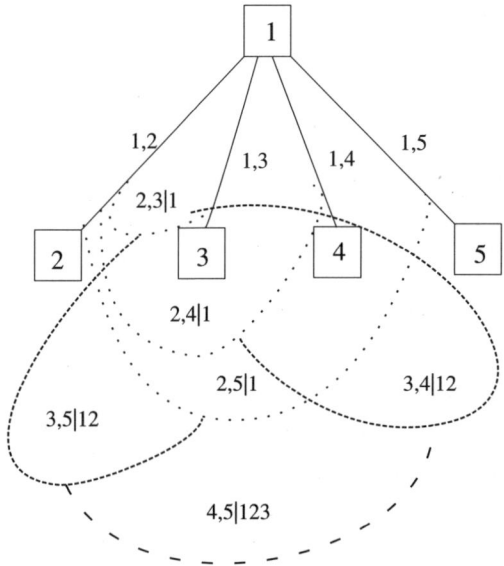

Figure 3.3. A C-vine on five variables with constraint sets.

Proposition 3.1. *Let $\mathcal{V} = \{T_1, \ldots, T_{n-1}\}$ be a regular vine, then*

(1) *the number of edges is $n(n-1)/2$,*
(2) *each conditioned set is a doubleton, each pair of variables occurs exactly once as a conditioned set,*
(3) *if two edges have the same conditioning set, then they are the same edge.*

Definition 3.3 (m-child; m-descendant). If node e is an element of node f, we say that e is an **m-child** of f; similarly, if e is reachable from f via the membership relation: $e \in e_1 \in \cdots \in f$, we say that e is an **m-descendant** of f.

Proposition 3.2. *For any node K of order $k > 0$ in a regular vine, if variable i is a member of the conditioned set of K, then i is a member of the conditioned set of exactly one of the m-children of K, and the conditioning set of an m-child of K is a subset of the conditioning set of K.*

The search for an optimal vine requires a method for enumerating and searching all vines. The number of regular vines grows very quickly. A closed formula for the number of regular vines on n elements was found in Morales-Nápoles et al.[44]:

Theorem 3.1.

(1) *For any regular vine on $n-1$ elements, the number of regular n-dimensional vines which extend this vine is 2^{n-3}.*
(2) *There are $\binom{n}{2} \times (n-2)! \times 2^{(n-2)(n-3)/2}$ labeled regular vines in total.*

Note that the number of extensions of a regular vine does not depend on the vine itself.

From Kurowicka and Cooke[36] (see also Chapter 7 in this volume), we have that for $n = 3$, all vines are in the same equivalence class, and for $n = 4$, all regular vines are either C-vines or D-vines. For $n \geq 5$, there are many vines that are neither C-vines nor D-vines. However, the C-vines and D-vines are boundary cases of the possible vines. An extension to non-regular vines is presented in Bedford and Cooke.[5]

We conclude this subsection with some examples to illustrate the notation. The examples consist of a C-vine for $n = 3$, a general C-vine, a general D-vine, a D-vine for $n = 4$, and a vine for $n = 5$ that is neither a C-vine nor a D-vine.

For a C-vine with $n = 3$, in T_1, $N_1 = \{1,2,3\}$ and $E_1 = \{\{1,2\};\{1,3\}\} = \{1,2;1,3\} = \{12;13\}$; then in T_2, $N_2 = E_1$, and $E_2 = \{[\{1,2\};\{1,3\}]\} = \{2,3|1\} = \{23|1\}$. The shorthand notation with fewer commas and braces is used for simplicity. For the edge $e = 23|1$ in T_2, $U_j^* = \{1,2\}$, $U_k^* = \{1,3\}$, the conditioning set is $D_e = \{1,2\} \cap \{1,3\} = \{1\}$, $C_{e,j} = \{1,2\}\setminus\{1\} = \{2\}$, $C_{e,k} = \{1,3\}\setminus\{1\} = \{3\}$, and the conditioned set is $C_{e,j} \cup C_{e,k} = \{2,3\}$.

For a general C-vine on n variables with standard indexing, $E_1 = \{1,i : i = 2,\ldots,n\}$, $E_2 = \{2,i|1 : i = 3,\ldots,n\}$, $\ldots$, $E_\ell = \{\ell,i|1,\ldots,i-1 : i = \ell+1,\ldots,n\}$, $\ldots$, $E_{n-1} = \{n-1,n|1,\ldots,n-2\}$, $T_1 = \{1,2,\ldots,n\}$ and $T_\ell = E_{\ell-1}$ for $\ell = 2,\ldots,n-1\}$. For an edge $e = [i_1,i_2|1,\ldots,i_1-1]$ with $1 \leq i_1 < i_2 \leq n$, the conditioning set is $D_e = \{1,\ldots,i_1-1\}$ and the conditioned set is $\{i_1,i_2\}$. If the indices $\{1,\ldots,n\}$ are permuted, the result is still a C-vine, since the C-vine is characterized by the degrees of the nodes for $T_1,\ldots,T_{n-1}$.

For a general D-vine on n variables with standard indexing, $E_1 = \{i,i+1 : i = 1,\ldots,n-1\}$, $E_2 = \{i,i+2|i+1 : i = 1,\ldots,n-2\}$, $\ldots$, $E_\ell = \{i,i+\ell|i+1,\ldots,i+\ell-1 : i = 1,\ldots,n-\ell\}$, $\ldots$, $E_{n-1} = \{1,n|2,\ldots,n-1\}$, $T_1 = \{1,2,\ldots,n\}$ and $T_\ell = E_{\ell-1}$ for $\ell = 2,\ldots,n-1\}$. For an edge $e = [i_1,i_2|i_1+1,\ldots,i_2-1]$ with $1 \leq i_1 < i_2 \leq n$, the conditioning set is

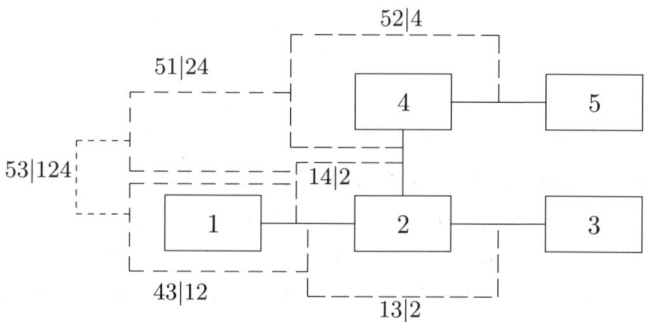

Figure 3.4. A regular vine on five variables which is neither a C-vine nor a D-vine with constraint sets.

$D_e = \{i_1 + 1, \ldots, i_2 - 1\}$ and the conditioned set is $\{i_1, i_2\}$. If the indices $\{1, \ldots, n\}$ are permuted, the result is still a D-vine.

Specific details in shorthand notation for the D-vine with $n = 4$ are: $N_1 = \{1, 2, 3, 4\}$, $E_1 = \{12; 23, 34\}$; then in T_2, $N_2 = E_1$ and $E_2 = \{13|2; 24|3\}$; finally, in T_3, $N_3 = E_2$ and $E_3 = \{14|23\}$. For the edge in T_3, the conditioning set is $D_e = \{2, 3\}$ and the conditioned set is $\{1, 4\}$.

An example of a five-dimensional regular vine that is neither a C-vine nor a D-vine is shown in Fig. 3.4. $E_1 = \{12; 23; 24; 45\}$, $E_2 = \{13|2; 14|2; 25|4\}$, $E_3 = \{34|12; 15|24\}$, $E_4 = \{35|124\}$.

3.3 Vine Types

Two main types of regular vines have been treated in the literature: vine copulae and partial correlation vine representations. Vine copulae or pair-copula constructions are obtained by assigning a bivariate copula to each edge in the vine. Similarly, a partial correlation vine representation of a correlation matrix is obtained by assigning a partial correlation to each edge in the vine. In this section, the two types of specifications are discussed.

3.3.1 Vine copula or pair-copula construction

A *bivariate copula vine specification* is called a *pair-copula construction*[1,2] or a *vine copula* (Section 4.4.2 of Kurowicka and Cooke[36]). It is obtained by assigning a bivariate copula C_e to each edge e in the union $E(\mathcal{V}) = E_1 \cup \cdots \cup E_{n-1}$ of the vine defined in the preceding subsection. The set of $\binom{n}{2}$ copulae is denoted by B. The elements of B can be chosen independently of each other (as long as they are bivariate copulae); this follows from Bedford and Cooke.[4]

In general, the form of the joint density of a regular vine copula with margins $F_1, \ldots, F_n$ is given by the following theorem:

Theorem 3.2 (Bedford and Cooke[4]). *Let $\mathcal{V} = (T_1, \ldots, T_{n-1})$ be a regular vine on n elements. For an edge $e \in E(\mathcal{V})$ with conditioned elements e_1, e_2 and conditioning set D_e, let the conditional copula and copula density be $C_{e_1,e_2|D_e}$ and $c_{e_1,e_2|D_e}$, respectively. Let the marginal distributions F_i with densities $f_i, i = 1, \ldots, n$ be given. Then the vine-dependent distribution is uniquely determined, and has a density given by*

$$f_{1\cdots n} = f_1 \cdots f_n \prod_{e \in E(\mathcal{V})} c_{e_1,e_2|D_e}(F_{e_1|D_e}, F_{e_2|D_e}). \tag{3.1}$$

Equation (3.1) shows that vine copulae have closed-form densities when $F_1, \ldots, F_n$ and the bivariate copulae in B are differentiable.

Note that C_e is a marginal bivariate copula for edges in T_1 and C_e is a conditional bivariate copula for edges in $T_2, \ldots, T_{n-1}$. For a C-vine, the set of bivariate copulae is denoted as $B = \{C_{i_1 i_2 | 1, \ldots, i_1 - 1} : 1 \leq i_1 < i_2 \leq n\} = \{C_{12}; \ldots; C_{1n}; C_{23|1}; \ldots; C_{2n|1}; \ldots, C_{n-1,n|1,\ldots,n-2}\}$. For a D-vine, the set of bivariate copulae is $B = \{C_{i_1 i_2 | i_1+1, \ldots, i_2-1} : 1 \leq i_1 < i_2 \leq n\} = \{C_{12}; \ldots; C_{n-1,n}; C_{13|2}; \ldots; C_{n-2,n|n-1}; \ldots, C_{1,n|2,\ldots,n-1}\}$. For the regular vine in Section 3.2 that is not a C-vine or a D-vine, the set of bivariate copulae is: $B = \{C_{12}; C_{23}; C_{24}; C_{45}; C_{13|2}; C_{14|2}; C_{25|4}; C_{34|12}; C_{15|24}; C_{35|124}\}$.

For applications, univariate margins $F_1, \ldots, F_n$ are specified or estimated, as well as the marginal or conditional copulae in B. The resulting multivariate distribution in the Fréchet class $\mathcal{F}(F_1, \ldots, F_n)$ has a form that can be shown recursively. We show the results for a C-vine with $n = 3$ and and a D-vine with $n = 4$.

First note that assuming F_1, F_2, C_{12} are differentiable with respective densities f_1, f_2, c_{12}, then $F_{12} = C_{12}(F_1, F_2)$ has density $f_{12} = c_{12}(F_1, F_2) f_1 f_2$ and conditional density $f_{2|1} = f_{12}/f_1 = c_{12}(F_1, F_2) f_2$.

For the C-vine with $n = 3$, the trivariate distribution comes from the specification $\{F_1, F_2, F_3, C_{12}, C_{13}, C_{23|1}\}$. The $(1,2)$ and $(2,3)$ margins are $F_{12} = C_{12}(F_1, F_2)$ and $F_{13} = C_{13}(F_1, F_3)$, from which conditional distribution $F_{2|1}, F_{3|1}$ can be obtained; then

$$F_{123}(x_1, x_2, x_3) = \int_{-\infty}^{x_1} C_{23|1}(F_{2|1}(x_2|z), F_{3|1}(x_3|z)) dF_1(z). \tag{3.2}$$

If F_i are differentiable with respective densities f_i, $i = 1, 2, 3$, and $C_{12}, C_{13}, C_{23|1}$ have densities $c_{12}, c_{13}, c_{23|1}$ respectively, then the conditional densities $f_{2|1}, f_{3|1}$ exist, and the mixed third-order derivative of (3.2) (as a special case of Theorem 3.2) is:

$$f_{123}(x_1, x_2, x_3) = c_{23|1}(F_{2|1}(x_2|x_1), F_{3|1}(x_3|x_1)) f_{2|1}(x_2|x_1) f_{3|1}(x_3|x_1) f_1(x_1)$$
$$= c_{23|1}(F_{2|1}(x_2|x_1), F_{3|1}(x_3|x_1)) c_{12}(F_1(x_1), F_2(x_2)) f_2(x_2)$$
$$\times c_{13}(F_1(x_1), F_3(x_3)) f_3(x_3) f_1(x_1)$$
$$= c_{12}(F_1(x_1), F_2(x_2)) c_{13}(F_1(x_1), F_3(x_3))$$
$$\times c_{23|1}(F_{2|1}(x_2|x_1), F_{3|1}(x_3|x_1)) \cdot \prod_{i=1}^{3} f_i(x_i).$$

For the D-vine with $n = 4$, the four-variate distribution comes from the specification $\{F_1, F_2, F_3, F_4, C_{12}, C_{23}, C_{34}, C_{13|2}, C_{24|3}, C_{14|23}\}$. The $(i, i+1)$ margins are $F_{i,i+1} = C_{i,i+1}(F_i, F_{i+1})$, F_{123} and F_{234} have expressions like (3.2), and then

$$F_{1234}(x_1, x_2, x_3, x_4) = \int_{-\infty}^{x_2} \int_{-\infty}^{x_3} C_{14|23}(F_{1|23}(x_1|z_2, z_3), F_{4|23}(x_4|z_2, z_3))$$
$$\times dF_{23}(z_2, z_3). \quad (3.3)$$

If F_i are differentiable with respective densities f_i, $i = 1, 2, 3, 4$, and C_e have densities c_e for edges e in this vine, then the mixed fourth-order derivative of (3.3) is:

$$f_{1234}(x_1, x_2, x_3, x_4) = c_{14|23}(F_{1|23}(x_1|x_2, x_3), F_{4|23}(x_4|x_2, x_3)) f_{1|23}(x_1|x_2, x_3)$$
$$\times f_{4|23}(x_4|x_2, x_3) f_{23}(x_2, x_3)$$
$$= c_{14|23}(F_{1|23}(x_1|x_2, x_3), F_{4|23}(x_4|x_2, x_3)) f_{123}(x_1, x_2, x_3)$$
$$\times f_{234}(x_2, x_3, x_4)/f_{23}(x_2, x_3)$$
$$= c_{14|23}(F_{1|23}(x_1|x_2, x_3), F_{4|23}(x_4|x_2, x_3))$$
$$\times c_{13|2}(F_{1|2}(x_1|x_2), F_{3|2}(x_3|x_2))$$
$$\times c_{24|3}(F_{2|3}(x_2|x_3), F_{4|3}(x_4|x_3))$$
$$\times c_{12}(F_1(x_1), F_2(x_2)) c_{23}(F_2(x_2), F_3(x_3))$$
$$\times c_{34}(F_3(x_3), F_4(x_4)) \cdot \prod_{i=1}^{4} f_i(x_i).$$

In applications of vine copulae to date, a parameter (vector) is associated with each $C_e \in B$, and then statistical inference can proceed with maximum likelihood; see Section 3.7.

Normal copulae When each bivariate copula C_e is a bivariate normal copula, then the resulting multivariate copula is a multivariate normal copula. For a multivariate normal copula represented as a vine, there is a correlation or partial correlation parameter associated with each C_e, and the parameters can be summarized into a *partial correlation vine*; see Section 3.3.2. Moreover, since the multivariate normal copula has the property that conditional correlations do not depend on the values of the conditioning variables, any multivariate normal copula has many representations as a vine copula. It can also be shown that the multivariate t_ν copulae are special cases of vine copulae.

Dependence properties The following dependence properties of vine copulae are shown in Joe[26] and Joe et al.[30]:

(1) Let edge e be in E_ℓ with $\ell > 1$ and let the conditioned set for e be $\{e_1, e_2\}$. If C_e is more concordant than C'_e, then the margin F_{e_1,e_2} is more concordant than F'_{e_1,e_2}.
(2) If C_e has upper (lower) tail dependence for all $e \in E_1$, and the remaining copulae have support on $[0,1]^2$, all bivariate margins of $F_{1\cdots n}(x_1,\ldots,x_n)$ have upper (lower) tail dependence.
(3) For parametric vine copulae with a parameter θ_e associated with C_e, a wide range of dependence is obtained if each $C_e(\cdot;\theta_e)$ can vary from the bivariate Fréchet lower bound to the Fréchet upper bound. Consider the Kendall tau triple $(\tau_{12}, \tau_{13}, \tau_{23})$ for $n=3$. For a three-dimensional vine copula, if $C_{23|1}$ is the conditional Fréchet upper (lower) bound copula, then τ_{23} achieves the maximum (minimum) possible, given τ_{12}, τ_{13}.

3.3.2 *Partial correlation vine*

In this section, we first give the definition of partial correlation. Then, we describe the partial correlation vine structure and, finally, we mention two applications of partial correlation vines.

3.3.2.1 *Partial correlation*

A partial correlation can be defined in terms of partial regression coefficients. Consider variable X_i with zero mean and standard deviation $\sigma_i = 1$,

$i = 1, \ldots, n$. Let the numbers $b_{ij;\{1,\ldots,n\}\setminus\{i,j\}}$ minimize

$$E\left[\left(X_i - \sum_{j:j\neq i} b_{ij;\{1,\ldots,n\}\setminus\{i,j\}} X_j\right)^2\right], \quad i = 1, \ldots, n.$$

Definition 3.4 (Partial correlation). The partial correlation of variables 1 and 2, given the remaining variables is:

$$\rho_{12;3,\ldots,n} = \mathrm{sgn}(b_{12;3,\ldots,n})(b_{12;3,\ldots,n} b_{21;3,\ldots,n})^{1/2}.$$

By permuting the indices, other partial correlations on n variables are defined.

Equivalently, we could define the partial correlation as

$$\rho_{12;3,\ldots,n} = -\frac{K_{12}}{\sqrt{K_{11} K_{22}}},$$

where K_{ij} denotes the (i,j) cofactor of the correlation matrix. The partial correlation $\rho_{12;3,\ldots,n}$ can be interpreted as the correlation between the orthogonal projections of X_1 and X_2 on the plane orthogonal to the space spanned by $X_3, \ldots, X_n$.

Partial correlations can be computed from correlations with the following recursive formula[52]:

$$\rho_{12;3,\ldots,n} = \frac{\rho_{12;3,\ldots,n-1} - \rho_{1n;3,\ldots,n-1} \cdot \rho_{2n;3,\ldots,n-1}}{\sqrt{1 - \rho_{1n;3,\ldots,n-1}^2}\sqrt{1 - \rho_{2n;3,\ldots,n-1}^2}}. \quad (3.4)$$

3.3.2.2 *Partial correlation vine*

A *partial correlation vine*,[5,34,41] which is a useful parametrization for a multivariate normal or elliptical distribution, is obtained by assigning a partial correlation ρ_e, with a value chosen arbitrarily in the interval $(-1, 1)$, to each edge e in the union $E(\mathcal{V}) = E_1 \cup \cdots \cup E_{n-1}$ of the vine defined in Section 3.2. Note that ρ_e is a correlation for edges in T_1 and ρ_e is a partial correlation for edges in $T_2, \ldots, T_{n-1}$. Theorem 3.3 in Bedford and Cooke[5] shows that a regular vine provides a bijective mapping from $(-1, 1)^{\binom{n}{2}}$ into the set of positive definite matrices with 1's on the diagonal.

Theorem 3.3. *For any regular vine on n elements, there is a one-to-one correspondence between the set of $n \times n$ positive definite correlation matrices and the set of partial correlation specifications for the vine.*

All assignments of the numbers between -1 and 1 to the edges of a partial correlation regular vine are consistent, and all correlation matrices can be obtained this way. Specific examples of partial correlation vines are the following: For a C-vine, the set of partial correlations is $\{\rho_{i_1 i_2; 1,\ldots,i_1-1} : 1 \leq i_1 < i_2 \leq n\} = \{\rho_{12},\ldots;\rho_{1n}, \rho_{23;1},\ldots,\rho_{2n;1},\ldots,\rho_{n-1,n;1,\ldots,n-2}\}$. For a D-vine, the set of partial correlations is $\{\rho_{i_1 i_2; i_1+1,\ldots,i_2-1} : 1 \leq i_1 < i_2 \leq n\} = \{\rho_{12},\ldots,\rho_{n-1,n}, \rho_{13;2},\ldots,\rho_{n-2,n;n-1},\ldots,\rho_{1,n;2,\ldots,n-1}\}$. For the regular vine in Section 3.2 that is not a C-vine or a D-vine, the set of partial correlations is $\{\rho_{12}, \rho_{23}, \rho_{24}, \rho_{45}, \rho_{13;2}, \rho_{14;2}, \rho_{25;4}, \rho_{34;12}, \rho_{15;24}, \rho_{35;124}\}$.

One verifies that the correlation between the ith and jth variables can be computed from the sub-vine generated by the constraint set of the edge whose conditioned set is $\{i,j\}$, using recursively Eq. (3.4) and the following lemma.[34]

Lemma 3.1. *If $z, x, y \in (-1, 1)$, then also $w \in (-1, 1)$, where*
$$w = z\sqrt{(1-x^2)(1-y^2)} + xy.$$

Thus, a regular vine may be seen as a way of picking out partial correlations which uniquely determine the correlation matrix and which are algebraically independent. The partial correlations in a partial correlation vine need not satisfy any algebraic constraint like positive definiteness. The "completion problem" for partial correlation vines is therefore trivial. An incomplete specification of a partial correlation vine may be extended to a complete specification by assigning arbitrary numbers in the $(-1,1)$ interval to the unspecified edges in the vine.

Partial correlation vines have another important property: the product of 1 minus the square of the partial correlations equals the determinant of the correlation matrix.

Theorem 3.4 (Kurowicka and Cooke[38]). *Let D be the determinant of the n-dimensional correlation matrix ($D > 0$). For any partial correlation vine,*
$$D = \prod_{e \in E(\mathcal{V})} (1 - \rho_{e_1, e_2; D_e}^2). \tag{3.5}$$

3.3.2.3 Applications

We mention two applications of partial correlation vines. One is the generation of random correlation matrices $\boldsymbol{R}$ that are uniform over the space of correlation matrices. Another is a reparametrization of statistical models where $\boldsymbol{R}$ is a parameter.

Random correlation matrices In Joe,[28] for the partial correlation D-vine, and in Lewandowski et al.,[41] for the general partial correlation vine, results and algorithms are given for generating a random correlation matrix $\boldsymbol{R}$ based on a partial correlation vine. This random correlation matrix generation is based on the property that the partial correlations in a regular vine are algebraically independent. By choosing the distributions of ρ_e to be appropriate beta distributions on $(-1, 1)$, Lewandowski et al.[41] have developed a method to obtain random correlation matrices with a uniform density or, more generally, a density proportional to $|\boldsymbol{R}|^{\eta-1}$ where $\eta > 0$.

Numerically, using the partial correlation C-vine is fastest but one might want to use a specific regular vine if there is an indexing of partial correlations of interest.

Reparametrization of statistical models Statistical models such as the multivariate probit model for ordinal data or the t-copula model have an $n \times n$ correlation matrix $\boldsymbol{R} = (\rho_{ij})$ as a parameter. To avoid checking the positive definiteness constraint in the middle of the numerical maximum likelihood iterations, the correlation matrix can be reparametrized via a partial correlation vine. The idea of reparametrizing the correlation matrix to $n-1$ correlations and $(n-1)(n-2)/2$ partial correlations (D-vine) was applied in Xu[51] as a way of allowing the correlation matrix to be a function of covariates. A more common way to deal with the positive definiteness constraint is to reparametrize via the lower triangular Cholesky matrix $\boldsymbol{A} = (a_{ij})$. The partial correlation C-vine might be a more interpretable parametrization. Note that if $\boldsymbol{R} = \boldsymbol{A}\boldsymbol{A}'$, then

$$a_{i1} = \rho_{1i}, \quad i = 1, \ldots, n,$$

$$a_{ij} = \rho_{ji;1\cdots j-1} \prod_{k=1}^{j-1} \sqrt{1 - \rho_{ki;1\cdots k-1}^2}, \quad j = 3, \ldots, n, \ i = j+1, \ldots, n,$$

$$a_{ii} = 1 - \sum_{k=1}^{i-1} a_{ik}^2, \quad i = 2, \ldots, n.$$

That is, each element of $\boldsymbol{A}$ that is below the diagonal is a function of partial correlations in the C-vine.

As shown in Section 5.2 of Kurowicka and Cooke,[36] with the use of expert judgement, it might be convenient to specify the conditional bivariate copulae by first assigning a constant conditional rank correlation to each edge of the vine. For $i = 1, \ldots, n-1$, with $e \in E_i$ having $\{j, k\}$

as the conditioned variables and D_e as the conditioning variables, we associate

$$r_{j,k|D_e}.$$

The resulting structure is called a *conditional rank correlation vine*.

In eliciting expert judgement on strengths of dependence, the conditional rank correlation vine avoids the constraints in a matrix of rank correlations. It is shown in Joe,[29] using conditional distributions in the form of D-vines, that in dimensions $n \geq 5$, the possible rank correlation matrices (or correlation matrices of dependent uniform random variables) is smaller than the set of all positive definite matrices with 1 on the diagonal.

3.4 Historical Origins

The first regular vine, *avant la lettre*, was introduced by Joe.[25] The motive was to extend the bivariate extreme-value copula to higher dimensions. Consider a multivariate survival function $G(z_1, \ldots, z_n) = \text{Prob}\{Z_1 > z_1, \ldots, Z_n > z_n\}$. If G is "min-stable", then it satisfies

$$G(tz_1, \ldots, tz_n) = e^{-A(tz_1, \ldots, tz_n)} = e^{-tA(z_1, \ldots, z_n)}. \tag{3.6}$$

As shown by Pickands (Galambos,[14] Chapter 5), the family of functions A satisfying this equation is infinite-dimensional. Joe's goal was to find finite-dimensional parametric subfamilies that would cover the whole family represented by (3.6). To this end, he introduced what would later be called the *D-vine*.

Joe[26] was interested in a class of n-variate distributions with given one-dimensional margins, and $n(n-1)$ dependence parameters, whereby $n-1$ parameters correspond to bivariate margins, and the others correspond to conditional bivariate margins. In the case of multivariate normal distributions, the parameters would be $n-1$ correlations and $(n-1)(n-2)/2$ partial correlations, which were noted to be algebraically independent in $(-1, 1)$. Implicit in this remark is the observation that partial correlations on what is now called the D-vine provide an algebraically independent parametrization of the set of positive definite correlation matrices.

One main idea[26,27] comes from the Fréchet class $\mathcal{F}(F_{iS}, F_{jS})$ where S is a set of indices of variables that does not contain i and j. That is, $\mathcal{F}(F_{iS}, F_{jS})$ is the class of distributions of cardinality $|S| + 2$ with the margin F_S in common.

If $S = \{k_1, \ldots, k_m\}$ with $m \geq 1$, a member of $\mathcal{F}(F_{iS}, F_{jS})$ has the form

$$\int_0^{x_{k_1}} \cdots \int_0^{x_{k_m}} F_{ij|S}(x_i, x_j | \boldsymbol{y}_S) \, dF_S(\boldsymbol{y}_S). \tag{3.7}$$

By Sklar's theorem, there are conditional copulae $\{C_{ij|S}(\cdot | \boldsymbol{y}_S)\}$ such that (3.7) is

$$\int_0^{x_{k_1}} \cdots \int_0^{x_{k_m}} C_{ij|S}(F_{i|S}(x_i|\boldsymbol{y}_S), F_{j|S}(x_y|\boldsymbol{y}_S)|\boldsymbol{y}_S) \, dF_S(\boldsymbol{y}_S). \tag{3.8}$$

By imitating multivariate Gaussian distributions, simpler distributions in $\mathcal{F}(F_{iS}, F_{jS})$ have a constant conditional copula: $C_{ij|S}(\cdot | \boldsymbol{y}_S) \equiv C_{ij|S}$ for all $\boldsymbol{y}_S$. By adding a dependence parameter, one can have a bivariate parametric copula family $C_{ij|S}(\cdot | \cdot; \boldsymbol{\theta})$. A wide range of conditional dependence obtains if $C_{ij|S}(\cdot | \cdot; \boldsymbol{\theta})$ interpolates the Fréchet upper bound, independence and the Fréchet lower bound.

Joe[26,27] applied the above idea of Fréchet classes recursively in a D-vine for $\mathcal{F}(F_{i,i+1,\ldots,j-1}, F_{j,i+1,\ldots,j-1})$ with $1 \leq i < j \leq d$. This was partly motivated by variables that might be indices in time or in a one-dimensional spatial direction. Properties of bivariate tail dependence, ordering by concordance, and range of dependences were obtained. The basic sampling strategy was also outlined.

An entirely different motivation underlays the first formal definition of vines in Cooke.[9] Uncertainty analyses of large risk models, such as those undertaken for the European Union and the US Nuclear Regulatory Commission for accidents at nuclear power plants, involve quantifying and propagating uncertainty over hundreds of variables.[16,19] Dependence information for such studies had been captured by *Markov trees*,[50] which are trees constructed with nodes as univariate random variables and edges as bivariate copulae. For n variables, there are at most $n-1$ edges for which dependence can be specified. New techniques at that time involved obtaining uncertainty distributions on modeling parameters by eliciting experts' uncertainties on other variables, which are predicted by the models. These uncertainty distributions are pulled back onto the model's parameters by a process known as probabilistic inversion.[15,36] The resulting distributions often displayed a dependence structure that could not be captured as a Markov tree (see Fig. 3.5).

This led to the invention of regular vines. Regular vines enable an additive decomposition of the mutual information that depends only on the expected mutual information of each edge. Making any conditional copula the conditionally independent copula lowers the mutual information.[9]

Figure 3.5. A simple Markov tree (left) and a vine (right) on three variables.

This remark shows that the minimal information completion of any partially specified regular vine is trivially found by making the unspecified conditional copulae conditionally independent. This situation compares favorably with the problem of completing a partially specified correlation matrix. If a partially specified regular vine has the property that no unspecified edge has specified m-parents, then the partial specification is called *m-saturated*. If we consider the indices in the conditioned sets of a partially specified regular vine, then placing an edge between two indices in the same conditioned set generates a graph. m-saturation is equivalent to the decomposability of this graph, which is equivalent to the graph being chordal and to the existence of a junction tree.[38] Bedford and Cooke[5] extend the result of Joe[26] and show that partial correlations in $(-1, 1)$ on the edges of any regular vine provide an algebraically independent parametrization of the positive definite correlation matrices, and introduce Cantor trees as a generalization of regular vines. Bedford and Cooke[4] give an explicit formula, factorizing any multivariate density in terms of (conditional) copula densities on any regular vine. This generalizes the Hammersley–Clifford theorem applied to Markov trees.[6]

3.5 Compatibility of Marginal Distributions

n-dimensional vine copulae are based on $\binom{n}{2}$ bivariate copulae which can be specified completely independently of each other. To do this, $n - 1$ of the bivariate copulae are bivariate margins and the remaining $(n-1)(n-2)/2$ are conditional copulae. In this section, we provide some results on sets of marginal distributions that can be compatible and hence provide some intuition for the definition of vines and vine copulae (pair-copula construction) in Section 3.2.

The Fréchet class $\mathcal{F}(F_j, 1 \leq j \leq n; F_{jk}, 1 \leq j < k \leq n)$ of given (continuous) univariate and bivariate margins is hard to study; there is no general result on when the set of $\binom{n}{2}$ bivariate margins or copulae are compatible with an n-variate distribution. Assuming that bivariate margins agree on the univariate margins, the maximal number of bivariate margins that can be compatible with no constraints is $n-1$. The maximum is attained if an acyclic condition is satisfied. Consider the Fréchet class $\mathcal{F}(F_j, 1 \leq j \leq n, F_{j_i k_i} : j_i < k_i, i = 1, \ldots, n-1)$ with $n-1$ distinct pairs. This class is non-empty for any choice of the $n-1$ bivariate margins if the graph with nodes $\{1, \ldots, n\}$ and edges $\{(j_i, k_i) : i = 1, \ldots, n-1\}$ has no cycles. This result follows the compatibility condition in Kellerer,[31] summarized in Section 3.7 of Joe.[27] It also follows from ideas presented below. If any additional bivariate margin $F_{j_n k_n}$ is added, then the graph will definitely have a cycle, and some choices of the $F_{j_i k_i}$ will lead to non-compatibility.

With the univariate margins fixed, let us illustrate the above results with bivariate copulae. Because we can permute indices of the variables without changing probabilistic properties, a simple way to get the $n-1$ bivariate margins is to have pairs

$$\{(1,2), (j_2, 3), \ldots, (j_{n-1}, n)\}, \quad j_i \in \{1, \ldots, i\}, \; i = 1, \ldots, n-1. \quad (3.9)$$

The first edge has nodes 1 and 2 and edge $(1,2)$. The second edge adds node 3 and connects to either node 1 or 2. The ith edge adds node i and connects to one of the nodes between 1 and $i-1$ inclusive. In this way, no cycle is formed, and the result is a tree after $n-1$ steps. If an nth edge is added without adding another node, there will definitely be a cycle (the reader can confirm this by drawing some diagrams).

Let us add the nth edge to get a cycle. By relabeling, we can assume that the edges of the cyclical subgraph are $\{(1,2), (2,3), \ldots, (m-1, m), (1, m)\}$ where $3 \leq m \leq n$. Consider copulae $C_{12}, C_{23}, \ldots, C_{m-1,m}, C_{1m}$. If $C_{12}, C_{23}, \ldots, C_{m-1,m}$ are co-monotonic (Fréchet upper bound) and C_{1m} is counter-monotonic (Fréchet lower bound), then this set of bivariate margins has no compatible n-variate or m-variate distribution.

We next show via an example why $n-1$ bivariate margins satisfying the tree condition (3.9) imply that there are no extra constraints for compatibility. Consider the Fréchet class of bivariate copulae $\mathcal{F}(C_{12}, C_{13}, C_{14}, C_{25})$ for $n=5$; the choice of bivariate margins satisfies (3.9). C_{12}, C_{13} can be specified completely independently because this is the same as specifying the univariate conditional distributions $\{C_{2|1}(\cdot|u_1), C_{3|1}(\cdot|u_1) : 0 < u_1 < 1\}$. For each u_1, $C_{2|1}, C_{3|1}$ can be coupled with a conditional copula, and from (3.7–3.8), one can construct all trivariate copulae with C_{12}, C_{13} as bivariate margins. The same statement holds for the pairs $\{C_{13}, C_{14}\}$. Hence, one can get distributions C_{123}, C_{134} with bivariate margins $\{C_{12}, C_{13}, C_{14}\}$. For the resulting C_{123}, C_{134}, one can build a four-variate copula C_{1234} via $\{C_{2|13}(\cdot; u_1, u_3), C_{4|13}(\cdot; u_1, u_3) : 0 < u_1, u_3 < 1\}$ and (3.7). After adding C_{25}, one can get a trivariate distribution C_{125} with bivariate margins $\{C_{12}, C_{25}\}$. By coupling the appropriate conditional distributions, one can get C_{1235} with C_{123}, C_{125} as trivariate margins. Finally, one can couple $C_{4|123}(\cdot|u_1, u_2, u_3)$ and $C_{5|123}(\cdot|u_1, u_2, u_3)$ in (3.7) to get a five-variate copula C_{12345}. Hence $\{C_{12}, C_{13}, C_{14}, C_{25}\}$ is a set of compatible copulae with no additional constraints.

The above example extends for any set of $n-1$ bivariate copulae with pairs of marginal indices of the form of (3.9). This explains the first tree of a vine. However, vines also provide conditions for bivariate conditional copulae. Working with bivariate conditional copulae is easier than studying conditions for the compatibility of trivariate (and higher-dimensional) margins. We next show that conditions for compatible trivariate margins are more complicated.

For trivariate margins, we take a subset of $n-2$ to consider compatibility. For $n = 5$, $n-2 = 3$ and there are three possible patterns of three trivariate margins from the full set of $10 = \binom{5}{3}$.

(a) Two indices appear in all three triplets: e.g., $\{(1,2,3),(1,2,4),(1,2,5)\}$: this is compatible with copulae for the univariate conditional distributions $3|12$, $4|12$ and $5|12$.
(b) Two of the three pairs intersects in two indices and one pair intersects in one index, e.g., $\{(1,2,3),(1,2,5),(1,3,4)\}$. The above construction shows that something like this will always be compatible.
(c) One of the three pairs intersects in two indices and the other two pairs intersect in one index, e.g., $\{(1,2,3),(2,3,4),(1,4,5)\}$. It is shown in Example 3.4 in Joe[27] that this set of trivariate margins does not satisfy the compatibility condition in Kellerer.[31]

In general, the condition to determine which sets of trivariate margins are always compatible is more complicated than the condition for bivariate margins given above. Vines are a way to specify a set of compatible bivariate margins and bivariate conditional distributions and they lead to compatible marginal distributions of higher dimensions. The condition for compatibility is straightforward to check, compared with something like Kellerer's condition. The examples in this section show why the definition of vines involves tree graphs and conditional distributions.

3.6 Sampling

We assume that variables $X_1, X_2, \ldots, X_n$ are uniform on $(0,1)$. Each edge in a regular vine may be associated with a conditional copula, that is, a conditional bivariate distribution with uniform margins. Given a conditional rank correlation vine as defined in Section 3.3.2.3, we choose a class of copulae indexed by correlation coefficients in the interval $(-1,1)$ and select the copulae with correlation corresponding to the conditional rank correlation assigned to the edge of the vine. A joint distribution satisfying the vine-copula specification can be constructed and sampled on the fly, and will preserve maximum entropy properties of the conditional bivariate distributions.[4,9]

The conditional rank correlation vine plus copula determines the whole joint distribution. There are two strategies for sampling such a distribution, which we term the *cumulative* and *density* approaches.

3.6.1 Sampling a D-vine

We first illustrate the cumulative approach with the distribution specified by a D-vine on four variables, D(1,2,3,4). Sample four independent variables distributed uniformly on interval $[0,1]$, U_1, U_2, U_3, U_4, and determine the values of correlated variables X_1, X_2, X_3, X_4 as follows:

(1) $x_1 = u_1$,
(2) $x_2 = F^{-1}_{r_{12};x_1}(u_2)$,
(3) $x_3 = F^{-1}_{r_{23};x_2}(F^{-1}_{r_{13|2};F_{r_{12};x_2}(x_1)}(u_3))$,
(4) $x_4 = F^{-1}_{r_{34};x_3}(F^{-1}_{r_{24|3};F_{r_{23};x_3}(x_2)}(F^{-1}_{r_{14|23};F_{r_{13|2};F_{r_{23};x_2}(x_3)}(F_{r_{12};x_2}(x_1))}(u_4)))$

where $F_{r_{ij|k};x_i}(X_j)$ denotes the cumulative distribution function of X_j, applied to X_j, given $X_i = x_i$ under the conditional copula with correlation $r_{ij|k}$. Notice that the D-vine sampling procedure uses conditional and inverse conditional distribution functions. A more general form of the above procedure simply refers to conditional cumulative distribution functions:

$$\begin{aligned} x_1 &= u_1, \\ x_2 &= F^{-1}_{2|1:x_1}(u_2), \\ x_3 &= F^{-1}_{3|2:x_2}(F^{-1}_{3|12:F_{1|2}(x_1)}(u_3)), \\ x_4 &= F^{-1}_{4|3:x_3}(F^{-1}_{4|23:F_{2|3}(x_2)}(F^{-1}_{4|123:F_{1|23}(x_1)}(u_4))). \end{aligned} \quad (3.10)$$

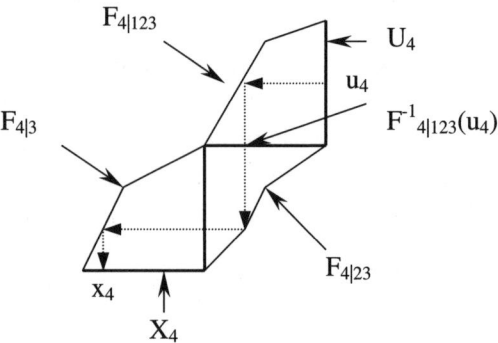

Figure 3.6. Staircase graph representation of D-vine sampling procedure.

Figure 3.6 depicts the sampling of X_4 in the D-vine with a so-called staircase graph. Following the dotted arrows, we start by sampling U_4 (realization u_4) and use this with the copula for the conditional rank correlation of $\{1,4\}$ given $\{2,3\}$ to find the argument of $F_{4|23}^{-1}$, etc. Notice that for the D-vine, values of $F_{2|3}$ and $F_{1|23}$ used to conditionalize copulae with correlations $r_{24|3}$ and $r_{14|23}$ to obtain $F_{4|23}$ and $F_{4|123}$, respectively, have to be calculated.

The staircase graph shows that if any of the cumulative conditional distributions in Fig. 3.6 is uniform, then the corresponding abscissa and ordinates can be identified. This corresponds to noting that the inverse cumulative function in (3.10) is the identity, which in turn corresponds to a conditional rank correlation being zero and the corresponding variables being conditionally independent. Notice that the conditional rank correlations can be chosen arbitrarily in the interval $[-1, 1]$; they need not be positive definite or satisfy any further algebraic constraint.

3.6.2 Sampling an arbitrary regular vine

The content of this section is based on Section 6.4.2 of Kurowicka and Cooke.[36] A regular vine on n nodes will have a single node in tree $n-1$. It suffices to show how to sample one of the conditioned variables in this node, say n, assuming we have sampled all the other variables. We proceed as follows:

(1) By Lemma 3.2, the variable n occurs in trees $1, \ldots, n-1$ exactly once as a conditioned variable. The variable with which it is conditioned in tree j is called its j-partner. We define an ordering for n as follows: index the

j-partner of variable n as variable j. We denote the conditional bivariate constraints corresponding to the partners of n as:

$$(n,1|\emptyset), (n,2|D_2^n), (n,3|D_3^n), \ldots, (n, n-1|D_{n-1}^n).$$

Again, by Lemma 3.2, variables $1, \ldots, n-1$ appear first as conditioned variables (to the left of "|") before appearing as conditioning variables (to the right of "|"). Also,

$$0 = \#D_1^n < \#D_2^n < \cdots < \#D_{n-1}^n = n - 2.$$

(2) Assuming that we have sampled all variables except n, sample one variable uniformly distributed on the interval $(0,1)$, denoted u_n. We use the general notation $F_{a|b:D}$ to denote $F_{a,b|D:F_{b|D}}$; that is the conditional distribution for $\{a,b|D\}$ conditional on a value of the cumulative conditional distribution $F_{b|D}$. Here, $\{a,b|D\}$ is the conditional bivariate constraint corresponding to a node in the vine.

(3) Sample x_n as follows:

$$x_n = F^{-1}_{n|1:D_1^n}(F^{-1}_{n|2:D_2^n}(\cdots(F^{-1}_{n|n-1:D_{n-1}^n}(u_n))\cdots)). \tag{3.11}$$

The innermost term of (3.11) is:

$$F^{-1}_{n|n-1:D_{n-1}^n} = F^{-1}_{n,n-1|D_{n-1}^n:F_{n-1|D_{n-1}^n}}$$

$$= F^{-1}_{n,n-1|D_{n-1}^n:F_{n-1,n-2|D_{n-2}^{n-1}:F_{n-2|D_{n-2}^{n-1}}}}.$$

See Chapter 7 in this volume for pseudocode for the regular vine.

3.6.3 *Density approach sampling*

When the vine-copula distribution is given as a density, the density approach to sampling may be used. Assume that the marginal distributions in (3.1) are uniform $[0,1]$. Then (3.1) can be rewritten as

$$\prod_{e \in E} c_{ij|D_e}(F_{i|D_e}(x_i), F_{j|D_e}(x_j)), \tag{3.12}$$

where, by uniformity, the density $f_i(x_i) = 1$. Expression (3.12) may be used to sample the vine distribution; namely, draw a large number of samples $(x_1, \ldots, x_n)$ uniformly, and then resample these with probability proportional to (3.12). This is less efficient than the general sampling algorithm given previously; however, it may be more convenient for conditionalization.

3.7 Parametric Inference for a Specific Pair-Copula Construction

Aas et al.[2] develop a maximum likelihood procedure to estimate parameters in copulae for D- and C-vines. The procedure can be extended to arbitrary regular vines but the algorithms are less transparent.

For notation to cover the C-vine, D-vine and other vines, let $C_{i_1 i_2 | m}(u_{i_1}, u_{i_2})$ denote the copula with conditioned set $\{i_1, i_2\}$ and conditioning set m. If $i_1 < i_2$, then $m = \{1, \ldots, i_1 - 1\}$ for the C-vine and $m = \{i_1 + 1, \ldots, i_2 - 1\}$ for the D-vine. For the partial derivatives with respect to u_j and u_{j+i}, we use the notation

$$C_{i_1 | i_2 : m}(u_{i_1} | u_{i_2}) = \frac{\partial C_{i_1 i_2 | m}}{\partial u_{i_2}}, \quad C_{i_2 | i_1 : m}(u_{i_2} | u_{i_1}) = \frac{\partial C_{i_1 i_2 | m}}{\partial u_{i_1}}.$$

The next illustration of notation is for the C-vine in (3.2). If this three-dimensional distribution is embedded in a C-vine of dimension four or more, then the conditional distribution $F_{3|12}$ is needed at the next stage, since (3.2) involves $C_{23|1}$. We use the notation $F_{3|12} = F_{3|2:1}$ to show that it depends on $C_{3|2:1}$. Differentiating (3.2) with respect to x_2, x_3, and then dividing by $f_{12}(x_1, x_2)$ leads to

$$F_{3|2:1}(x_3 | x_2, x_1) = C_{3|2:1}(F_{3|1}(x_3 | x_1) | F_{2|1}(x_2 | x_1)).$$

Expressions like this must be computed for the likelihood of a C-vine (more generally, a regular vine).

Note that when estimating the parameters here, we assume that the conditional bivariate copulae are constant over the values of the conditioning variables; see Hobæk Haff et al.[21] for examples that relate to this assumption. In the general representation of any multivariate distribution in (3.1) or (3.8), the conditional bivariate copula can vary over the values of the conditioning variables.

Assume that we observe n variables at T time points, or more generally a random sample of size T. Let $\boldsymbol{x}_i = (x_{i,1}, \ldots, x_{i,T})$, $i = 1, \ldots, n$, denote the ith observation vector in the data set. First, we assume for simplicity that the T observations of each variable are independent over time. This is not a limiting assumption, since in the presence of temporal dependence, univariate time-series models can be fitted to the margins and the analysis can henceforth proceed with the residuals.

It is important to emphasize that unless the margins are known (which they never are in practice), the estimation method presented below must rely on the normalized ranks of the data, or on a two-stage procedure where univariate margins have been estimated first and then transformed to uniform. Normalized ranks are only approximately uniform and independent, meaning that what is being maximized is a pseudo-likelihood. A two-stage procedure is better if inferences on tail probabilities are needed; the theory of estimating equations applies for the inference in this case.

3.7.1 Inference for a C-vine

In this subsection, we provide an algorithm for computing the log-likelihood of a parametric C-vine where there is a parameter $\theta_{j,i}$ associated with the bivariate copula $C_{j,j+i|1\cdots j-1}(u_j, u_{j+i})$, for $i = 1, \ldots, n-j$, $j = 1, \ldots, n-1$. Here j is the index for the tree level of the vine.

Further, let $\Theta_{j,i}$ be the set of parameters in the copula density $c_{j,j+i|1,\ldots,j-1}(F_{j|1:2\cdots j-1}, F_{j+i|1:2\cdots j-1})$. Note that $F_{j+i|1:2\cdots j-1}$ depends recursively on $\theta_{\ell,k}$, $k = 1, \ldots, j - \ell$ and $j + i - \ell$, $\ell = 1, \ldots, j - 1$.

For the canonical vine, the log-likelihood (for the copula parameters, assuming univariate margins have been estimated or transformed to uniform) is given by

$$\sum_{j=1}^{n-1}\sum_{i=1}^{n-j}\sum_{t=1}^{T} \log[c_{j,j+i|1\cdots j-1}\{F_{j|j-1:1\cdots j-2}(x_{j,t}|\boldsymbol{x}_t^{(j-1)}),$$

$$F_{j+i|j-1:1\cdots j-2}(x_{j+i,t}|\boldsymbol{x}_t^{(j-1)})\}] \quad (3.13)$$

where $\boldsymbol{x}_t^{(j-1)} = (x_{1,t}, \ldots, x_{j-1,t})$. For each copula in the sum (3.13) there is at least one parameter to be determined. The number depends on which copula type is used. The log-likelihood must be numerically maximized over all parameters. If parametric univariate margins are also estimated, say, $f_i(\cdot; \alpha_i)$, $i = 1, \ldots, n$, then the added contribution to (3.13) is

$$\sum_{i=1}^{n}\sum_{t=1}^{T} \log f_i(x_{i,t}; \alpha_i).$$

Algorithm 3.1 evaluates the likelihood for the canonical vine. The outer for-loop corresponds to the outer sum in (3.13), corresponding to the tree level of the vine. This for-loop consists in turn of two other for-loops. The first of these corresponds to the sum over i in (3.13). In the other, the conditional distribution functions needed for the next run of the outer for-loop are computed. In the algorithm,

$$L_{j,j+i}(\boldsymbol{y}, \boldsymbol{v}, \Theta) = \sum_{t=1}^{T} \log\{c_{j,j+i|1\cdots j-1}(y_t, v_t, \Theta)\} \quad (3.14)$$

is the contribution to the log-likelihood from the copula $c_{j,j+i|1\cdots j-1}$.

Algorithm 3.1.

1: log-likelihood $\leftarrow 0$
2: **for** $i \leftarrow 1, \ldots, n$ **do**
3: $\quad v_{0,i} \leftarrow x_i$ (vectorized over t)
4: **end for**
5: **for** $j \leftarrow 1, \ldots, n-1$ **do** (tree level j)
6: $\quad$ **for** $i \leftarrow 1, \ldots, n-j$ **do**
7: $\quad\quad$ log-likelihood $\leftarrow$ log-likelihood $+ L_{j,j+i}(v_{j-1,1}, v_{j-1,i+1}, \Theta_{j,i})$
8: $\quad$ **end for**
9: $\quad$ **if** $j == n-1$ **then**
10: $\quad\quad$ Stop
11: $\quad$ **end if**
12: $\quad$ **for** $i \leftarrow 1, \ldots, n-j$ **do**
13: $\quad\quad v_{j,i} \leftarrow C_{j+i|j:1\cdots j-1}(v_{j-1,i+1}|v_{j-1,1}; \Theta_{j,i})$ (vectorized over t)
14: $\quad$ **end for**
15: **end for**

Starting values of the parameters needed in the numerical maximization of the log-likelihood may be determined as follows:

(a) Estimate the parameters of the copulae in tree 1 from the original data.
(b) Compute observations (i.e., conditional distribution functions) for tree 2 using the copula parameters from tree 1 and the conditional distributions.
(c) Estimate the parameters of the copulae in tree 2 using the observations from (b).
(d) Compute observations for tree 3 using the copula parameters at level 2 and the conditional distributions.
(e) Estimate the parameters of the copulae in tree 3 using the observations from (d).
(f) etc.

Note that each estimation here is easy to perform, since the data set is only of dimension 2 in each step.

3.7.2 Inference for a D-vine

Similar to the preceding subsection, for the D-vine, the log-likelihood is given by

$$\sum_{j=1}^{n-1}\sum_{i=1}^{n-j}\sum_{t=1}^{T} \log[c_{i,i+j|i+1\cdots i+j-1}\{F_{i|i+j-1:i+1\cdots i+j-2}(x_{i,t}|\boldsymbol{x}_t^{(i,j-1)}),$$

$$F_{i+j|i+1:i+2\cdots i+j-1}(x_{i+j,t}|\boldsymbol{x}_t^{(i,j-1)})\}],$$

where $\boldsymbol{x}_t^{(i,j-1)} = (x_{i+1,t},\ldots,x_{i+j-1,t})$. The D-vine log-likelihood must also be numerically optimized. Algorithm 3.2 evaluates the likelihood. $\Theta_{j,i}$ is the set of parameters of copula density $c_{i,i+j|i+1,\ldots,i+j-1}$ ($F_{i|i+j-1:i+1\cdots i+j-2}, F_{i+j|i+1:i+2\cdots i+j-1}$). Note that the algorithm requires $2(n-j-1)$ conditional distributions at step j for $j = 1,\ldots,n-1$. For $j=1$, $C_{i|i+1}, C_{i+1|i}$, $i = 1,\ldots,n-1$, are all needed except for $C_{2|1}$ and $C_{n-1|n}$. A similar pattern holds for $j > 1$. In the notation in Algorithm 3.2, $v'_{j,i}$ is used in the tree level j when the conditional distribution is $C_{i|i+j:i+1\cdots i+j-1}$ and $v_{j,i}$ is used when the conditional distribution is $C_{i+j|i:i+1\cdots i+j-1}$.

Similar to the C-vine, in the D-vine algorithm,

$$L_{i,i+j}(\boldsymbol{y},\boldsymbol{v},\Theta) = \sum_{t=1}^{T} \log\{c_{i,i+j|i+1\cdots i+j-1}(y_t,v_t,\Theta)\}$$

is the contribution to the log-likelihood from the copula $c_{i,i+j|i+1\cdots i+j-1}$.

Algorithm 3.2.

1: log-likelihood $\leftarrow 0$
2: **for** $i \leftarrow 1,\ldots,n$ **do**
3:     $\boldsymbol{v}_{0,i} \leftarrow \boldsymbol{x}_i$ (vectorized over t)
4: **end for**
5: **for** $i \leftarrow 1,\ldots,n-1$ **do**
6:     log-likelihood $\leftarrow$ log-likelihood$+ L_{i,1+i}(\boldsymbol{v}_{0,i},\boldsymbol{v}_{0,i+1},\Theta_{1,i})$
7: **end for**
8: $\boldsymbol{v}'_{1,1} \leftarrow C_{1|2}(\boldsymbol{v}_{0,1}|\boldsymbol{v}_{0,2};\Theta_{1,1})$ (vectorized over t; similarly below)
9: **for** $k \leftarrow 1,\ldots,n-3$ **do**
10:     $\boldsymbol{v}_{1,k+1} \leftarrow C_{k+2|k+1}(\boldsymbol{v}_{0,k+2}|\boldsymbol{v}_{0,k+1};\Theta_{1,k+1})$
11:     $\boldsymbol{v}'_{1,k+1} \leftarrow C_{k+1|k+2}(\boldsymbol{v}_{0,k+1}|\boldsymbol{v}_{0,k+2};\Theta_{1,k+1})$

12: **end for**
13: $v_{1,n-1} \leftarrow C_{n|n-1}(v_{0,n}|v_{0,n-1}; \Theta_{1,n-1})$
14: **for** $j \leftarrow 2, \ldots, n-1$ **do** (tree level j)
15:     **for** $i \leftarrow 1, \ldots, n-j$ **do**
16:        log-likelihood $\leftarrow$ log-likelihood$+ L_{i,i+j}(v'_{j-1,i}, v_{j-1,i+1}, \Theta_{j,i})$
17:     **end for**
18:     **if** $j == n-1$ **then**
19:        Stop
20:     **end if**
21:     $v'_{j,1} \leftarrow C_{1|j+1:2\cdots j}(v'_{j-1,1}|v_{j-1,2}; \Theta_{j,1})$
22:     **if** $n > 4$ **then**
23:        **for** $i \leftarrow 1, 2, \ldots, n-j-2$ **do**
24:           $v_{j,i+1} \leftarrow C_{i+j+1|i+1:i+2\cdots i+j}(v_{j-1,i+2}|v'_{j-1,i+1}; \Theta_{j,i+1})$
25:           $v'_{j,i+1} \leftarrow C_{i+1|i+j+1:i+2\cdots i+j}(v'_{j-1,i+1}|v_{j-1,i+2}; \Theta_{j,i+1})$
26:        **end for**
27:     **end if**
28:     $v_{j,n-j} \leftarrow C_{n|n-j:n-j+1\cdots n-1}(v_{j-1,n-j+1}|v'_{j-1,n-j}; \Theta_{j,n-j})$
29: **end for**

Note that, similar to other algorithms for C-vines and D-vines, the D-vine algorithm for the likelihood calculation is more complicated than that for the C-vine. Other comments for the C-vine inference also apply to D-vine inference.

3.8 Model Inference

Model inference relates to the problem of choosing a regular vine to model a multivariate data set. If the conditional copulae are not constant, then any regular vine can be used to describe any multivariate distribution. Following Joe,[26] the motive underlying the vine-copula approach to modeling is to have a flexible low parameter set of models. In the first instance, this has led to the restriction to constant conditional copulae. When a joint distribution is defined by one particular regular vine with constant conditional copulae, these copulae will not in general remain constant when a different regular vine is used.

In Section 3.7, we described how to do inference for some specific pair-copula decompositions. However, this is only a part of the full estimation problem. Full inference for a pair-copula decomposition should in principle consider (a) the selection of a regular vine, (b) the choice of (conditional) copula types and (c) the estimation of the copula parameters. For smaller dimensions (say three and four), one may estimate the parameters of all possible factorizations using the procedure described in Section 3.7 and compare the resulting log-likelihoods, Akaike information criterion (AIC) values or out-of-sample predictions. This is in practice infeasible for higher dimensions, in view of Theorem 3.1. Heuristic strategies are required to choose which decompositions to investigate. In this section, we review two approaches that have been suggested for choosing the "best" regular vine; the first is a modified version of the sequential estimation procedure outlined in Section 3.7 while the other is based on the mutual information.

3.8.1 *Sequential selection*

In this approach, one first has to decide whether to use a C- or D-vine. D-vines may be more appropriate than C-vines in situations where a distinguished variable of maximal degree at each echelon cannot readily be identified. The next step is to decide the order of the variables. One possibility that has turned out to be promising in practice is to base this decision on the strength of dependence between the variables, ordering the variables such that the copulae to be fitted in tree 1 in the decomposition are those associated with the strongest dependence.

Given data and an assumed pair-copula decomposition, it is necessary to specify the parametric shape of each pair-copula. For example, for the decomposition in Section 3.7, we need to decide which copula type to use for $C_{12}(\cdot,\cdot)$, $C_{23}(\cdot,\cdot)$ and $C_{13|2}(\cdot,\cdot)$. The pair-copulae do not have to belong to the same family. The resulting multivariate distribution will be valid if we choose for each pair of variables the parametric copula that best fits the data. If we choose not to stay in one predefined class, we need a way of determining which copula to use for each pair of (transformed) observations. We propose to use a modified version of the sequential estimation procedure outlined in Section 3.7:

(1) Determine which copula types to use in tree 1 by plotting the original data, and checking for tail dependence or asymmetries (these are the patterns that make the multivariate normal copula inadequate).

(2) Estimate the parameters of the selected copulae using the original data.
(3) Transform observations as required for tree 2, using the copula parameters from tree 1 and the conditional functions in Section 3.7.
(4) Determine which copula types to use in tree 2 in the same way as in tree 1.
(5) Iterate.

The observations used to select the copulae at a specific level depend on the specific pair-copulae chosen up-stream in the decomposition. This selection mechanism does not guarantee a globally optimal fit. Having determined the appropriate parametric shapes for each copulae, one may use the procedures in Section 3.7 to estimate their parameters.

3.8.2 *Information-based model inference*

A different approach to model learning inspired by Whittaker[50] was developed in Kurowicka and Cooke,[36] based on the factorization of the determinant in Theorem 3.4. We sketch here a more general approach based on the mutual information. Following Joe,[23,24] the mutual information is taken as a general measure of dependence. The strategy is to choose a regular vine which captures the mutual information in a small number of conditional bivariate terms, and to find a copula which renders these mutual information values. Before describing the approach, we give some definitions.

3.8.2.1 *Definitions and theorems*

Definition 3.5 (Relative information, mutual information). Let f and g be densities on $\mathbb{R}^n$ with f absolutely continuous with respect to g;

- the **relative information** of f with respect to g is

$$I(f|g) = \int_1 \cdots \int_n f(x_1, \ldots, x_n) \ln\left(\frac{f(x_1, \ldots, x_n)}{g(x_1, \ldots, x_n)}\right) dx_1 \ldots dx_n$$

- the **mutual information** of f is

$$MI(f) = I\left(f \middle| \prod_{i=i}^{n} f_i\right) \quad (3.15)$$

where f_i is the ith univariate marginal density of f and $\Pi_{i=1}^{n} f(x_1, \ldots, x_n)$ is the independent distribution with univariate margins $\{f_i\}$.

Relative information is also called the Kullback–Leibler information and the directed divergence. The mutual information is also called the information proper. The mutual information will be used to capture general dependence in a set of multivariate data. We do not possess something like an "empirical mutual information". It must rather be estimated with kernel estimators, as suggested in Joe.[24] For some copulae, the mutual information can be expressed in closed form[36]:

Theorem 3.5. *Let g be the elliptical copula with correlation ρ, then the mutual information of g is*

$$1 + \ln 2 + \ln(\pi\sqrt{1-\rho^2}\,).$$

Let h be the diagonal band copula with vertical bandwidth parameter $1-\alpha$, then the mutual information of h is

$$-\ln(2^{|\alpha|}(1-|\alpha|)).$$

Note that the mutual information of the elliptical copula with zero correlation is not zero, reflecting the fact that zero correlation in this case does not entail independence.

Theorem 3.6 (Whittaker[50]). *Let f be a joint normal density with mean vector zero, then*

$$MI(f) = -\tfrac{1}{2}\ln(D),$$

where D is the determinant of the correlation matrix.

For a bivariate normal, Theorem 3.6 says that $MI(f) = -\tfrac{1}{2}\ln(1-\rho^2)$. Substituting the appropriate conditional bivariate normal distributions on the right-hand side of (3.16), we find $MI(f) = -\tfrac{1}{2}\sum_{e\in E(\mathcal{V})} \ln(1-\rho^2_{e_1,e_2;D_e})$, which agrees with Theorem 3.4.

The determinant of a correlation matrix indicates the "amount of linearity" in a joint distribution. It takes the value 1 if the variables are uncorrelated, and the value zero if there is a linear dependence. Theorem 3.6 suggests that

$$e^{-2MI(f)}$$

is the appropriate generalization of the determinant to capture general dependence.

Proposition 3.3. $e^{-2MI(f)} = 1$ *if and only if $f = \Pi f_i$ and $e^{-2MI(f)} = 0$ if f has positive mass on a set of Πf_i measuring zero.*

Theorem 3.7 (Cooke,[9] Bedford and Cooke[5]). *Let g be an n-dimensional density satisfying the bivariate vine specification $(F, \mathcal{V}, B)$ with density g and one-dimensional marginal densities $g_1, \ldots, g_n$; then*

$$I\left(g \middle| \prod_{i=1}^{n} g_i\right) = \sum_{e \in E(\mathcal{V})} E_{D_e} I(g_{e_1, e_2 | D_e} | g_{e_1 | D_e} \cdot g_{e_2 | D_e}). \tag{3.16}$$

If D_e is vacuous, then by definition

$$E_{D_e} I(g_{e_1, e_2 | D_e} | g_{e_1 | D_e} \cdot g_{e_2 | D_e}) = I(g_{e_1, e_2} | g_{e_1} \cdot g_{e_2}).$$

3.8.2.2 Strategy for model inference

Theorem 3.7 may be rewritten as

$$MI(f) = \sum_{\{i,j | K(ij)\} \in \mathcal{V}} b_{ij; K(ij)}, \tag{3.17}$$

where $K(ij)$ is the conditioning set for the node in $\mathcal{V}$ with conditioned set $\{i, j\}$. The terms $b_{ij; K(ij)}$ will depend on the regular vine which we choose to represent the dependence structure; however, the sum of these terms must satisfy (3.17). We seek a regular vine for which the terms $b_{ij; K(ij)}$ in (3.17) are as "spread out" as possible. In other words, we wish to capture the total dependence $MI(f)$ in a small number of terms, with the remaining terms being close to zero. This concept is made precise with the notion of *majorization*.[42]

Definition 3.6. *Let $\boldsymbol{x}, \boldsymbol{y} \in \mathbb{R}^n$ be such that $\sum_{i=1}^{n} x_i = \sum_{i=1}^{n} y_i$; then $\boldsymbol{x}$ majorizes $\boldsymbol{y}$ if for all k; $k = 1, \ldots, n$*

$$\sum_{j=1}^{k} x_{(j)} \leq \sum_{j=1}^{k} y_{(j)}, \tag{3.18}$$

where $x_{(1)} \leq \cdots \leq x_{(j)} \leq \cdots \leq x_{(n)}$ are the increasing arrangement of the components of $\boldsymbol{x}$, and similarly for $\{y_{(j)}\}$ and $\boldsymbol{y}$.

In view of (3.17) the model inference problem may be cast as the problem of finding a regular vine whose terms $b_{ij; K(ij)}$ are non-dominated in the sense of majorization. In that case, setting the smallest mutual informations equal to zero will change the overall mutual information as little as possible. Pairs of variables whose (conditional) mutual information is zero are (conditionally) independent. Finding non-dominated solutions may be difficult but a necessary condition for non-dominance can be found by maximizing any Schur convex function.

Definition 3.7. A function $\phi : \mathbb{R}^n \to \mathbb{R}$ is Schur convex if $\phi(\boldsymbol{x}) \geq \phi(\boldsymbol{y})$ whenever $\boldsymbol{x}$ majorizes $\boldsymbol{y}$.

Schur convex functions have been studied extensively. A sufficient condition for Schur convexity is given by Marshall and Olkin.[42]

Proposition 3.4. *If $\phi : \mathbb{R}^k \to \mathbb{R}$ may be written as $\phi(\boldsymbol{x}) = \sum \varphi(x_i)$ with φ convex, then ϕ is Schur convex.*

Vine Inference Strategy: The following strategy for model inference suggests itself:

(1) Choose a Schur convex function $\phi : \mathbb{R}^{n(n-1)/2} \to \mathbb{R}$;
(2) Find a regular vine $\mathcal{V}(n)$ whose vector $b_{ij;K(ij)}$ maximizes ϕ;
(3) Set the mutual information in $\mathcal{V}(n)$ equal to zero for which the terms $b_{ij;K(ij)}$ are smallest;
(4) Associate copulae with the nodes in the vine, such that the non-zero mutual information values are preserved.

A different strategy for model inference is proposed in Chapter 11.

3.9 Applications

This section references applications in the wider sense, including uses of vine-copula representations of multivariate distributions for mathematical and modeling purposes, as well as applications to analysis of multivariate data.

3.9.1 *Multivariate data analysis*

Due to their high flexibility and simple structure, pair-copula constructions/vines are becoming increasingly popular for constructing continuous multivariate distributions. While built exclusively from pair-copulae, they can model a wide range of complex dependencies. Lately, a number of publications on applications of pair-copula constructions have appeared in the literature. Most of the publications treat financial applications[1,2,8,11,13,20,43,48] while Kolbjornsen and Stien[32] present a non-parametric petroleum-related application of pair-copula constructions. For more details on some of these applications, see the application chapters (beginning with Chapter 12) of this book.

The studies of Aas and Berg[1] and Fischer et al.[13] also compare pair-copula constructions with other multivariate models, e.g., n-dimensional parametric copulae and hierarchical Archimedean constructions,[27] and conclude with the superiority of the pair-copula constructions. In Chapter 15 of this book, a short version of Aas and Berg[1] is given.

Some more applications of vine copulae are as follows. Biller[7] uses vine copulae for copula-based multivariate time-series input models, and compares them with Vector-Autoregressive-To-Anything (VARTA). De Michele et al.[12] and Salvadori et al.[47] have application to a multivariate model of sea storms.

3.9.2 Non-parametric Bayesian belief nets

Bayesian belief nets[10,22,46,49] (*BBNs*) are directed acyclic graphs. The nodes of the graph represent univariate random variables, which can be discrete or continuous, and the arcs represent directed influences. BBNs provide a compact representation of high-dimensional uncertainty distributions over a set of variables $(X_1, \ldots, X_n)$ and encode the probability density of these variables by specifying a set of conditional independence statements in the form of an acyclic directed graph and a set of probability functions. In their most popular form, BBNs were introduced in the 1980s as a knowledge representation formalism to encode and use the information acquired from human experts in automated reasoning systems to perform diagnostics and predictions.[46]

Until recently, most BBNs were discrete. Moreover, there were only two ways of dealing with continuous BBNs. One was to discretize the continuous variables and work with the corresponding discrete model, while the other was to assume joint normality. Both these methods have serious drawbacks. Therefore, non-parametric BBNs (NPBBNs) were introduced in Kurowicka and Cooke[35] and extended in Hanea.[17] In the NPBBNs, nodes are associated with arbitrary continuous invertible distributions and arcs with (conditional) rank correlations that are realized by a chosen copula. No joint distribution is assumed, which makes this BBN non-parametric. Non-parametric BBNs have seen several applications to date, the most notable being a very large model for civil aviation transport safety.[3] There is a close relationship between regular vines and NPBBNs. Chapter 14 of this book provides some insights into the differences and similarities between the two types of models.

In a BBN, the arcs of a directed graph can be associated with conditional copula, where the conditioned variables are the source and sink of the arc, and the conditioning variables are a subset of the other parents of the sink node. These conditional copulae, together with the one-dimensional marginal distributions and the conditional independence statements implied by the BBN graph uniquely determine the joint distribution, and every such specification is consistent.[35,36] This requires a copula type for which zero correlation implies independence. The proof pivots on representing the parents of a child node as a D-vine. When the number of non-independent conditional copulae is not too large, BBNs provide a much more perspicuous representation of the dependence structure than regular vines. In a regular vine all edges must be drawn, even if the conditional copula is independent.

References

1. Aas K. and Berg D. (2009). Models for construction of multivariate dependence: A comparison study. *European Journal of Finance*, 15:639–659.
2. Aas K., Czado C., Frigessi A. and Bakken H. (2009). Pair-copula constructions of multiple dependence. *Insurance: Mathematics and Economics*, 44(2):182–198.
3. Ale B.J.M., Bellamy L.J., van der Boom R., Cooper J., Cooke R.M., Goossens L.H.J., Hale A.R., Kurowicka D., Morales O., Roelen A.L.C. and Spouge J. (2009). Further development of a causal model for air transport safety (CATS): Building the mathematical heart. *Reliability Engineering and System Safety Journal*, 94(9):1433–1441.
4. Bedford T.J. and Cooke R.M. (2001). Probability density decomposition for conditionally dependent random variables modeled by vines. *Annals of Mathematics and Artificial Intelligence*, 32:245–268.
5. Bedford T.J. and Cooke R.M. (2002). Vines: A new graphical model for dependent random variables. *Annals of Statistics*, 30(4):1031–1068.
6. Besag J. (1974). Spatial interaction and the statistical analysis of lattice systems. *Journal of the Royal Statistical Society Series B*, 34:192–236.
7. Biller B. (2009). Copula-based multivariate input models for stochastic simulation. *Operations Research*, 57:878–892.
8. Chollete L., Heinen A. and Valdesogo A. (2009). Modeling international financial returns with a multivariate regime switching copula. *Journal of Financial Econometrics*, 7(4):437–480.
9. Cooke R.M. (1997). Markov and entropy properties of tree and vines-dependent variables. In *Proceedings of the ASA Section of Bayesian Statistical Science*. American Statistical Association, Washington.
10. Cowell R.G., Dawid A.P., Lauritzen S.L. and Spiegelhalter D.J. (1999). *Probabilistic Networks and Expert Systems*, Statistics for Engineering and Information Sciences. Springer-Verlag, New York.
11. Czado C., Min A., Baumann T. and Dakovic R. (2009). Pair-copula constructions for modeling exchange rate dependence. Technical report, Technische Universität München.
12. De Michele C., Salvadori G., Passoni G. and Vezzoli R. (2007). A multivariate model of sea storms using copulas. *Coastal Engineering*, 54:734–751.
13. Fischer M., Köck C., Schlüter S. and Weigert F. (2009). Multivariate copula models at work. *Quantitative Finance*, 9(7):839–854.

14. Galambos J. (1987). *The Asymptotic Theory of Extreme Order Statistics*. Kreiger, Malabar, FL.
15. Goossens L.H.J., Kraan B.C., Cooke R.M., Jones J.A., Brown J., Ehrhardt J., Fischer F. and Hasemann I. (2001). Methodology and processing techniques. Directorate-General for Research EUR 18827 EN, European Commission, Luxembourg.
16. Goossens L.H.J., Harper F.T., Kraan B.C.P. and Métivier H. (2000). Expert judgement for a probabilistic accident consequence uncertainty analysis. *Radiation Protection Dosimetry*, 90(3):295–301.
17. Hanea A.M. (2008). Algorithms for non-parametric Bayesian belief nets. PhD thesis, Delft Institute of Applied Mathematics, Delft University of Technology.
18. Hanea A.M., Kurowicka D., Cooke R.M. and Ababei D.A. (2010). Mining and visualising ordinal data with non-parametric continuous BBNs. *Computational Statistics and Data Analysis*, 54:668–687.
19. Harper F., Goossens L.H.J., Cooke R.M., Hora S., Young M., Pasler-Sauer J., Miller L., Kraan B.C.P., Lui C., McKay M., Helton J. and Jones A. (1994). Joint USNRC CEC consequence uncertainty study: Summary of objectives, approach, application, and results for the dispersion and deposition uncertainty assessment. Technical Report VOL. III, NUREG/CR-6244, EUR 15755 EN, SAND94-1453.
20. Heinen A. and Valdesogo A. (2009). Asymmetric CAPM dependence for large dimensions: The canonical vine autoregressive model. Technical report.
21. Hobaek Haff I., Aas K. and Frigessi A. (2010). On the simplified pair-copula construction — Simply useful or too simplistic? *Journal of Multivariate Analysis*, 101: 1296–1310.
22. Jensen F.V. (2001). *Bayesian Networks and Decision Graphs*. Springer-Verlag, New York.
23. Joe H. (1989). Relative entropy measures of multivariate dependence. *Journal of the American Statistical Association*, 84(405):157–164.
24. Joe H. (1993). Multivariate dependence measures and data analysis. *Computational Statistics and Data Analysis*, 16:279–297.
25. Joe H. (1994). Multivariate extreme-value distributions with applications in environmental data. *The Canadian Journal of Statistics*, 22:47–64.
26. Joe H. (1996). Families of m-variate distributions with given margins and $m(m-1)/2$ bivariate dependence parameters. In L. Rüschendorf, B. Schweizer and M.D. Taylor (eds.), *Distributions with Fixed Marginals and Related Topics*, Vol. 28, pp. 120–141. IMS Lecture Notes.
27. Joe H. (1997). *Multivariate Models and Dependence Concepts*. Chapman & Hall, London.
28. Joe H. (2005). Generating random correlation matrices based on partial correlations. *Journal of Multivariate Analysis*, 97:2177–2189.
29. Joe H. (2006). Range of correlation matrices for dependent random variables with given marginal distributions. In *Advances in Distribution Theory, Order Statistics and Inference, in Honor of Barry Arnold*, N. Balakrishnan, E. Castillo and J.M. Sarabia (eds.), pp. 125–142. Birkhauser, Boston.
30. Joe H., Li H. and Nikoloulopoulos A.K. (2010). Tail dependence functions and vine copulas. *Journal of Multivariate Analysis*, 101:252–270.
31. Kellerer H.G. (1964) Verteilungsfunktionen mit gegebenen Marginal-verteilungen. *Zeitschrift für Wahrscheinlichkeitstheorie und verwandte Gebiete*, 3:247–270.

32. Kolbjornsen O. and Stien M. (2008). The D-vine creation of non-Gaussian random fields. *Proceedings of the Eighth International Geostatistics Congress.*
33. Kraan B.C.P. and Cooke R.M. (2000). Processing expert judgements in accident consequence modeling. *Radiation Protection Dosimetry*, 90(3).
34. Kurowicka D. and Cooke R.M. (2003). A parametrization of positive definite matrices in terms of partial correlation vines. *Linear Algebra and its Applications*, 372: 225–251.
35. Kurowicka D. and Cooke R.M. (2004). Distribution-free continuous Bayesian belief nets. *Proceedings of Mathematical Methods in Reliability Conference.* Sante Fe, New Mexico, USA.
36. Kurowicka D. and Cooke R.M. (2006). *Uncertainty Analysis with High Dimensional Dependence Modelling.* Wiley, Chichester.
37. Kurowicka D., Cooke R.M. and Callies U. (2007). Vines inference. *Brazilian Journal of Probability and Statistics*, 21(1):103–120.
38. Kurowicka D. and Cooke R.M. (2006). Completion problem with partial correlation vines. *Linear Algebra and Its Applications*, 418(1):188–200.
39. Kurowicka D. and Cooke R.M. (2007). Sampling algorithms for generating joint uniform distributions using the vine-copula method. *Computational Statistics and Data Analysis*, 51:2889–2906.
40. Lewandowski D. (2008). *High Dimensional Dependence. Copulae, Sensitivity, Sampling.* PhD thesis, Delft Institute of Applied Mathematics, Delft University of Technology.
41. Lewandowski D., Kurowicka D. and Joe H. (2009). Generating random correlation matrices based on vines and extended onion method. *Journal of Multivariate Analysis*, 100:1989–2001.
42. Marshall A.W. and Olkin I. (1979). *Inequalities: Theory of Majorization and its Applications.* Springer, New York.
43. Min A. and Czado C. (2010). Bayesian inference for multivariate copulas using pair-copula constructions. *Journal of Financial Econometrics.* In press.
44. Morales-Nápoles O., Cooke R.M. and Kurowicka D. (2008). The number of vines and regular vines on n nodes. Submitted to *Discrete Applied Mathematics.*
45. Nelsen R.B. (2006). *An Introduction to Copulas*, 2nd ed. Springer, New York.
46. Pearl J. (1988). *Probabilistic Reasoning in Intelligent Systems: Networks of Plausible Inference.* Morgan Kaufman Publishers, San Mateo.
47. Salvadori G., De Michele C., Kottegoda N.T. and Rosso R. (2007). *Extremes in Nature. An Approach Using Copulas.* Volume 56 of Water Science and Technology Library. Springer, Dordrecht.
48. Schirmacher D. and Schirmacher E. (2008). Multivariate dependence modeling using pair-copulas. Technical report, presented at the 2008 ERM Symposium, Chicago.
49. Shachter R.D. and Kenley C.R. (1989). Gaussian influence diagrams. *Management Science*, 35(5):527–550.
50. Whittaker J. (1990). *Graphical Models in Applied Multivariate Statistics.* Wiley, Chichester.
51. Xu J.J. (1996). *Statistical Modelling and Inference for Multivariate and Longitudinal Discrete Response Data.* PhD thesis, Department of Statistics, University of British Columbia.
52. Yule G.U. and Kendall M.G. (1965). *An Introduction to the Theory of Statistics*, 14th ed. Charles Griffin & Co, Belmont, California.

CHAPTER 4

Sampling Count Variables with Specified Pearson Correlation: A Comparison Between a Naive and a C-Vine Sampling Approach

Vinzenz Erhardt* and Claudia Czado

Technische Universität München, Zentrum Mathematik
Boltzmannstr. 3, 85747 Garching, Germany
*erhardt@ma.tum.de

Erhardt and Czado[11] suggest an approximative method for sampling high-dimensional count random variables with a specified Pearson correlation. They utilize Gaussian copulae for the construction of multivariate discrete distributions. A major task is to determine the appropriate copula parameters for the achievement of a specified target correlation. Erhardt and Czado[11] develop an optimization routine to determine these copula parameters sequentially. Thereby, they use pair-copula decompositions of n-dimensional distributions, i.e., a decomposition consisting only of bivariate copula with one parameter each. C-vines, a graphical tool to organize such pair-copula decompositions, are used to select a possible decomposition. In the paper mentioned, the approach was compared to the NORTA method for discrete margins described in Ref. 2. Here, we will compare it to a widely used naive sampling approach for an even larger variety of marginal distributions such as the Poisson, generalized Poisson, negative binomial and zero-inflated generalized Poisson distributions.

4.1	Introduction	74
4.2	Copulae and Multivariate Distributions	75
4.3	Naive Sampling with Illustration to GP Count Data	78
4.4	Simulation Study	80
4.5	Summary and Discussion	86
	References	86

4.1 Introduction

Erhardt and Czado[11] suggest a method for approximately sampling high-dimensional count variables with a prespecified Pearson correlation. The goal of this chapter is to sample from count random variables (rv's) $Y_1, \ldots, Y_n$ with $Y_i \sim F_i$ (e.g. Poisson), $i = 1, \ldots, n$ with prespecified corr($\boldsymbol{Y}$) = $\boldsymbol{\Sigma}^Y$, with (i,j)th element $\boldsymbol{\Sigma}^Y_{ij} = \rho_{ij}$ and $\rho_{ii} = 1$. Genest and Neslehova[13] review several facts about copulae linked to discrete margins specifically for rank-based dependence measures. Multivariate discrete distributions discussed in the literature have several shortcomings which we consider now. Kawamura[18] defines a multivariate Poisson distribution which can be obtained as a limiting case of a multivariate binomial distribution. Since these multivariate Poisson models only allow for a single common correlation parameter $\rho_{ij} = \rho$,[17] we are able to construct a model which allows for individual correlations between each pair of variables. However, these pairwise correlations are required to satisfy the positive definiteness constraint. According to Tsiamyrtzis and Karlis,[29] the usefulness of multivariate discrete models is limited since calculating the required probabilities is difficult. Therefore, they suggest algorithms that calculate the joint probabilities in a more efficient way, thus reducing the computational time. A multivariate negative binomial distribution has been discussed, for example, by Kopocinski.[19] A multivariate generalization of the generalized Poisson distribution (see Ref. 9) capable of modeling only exchangeable covariance structures has been developed by Vernic[30] and applied to the insurance field.

In the sampling approach of Erhardt and Czado,[11] dependency is modeled using a pair-copula decomposition of a general multivariate distribution. The graphical tool for organizing such decompositions is called a regular vine and goes back to the work on vines by Refs. 3, 4, 15 and 5. Aas et al.[1] propose a new method to perform inference of such pair-copula decompositions. In particular, the approach of Erhardt and Czado[11] is based on the Gaussian copula and a C-vine decomposition. The idea is to use a conditional sampling approach where conditional cdfs and quantiles are defined via a pair-copula construction. Here the bivariate copulae have only one parameter each, therefore a root-finding routine such as bisection can be utilized to sequentially determine optimal parameters for each pair-copula. They compare their approach to a widely used naive sampling approach.

An approximate method for sampling correlated continuous random variables from partially specified distributions has been introduced by Lurie and Goldberg.[21] This method is an enhancement of an approach by Li and Hammond[20] and is based on the multivariate normal distribution. Their approach optimizes the set of parameters such that the empirical correlations come close to the target correlations according to some distance measure, therefore the empirical and target correlations will closely match, if not agree. Whereas Erhardt and Czado[11] compare their sampling approach to the NORTA method, in this chapter, we will compare it to a "naive" sampling method often used. The NORTA method ("NORmal To Anything", see Refs. 6–8) is based on the work of Mardia[22] and Li and Hammond.[20] The naive sampling method assumes that the Gaussian copula parameters specifying the underlying multivariate distribution of the desired margins coincide with the target correlation parameters. The contribution of this chapter will be two-fold. The simulation study in Ref. 11 will be completed by considering also the generalized Poisson distribution and the zero-inflated generalized Poisson distribution. Since the presence of zero-inflation causes the margins to be even more discrete we are interested in investigating the influence of zero-inflation on the sampling results. Second, investigating the results of the naive approach will quantify the impact of this simplifying assumption.

This chapter is organized as follows: In Section 4.2, we will review some basic properties of multivariate distributions and copulae and will also review the concept of partial correlations, which the approach is based on. We will summarize the naive sampling method in Section 4.3. For generalized Poisson data in dimension 8, we will compare the C-vine sampling approach to the naive sampling method. An extensive simulation study comparing the two approaches is given in Section 4.4. We conclude with a summary and discussion in Section 4.5.

4.2 Copulae and Multivariate Distributions

Marginal distributions considered in this chapter will be the Poisson, generalized Poisson (GP), zero-inflated generalized Poisson (ZIGP) and the negative binomial (NB) distributions. Similar to the NB distribution, the GP distribution introduced in Ref. 9 can model overdispersion with respect to the Poisson model. Its advantage over the NB distribution is that the overdispersion factor in the GP case depends on one additional parameter φ whereas in the NB case it depends on an additional parameter as well as the mean parameter. A second advantage of the GP distribution is that for $\varphi = 1$ it reduces to the Poisson distribution. The ZIGP distribution is obtained by a mixing between the zero and the GP distribution. The probability of the mixing variable is an additional zero-inflation parameter ω, i.e., for $\omega = 0$ the distribution simplifies to the GP distribution.

Table 4.1. Probability mass functions of the Poisson, GP, ZIGP and NB distributions together with their means and variances.

	$P(Y=y)$
Poisson	$\frac{\mu^y}{y!}e^{-\mu}$ $E(Y) = \mu$, $\text{Var}(Y) = \mu$
GP	$\frac{\mu(\mu+(\varphi-1)y)^{y-1}}{y!}\varphi^{-y}e^{-\frac{1}{\varphi}(\mu+(\varphi-1)y)}$, where $\varphi > \max(\frac{1}{2}, 1 - \frac{\mu}{m})$ and m is the largest natural number with $\mu + m(\varphi - 1) > 0$, if $\varphi < 1$. $E(Y) = \mu$, $\text{Var}(Y) = E(Y)\varphi^2$
ZIGP	$\mathbf{1}_{\{y=0\}}\left[\omega + (1-\omega)e^{-\frac{\mu}{\varphi}}\right]$ $+\mathbf{1}_{\{y>0\}}\left[(1-\omega)\frac{\mu(\mu+(\varphi-1)y)^{y-1}}{y!}\varphi^{-y}e^{-\frac{1}{\varphi}(\mu+(\varphi-1)y)}\right]$ where in the case of $\varphi < 1$ the same condition as in the GP case must hold. $E(Y) = (1-\omega)\mu$, $\text{Var}(Y) = E(Y)\left(\varphi^2 + \mu\omega\right)$
NB	$\frac{\Gamma(y+\Psi)}{\Gamma(\Psi)y!}\left(\frac{\Psi}{\mu+\Psi}\right)^{\Psi}\left(\frac{\mu}{\mu+\Psi}\right)^{y}$ $E(Y) = \mu$, $\text{Var}(Y) = \mu(1 + \frac{\mu}{\Psi})$

Excess zeros can be regarded as a second source of zero-inflation. In order to allow for a comparison between these two distributions, we choose the mean parametrization for all of the distributions. Their probability mass function (pmf) together with means and variances are given in Table 4.1.

We will use copulae to obtain multivariate count distributions with marginal counts as specified above. An n-dimensional copula C_n is a multivariate cdf $C_n : [0,1]^n \to [0,1]$ whose univariate margins are uniform on $[0,1]$, i.e., $C_n(1, \ldots, 1, u_i, 1, \ldots, 1) = u_i \;\; \forall i \in \{1, \ldots, n\}$. For n continuous rv's $\mathbf{Y} := (Y_1, \ldots, Y_n)'$ with marginal distributions $F_1, \ldots, F_n$, the rv $F_i(Y_i)$ is uniform on $[0,1]$. Sklar[26] shows that while F_i reflects the marginal distribution of Y_i, C_n reflects the dependence, i.e.

$$F_{\mathbf{Y}}(y_1, \ldots, y_n) = C_n(F_1(y_1|\boldsymbol{\theta}_1), \ldots, F_n(y_n|\boldsymbol{\theta}_n)|\boldsymbol{\tau}), \tag{4.1}$$

where $\boldsymbol{\tau}$ are the corresponding copula parameters. Hence, for a multivariate cdf of $\mathbf{Y}$, there always is a copula C_n separating the dependence structure from the marginal distributions. However, C_n is unique only for continuous margins. Vice versa, a multivariate cdf can be constructed by virtue of (4.1) from n marginal distributions using an n-dimensional copula C_n. The sampling approach by Erhardt and Czado[11] is based on Gaussian copulae. For a more detailed introduction to copulae including the Gaussian copula, see for instance Refs. 16, 23 or Ref. 10. Copulae with discrete margins are discussed, for example, by.[27]

Definition 4.1 (Gaussian copula). The n-dimensional Gaussian copula is a function $C_n : [0,1]^n \to [0,1]$ with

$$C_n(u_1, \ldots, u_n | \mathbf{\Sigma}^Z) := \Phi_n \left(\Phi^{-1}(u_1), \ldots, \Phi^{-1}(u_n) | \mathbf{\Sigma}^Z \right), \quad (4.2)$$

where $\Phi_n(\cdot | \mathbf{\Sigma}^Z)$ is the cdf of the n-dimensional normal distribution with mean $\boldsymbol{\mu} = \mathbf{0}_n$ and covariance $\mathbf{\Sigma}^Z$ and $\Phi^{-1}(\cdot)$ is the univariate standard normal quantile function.

In the special case of $n = 2$, we write $C_2(u_1, u_2 | \tau_{12}) = \Phi_2(\Phi^{-1}(u_1), \Phi^{-1}(u_2) | \tau_{12})$ instead of (4.2). The n-dimensional Gaussian copula density is

$$c_n(u_1, \ldots, u_n | \mathbf{\Sigma}^Z) = \phi_n \left(\Phi^{-1}(u_1), \ldots, \Phi^{-1}(u_n) | \mathbf{\Sigma}^Z \right) \prod_{i=1}^n \frac{1}{\phi(\Phi^{-1}(u_i))},$$

with ϕ_n being the n-dimensional normal pdf with mean $\boldsymbol{\mu} = \mathbf{0}_n$ and covariance $\mathbf{\Sigma}^Z$.

Erhardt and Czado[11] stress that for a joint distribution of count margins defined by a Gaussian copula, there are three levels of correlated random variables:

(1) **Multivariate normal level**: $(Z_1, \ldots, Z_n) \sim N_n \left(\mathbf{0}, \mathbf{\Sigma}^Z \right)$, where the $(i,j)^{\text{th}}$ element of $\mathbf{\Sigma}^Z$ will be denoted by τ_{ij}. We refer to τ_{ij} as the "association parameter".
(2) **Uniform level**: $U_1, \ldots, U_n \sim unif(0,1)$, $U_i := \Phi(Z_i)$, $i = 1, \ldots, n$. The joint cdf $G(u_1, \ldots, u_n) = C_n(u_1, \ldots, u_n | \mathbf{\Sigma}^Z)$ is defined by the Gaussian copula cdf with association parameters $\mathbf{\Sigma}^Z$.
(3) **Count level**: $\mathbf{Y} := (Y_1, \ldots, Y_n)'$ are counts, where $Y_i := F_i^{-1}(U_i | \boldsymbol{\theta}_i)$, $i = 1, \ldots, n$ and $\boldsymbol{\theta}_i$ are the parameters of margin i. Further, $F_i^{-1}(U_i | \boldsymbol{\theta}_i)$ is the pseudo-inverse of F_i at U_i. The joint cdf is $F(y_1, \ldots, y_n | \boldsymbol{\theta}_1, \ldots, \boldsymbol{\theta}_n) = C_n(F_1(y_1 | \boldsymbol{\theta}_1), \ldots, F_n(y_n | \boldsymbol{\theta}_n) | \mathbf{\Sigma}^Z)$. For $Y_1, \ldots, Y_n$ with $Y_i \sim F_i$, $i = 1, \ldots, n$, $\text{corr}(\mathbf{Y}) =: \mathbf{\Sigma}^Y$, where $\Sigma_{ij}^Y = \rho_{ij}$ and $\rho_{ii} = 1$.

They argue that the main problem of sampling from such a copula specification is that $\text{corr}(Z_i, Z_j) \neq \text{corr}(U_i, U_j) \neq \text{corr}(Y_i, Y_j)$.

An important concept for the sampling approach in Ref. 11 is partial correlations. Here we review an important property of partial correlations since it will be needed in the simulation study in this chapter. Partial correlation is the correlation between two variables while controlling for a third variable or more. Let $\mathbf{W}$ be a standardized n-dimensional random vector, where we partition $\mathbf{W} = (W_1, W_2, \mathbf{W}_3')'$, and $\mathbf{W}_3 = (W_3, \ldots, W_n)'$

is an $(n-2)$-dimensional random vector. Mean and correlation matrix are $\boldsymbol{\mu} = (\mu_1, \mu_2, \boldsymbol{\mu}_3')'$ and

$$\boldsymbol{\Sigma} = \begin{pmatrix} \sigma_{11} & \sigma_{12} & \boldsymbol{\sigma}_{13}' \\ \sigma_{12} & \sigma_{22} & \boldsymbol{\sigma}_{23}' \\ \boldsymbol{\sigma}_{13} & \boldsymbol{\sigma}_{23} & \boldsymbol{\Sigma}_{33} \end{pmatrix}, \quad \boldsymbol{\Sigma}^{-1} =: \begin{pmatrix} \sigma^{11} & \sigma^{12} & \boldsymbol{\sigma}^{13'} \\ \sigma^{12} & \sigma^{22} & \boldsymbol{\sigma}^{23'} \\ \boldsymbol{\sigma}^{13} & \boldsymbol{\sigma}^{23} & \boldsymbol{\Sigma}^{33} \end{pmatrix}.$$

According to Srivastava and Khatri[28] (p. 53), the partial correlation between W_1 and W_2 with $\boldsymbol{W}_3$ denoted by $\rho_{12;3:n}$ is defined as the correlation between $W_1 - \boldsymbol{\sigma}_{13}'\boldsymbol{\Sigma}_{33}^{-1}\boldsymbol{W}_3$ and $W_2 - \boldsymbol{\sigma}_{23}'\boldsymbol{\Sigma}_{33}^{-1}\boldsymbol{W}_3$, which is the correlation between W_1 and W_2 after eliminating the best linear effects of $\boldsymbol{W}_3$ from both variables. It can be calculated as $\rho_{12;3:n} = \frac{-\sigma^{12}}{\sqrt{\sigma^{11}\sigma^{22}}}$. An important property of partial correlations is a recursive formula (see e.g., Ref. 24): for $I := \{1, \ldots, n\}$ and for any subset $I^* \subseteq I$, which contains at least i, j and k,

$$\rho_{ij;I^*\setminus\{i,j\}} = \frac{\rho_{ij;I^*\setminus\{i,j,k\}} - \rho_{ik;I^*\setminus\{i,j,k\}} \cdot \rho_{jk;I^*\setminus\{i,j,k\}}}{\sqrt{(1-\rho_{ik;I^*\setminus\{i,j,k\}}^2)(1-\rho_{jk;I^*\setminus\{i,j,k\}}^2)}}, \quad (4.3)$$

i.e., partial correlations of order $(n-2)$ can be calculated from those of order $(n-3)$.

4.3 Naive Sampling with Illustration to GP Count Data

In this section, we will compare our sampling approach to a naive approach of sampling count random variables. The naive approach is to use our desired target correlation $\boldsymbol{\Sigma}^Y$ and generate for a sample of N subjects n-dimensional multivariate normal random vectors with covariance $\boldsymbol{\Sigma}^Y$, i.e. $\boldsymbol{Z}_k \sim N_n(\boldsymbol{0}, \boldsymbol{\Sigma}^Y)$, $k = 1, \ldots, N$. Next, we transform the sample $\boldsymbol{z}_k = (z_{k1}, \ldots, z_{kn})'$ to the uniform level $\boldsymbol{u}_k := (\Phi(z_{k1}), \ldots, \Phi(z_{kn}))'$, $k = 1, \ldots, N$ and determine the sample correlation $\hat{\boldsymbol{\Sigma}}^U$ of $\{\boldsymbol{u}_k, k = 1, \ldots, N\}$. Then we generate outcomes according to the generalized Poisson distribution (see Table 4.1) with cdf F_i by determining the quantiles of the GP distribution with mean μ_i and variance $\mu_i\varphi_i^2$ at u_{ki}, $k = 1, \ldots, N$, $i = 1, \ldots, n$, i.e., $y_{ki}^{naive} := F_i^{-1}(u_{ki}|\mu_i, \varphi_i)$, and $\boldsymbol{y}_k^{naive} := (y_{k1}^{naive}, \ldots, y_{kn}^{naive})'$. The sample correlation of $\{\boldsymbol{y}_k^{naive}, k = 1, \ldots, N\}$ will be denoted by $\hat{\boldsymbol{\Sigma}}^{Y^{naive}}$.

For $n = 8$ and $N = 100000$, we use as a target correlation matrix an exchangeable structure, i.e., $\boldsymbol{\Sigma}^Y = (\rho_{ij})$ with $\rho_{ij} = 0.6$ $\forall i \neq j$ and $\rho_{ii} = 1$. Marginal means of the eight-dimensional GP distribution were set to $\boldsymbol{\mu} := (4, 25, 120, 2, 28, 7, 27, 5)'$, dispersion parameters to

$\varphi := (1.5, 3.5, 2, 2.5, 2, 3, 1.5, 2.5)'$. The empirical correlation matrix $\hat{\Sigma}^U$ is determined to be

$$\hat{\Sigma}^U = \begin{pmatrix} 1.0000, 0.5814, 0.5836, 0.5799, 0.5812, 0.5815, 0.5821, 0.5807 \\ 0.5814, 1.0000, 0.5849, 0.5841, 0.5837, 0.5855, 0.5837, 0.5821 \\ 0.5836, 0.5849, 1.0000, 0.5839, 0.5840, 0.5819, 0.5832, 0.5853 \\ 0.5799, 0.5841, 0.5839, 1.0000, 0.5809, 0.5829, 0.5842, 0.5831 \\ 0.5812, 0.5837, 0.5840, 0.5809, 1.0000, 0.5827, 0.5804, 0.5818 \\ 0.5815, 0.5855, 0.5819, 0.5829, 0.5827, 1.0000, 0.5839, 0.5822 \\ 0.5821, 0.5837, 0.5832, 0.5842, 0.5804, 0.5839, 1.0000, 0.5848 \\ 0.5807, 0.5821, 0.5853, 0.5831, 0.5818, 0.5822, 0.5848, 1.0000 \end{pmatrix},$$

where the average absolute deviation of all off-diagonal elements from Σ^Y is 0.0172. Naively transforming the obtained uniform variables to the count level gives us a sample of count variables whose empirical correlation matrix is calculated to be

$$\hat{\Sigma}^{Y^{naive}} = \begin{pmatrix} 1.0000, 0.5711, 0.5788, 0.5036, 0.5791, 0.5386, 0.5808, 0.5473 \\ 0.5711, 1.0000, 0.5717, 0.5062, 0.5777, 0.5497, 0.5727, 0.5491 \\ 0.5788, 0.5717, 1.0000, 0.4781, 0.5956, 0.5298, 0.5979, 0.5400 \\ 0.5036, 0.5062, 0.4781, 1.0000, 0.4909, 0.5007, 0.4850, 0.5069 \\ 0.5791, 0.5777, 0.5956, 0.4909, 1.0000, 0.5402, 0.5919, 0.5452 \\ 0.5386, 0.5497, 0.5298, 0.5007, 0.5402, 1.0000, 0.5361, 0.5360 \\ 0.5808, 0.5727, 0.5979, 0.4850, 0.5919, 0.5361, 1.0000, 0.5429 \\ 0.5473, 0.5491, 0.5400, 0.5069, 0.5452, 0.5360, 0.5429, 1.0000 \end{pmatrix}.$$

The off-diagonal average absolute deviation is 0.0556. If we, however, use our approach for sampling correlated GP variables, we get

$$\hat{\Sigma}^Y = \begin{pmatrix} 1.0000, 0.5938, 0.5984, 0.6022, 0.5992, 0.5921, 0.5975, 0.5953 \\ 0.5938, 1.0000, 0.5977, 0.6019, 0.6030, 0.5989, 0.6012, 0.6072 \\ 0.5984, 0.5977, 1.0000, 0.5589, 0.6161, 0.5828, 0.6243, 0.5898 \\ 0.6022, 0.6019, 0.5589, 1.0000, 0.5721, 0.6317, 0.5632, 0.6405 \\ 0.5992, 0.6030, 0.6161, 0.5721, 1.0000, 0.5948, 0.6249, 0.5985 \\ 0.5921, 0.5989, 0.5828, 0.6317, 0.5948, 1.0000, 0.5930, 0.6301 \\ 0.5975, 0.6012, 0.6243, 0.5632, 0.6249, 0.5930, 1.0000, 0.6065 \\ 0.5953, 0.6072, 0.5898, 0.6405, 0.5985, 0.6301, 0.6065, 1.0000 \end{pmatrix},$$

where the off-diagonal absolute deviations have an average value of 0.0130.

4.4 Simulation Study

In this section, we want to perform a systematic comparison of the small sample performance of the two sampling approaches for a correlated count random vector $\boldsymbol{Y} = (Y_1, \ldots, Y_n)$ with target correlation $\rho_{ij} = \text{corr}(Y_i, Y_j)$, $1 \leq i < j \leq n$. We consider two methods for measuring the performance of the approaches. The description of these measures and of the specification of the simulation settings are given in detail in Section 6 of Erhardt and Czado.[11]

Relative bias with respect to target correlation

In R independent replications, we generate an N-dimensional i.i.d. sample of $\boldsymbol{Y}$. For $\boldsymbol{y}_i^r := (y_{1i}^r, \ldots, y_{Ni}^r)'$, $i = 1, \ldots, n$, $r = 1, \ldots, R$, let $\hat{\rho}_{ij}^r$ be the empirical correlation coefficient based on $\boldsymbol{y}_i^r$ and $\boldsymbol{y}_j^r$. Then the estimated relative bias is $\widehat{rb}_{ij} := \frac{1}{R} \sum_{r=1}^{R} \frac{\hat{\rho}_{ij}^r}{\rho_{ij}} - 1$, where ρ_{ij} is the target correlation. These estimated biases will be dependent, therefore we will consider the maximal estimated relative bias $MAXRB := \max_{1 \leq i < j \leq n} \widehat{rb}_{ij}$ as an overall measure for all $1 \leq i < j \leq n$.

Average number of acceptance of specified correlation

We would like to test

$$H_0 : \rho_{ij} = \rho_{ij}^0 \quad \forall 1 \leq i < j \leq n \text{ versus } H_1 : \text{not } H_0, \quad (4.4)$$

where ρ_{ij}^0 is the target correlation. This composite test consists of $\frac{n(n-1)}{2}$ individual tests, i.e., we reject H_0 if for some (i, j)

$$H_0^{ij} : \rho_{ij} \neq \rho_{ij}^0 \text{ versus } H_1^{ij} : \rho_{ij} = \rho_{ij}^0 \quad (4.5)$$

cannot be rejected. Thus, we are dealing with a multiple testing problem. The classic way to account for this is to use the Bonferroni correction (see Ref. 25) where the overall α level test for (4.4) is obtained by performing $\frac{n(n-1)}{2}$ individual tests (4.5) based on level α_c with $\alpha_c = \frac{\alpha}{n(n-1)/2}$. Further, since the distribution of $\hat{\rho}_{ij}^r$ is unknown, we use the Fisher z-transform to $\mathcal{R}$ by defining $\hat{z}_{ij}^r := \tanh^{-1}(\hat{\rho}_{ij}^r)$ and $z_{ij}^0 := \tanh^{-1}(\rho_{ij}^0)$. Then according to Ref. 12 an asymptotic α_c-level test for (4.5) is given by

$$\text{Reject } H_0^{ij} : \rho_{ij} \neq \rho_{ij}^0 \Leftrightarrow \frac{|\hat{z}_{ij}^r - z_{ij}^0|}{1/\sqrt{N-3}} \leq q_{\alpha_c},$$

where q_{α_c} is the $(1 - \alpha_c)$ quantile of a standard normal distribution. If an $i < j$ exists such that $H_0^{ij} : \rho_{ij} \neq \rho_{ij}^0$ is not rejected on level α_c, reject

$H_0 : \rho_{ij} = \rho_{ij}^0 \; \forall \, 1 \leq i < j \leq n$ at level α. We set ACC_α as the percentage of acceptances of H_0 at level α among the R replications.

The number of replications in our simulation study is $R = 1000$, N is now chosen to be 500. We consider the four distributions introduced in Section 4.2. Marginal parameters $\boldsymbol{\theta}_i$ are μ_i in the Poisson case, (μ_i, φ_i) in the GP case, $(\mu_i, \varphi_i, \omega_i)$ in the ZIGP case and (μ_i, ψ_i) in the NB case. Variances $Var(Y_{ki}^r)$ will be equal in the GP and NB case if we set $\varphi_i^2 = 1 + \frac{\mu_i}{\psi_i}$ or equivalently $\psi_i = \frac{\mu_i}{\varphi_i^2 - 1}$. According to Table 4.1, a high ψ_i corresponds to low overdispersion and vice versa.

(1) First, we investigated the influence of the dimension n and the size of the correlation in an exchangeable target correlation structure, i.e. $\rho_{ij} = \rho$. The settings were $\rho \in \{0.1, 0.5, 0.9\}$, $n \in \{2, 5, 10\}$. Medium-sized marginal parameters according to Table 4.2 were used. Results are summarized in Table 4.4.

(2) For the exchangeable target correlation structure, we looked at the influence of the marginal parameters. Here, $\rho = 0.5$ and $n = 5$ were fixed. For $\boldsymbol{\mu}$, $\boldsymbol{\varphi}$ and $\boldsymbol{\omega}$, sets of small values (S) were compared to sets of larger (L) values. Again, for $\boldsymbol{\psi}^S := (\psi_1^S, \ldots, \psi_n^S)$ and $\boldsymbol{\psi}^L := (\psi_1^L, \ldots, \psi_n^L)$, the

Table 4.2. Marginal parameter choices for $n = 2$, 5 and 10 and exchangeable correlation structure for different marginal distributions (marginal variances for GP and NB margins are chosen to be equal).

	T	Parameters
Poi	2	$\boldsymbol{\mu} := (10, 15)'$
	5	$\boldsymbol{\mu} := (10, 15, 12, 20, 28)'$
	10	$\boldsymbol{\mu} := (10, 15, 12, 20, 28, 17, 27, 13, 19, 25)'$
GP		$\boldsymbol{\mu}$ as in Poisson case
	2	$\boldsymbol{\varphi} := (1.5, 3.5)'$
	5	$\boldsymbol{\varphi} := (1.5, 3.5, 1.5, 2, 2.5)'$
	10	$\boldsymbol{\varphi} := (1.5, 3.5, 1.5, 2, 2.5, 2, 3, 1.5, 1.5, 2.5)'$
ZIGP		$\boldsymbol{\mu}$ and $\boldsymbol{\varphi}$ as in GP case
	2	$\boldsymbol{\omega} := (0.25, 0.15)'$
	5	$\boldsymbol{\omega} := (0.25, 0.15, 0.10, 0.3, 0.2)'$
	10	$\boldsymbol{\omega} := (0.25, 0.15, 0.10, 0.3, 0.2, 0.17, 0.24, 0.24, 0.2, 0.15)'$
NB		$\boldsymbol{\mu}$ as in Poisson case
	2	$\boldsymbol{\psi} := (8, 1\frac{1}{3})'$
	5	$\boldsymbol{\psi} := (8, 1\frac{1}{3}, 9.6, 6\frac{2}{3}, 5\frac{1}{3})'$
	10	$\boldsymbol{\psi} := (8, 1\frac{1}{3}, 9.6, 6\frac{2}{3}, 5\frac{1}{3}, 5\frac{2}{3}, 3.375, 10.4, 15.2, 4.762)'$

Table 4.3. Marginal parameter choices for investigating the influence of marginal parameter sizes ($\psi^S(\mu)$ corresponds to large overdispersion, $\psi^L(\mu)$ to small overdispersion).

Small	Large
$\boldsymbol{\mu}^S := (1, 3, 2, 2, 1.5)'$	$\boldsymbol{\mu}^L := (30, 20, 35, 50, 25)'$
$\boldsymbol{\varphi}^S := (1.1, 2.5, 1.5, 3, 2)'$	$\boldsymbol{\varphi}^L := (6, 5, 3, 4, 4.5)'$
$\boldsymbol{\omega}^S := (0.05, 0.1, 0.05, 0.08, 0.07)'$	$\boldsymbol{\omega}^L := (0.25, 0.2, 0.35, 0.15, 0.4)'$
$\boldsymbol{\psi}^S(\boldsymbol{\mu}^S) := (4.76, 0.57, 1.6, 0.25, 0.5)'$	$\boldsymbol{\psi}^L(\boldsymbol{\mu}^S) := (0.03, 0.13, 0.25, 0.13, 0.08)'$
$\boldsymbol{\psi}^S(\boldsymbol{\mu}^L) := (142.9, 3.810, 28, 6.25, 8.33)'$	$\boldsymbol{\psi}^L(\boldsymbol{\mu}^L) := (0.86, 0.83, 4.38, 3.33, 1.30)'$

entries were calculated according to $\psi_i^S(\mu_i) = \frac{\mu_i}{(\varphi_i^S)^2 - 1}$ and $\psi_i^L(\mu_i) = \frac{\mu_i}{(\varphi_i^L)^2 - 1}$, respectively, where μ_i could either be μ_i^S or μ_i^L (see Table 4.3). Results can be found in Table 4.5.

(3) Finally, AR(1) and unstructured target correlations were investigated (Table 4.6).

AR(1) and unstructured correlation matrices:

For $R = 1000$ replications, $N = 500$ and $n = 5$, we investigated as target correlation also AR(1) and unstructured correlation matrices, i.e., for the AR(1) case we used $\boldsymbol{\Sigma}^Y = (\rho_{ij})$ with $\rho_{ij} = 0.7^{|i-j|} \ \forall \, i \neq j$ and $\rho_{ii} = 1$. In order to obtain unstructured correlation matrices, we generated a sample of $R = 1000$ unstructured partial correlations fully specifying a C-vine decomposition. Then we calculated the corresponding correlation matrix from them using the recursive expression (4.3). Note that not all correlations can be sampled. For very high and very low target correlations and especially for low marginal means in i and/or j, $\tau_{ij}(\boldsymbol{\Sigma}^Y|\boldsymbol{\theta})$ might not exist. We did not discard the simulation in these replications but used the result generated from the closest association parameters obtained in the bisection step when no further optimization could be achieved. We briefly interpret the obtained results.

Influence of the choice of ρ:

According to Table 4.4, the higher the target correlation chosen, the smaller $ACC_{0.05}$ was and hence the worse the approximations became. The maximal estimated relative bias, however, shrinks. This is due to the standardization by the true correlation parameters.

Table 4.4. Maximal estimated relative bias ($MAXRB$) and proportion of tests which accepted target correlation ($ACC_{0.05}$) based on $R = 1000$ replications of $N = 500$ samples of size n for exchangeable target correlation ρ and different count margins and parameters as in Table 4.2 (bold: C-vine sampling, italics: naive sampling).

		Poisson		GP		ZIGP		NB	
ρ	n	$MAXRB$	$ACC_{0.05}$	$MAXRB$	$ACC_{0.05}$	$MAXRB$	$ACC_{0.05}$	$MAXRB$	$ACC_{0.05}$
0.1	2	**0.0018**	**1.000**	**0.0036**	**1.000**	**0.0011**	**1.000**	**0.0004**	**1.000**
		0.0236	*0.938*	*0.0859*	*0.935*	*0.1275*	*0.929*	*0.0905*	*0.944*
	5	**0.0372**	**1.000**	**0.0191**	**1.000**	**0.0299**	**1.000**	**0.0279**	**1.000**
		0.0338	*0.959*	*0.1446*	*0.933*	*0.1511*	*0.936*	*0.1037*	*0.937*
	10	**0.1068**	**1.000**	**0.0659**	**1.000**	**0.0703**	**1.000**	**0.0735**	**1.000**
		0.0350	*0.940*	*0.1295*	*0.932*	*0.1311*	*0.937*	*0.1091*	*0.947*
0.5	2	**0.0002**	**1.000**	**0.0001**	**1.000**	**0.0005**	**1.000**	**0.0000**	**1.000**
		0.0119	*0.951*	*0.0776*	*0.770*	*0.0939*	*0.708*	*0.0619*	*0.826*
	5	**0.0191**	**0.995**	**0.0110**	**0.992**	**0.0176**	**0.998**	**0.0083**	**0.996**
		0.0114	*0.952*	*0.0774*	*0.764*	*0.1004*	*0.709*	*0.0589*	*0.836*
	10	**0.0309**	**1.000**	**0.0119**	**0.998**	**0.0231**	**0.998**	**0.0091**	**0.999**
		0.0093	*0.955*	*0.0748*	*0.792*	*0.1242*	*0.731*	*0.0615*	*0.850*
0.9	2	**0.0006**	**1.000**	**0.0006**	**1.000**	**0.0003**	**1.000**	**0.0004**	**1.000**
		0.0077	*0.877*	*0.0456*	*0.038*	*0.0699*	*0.000*	*0.0323*	*0.162*
	5	**0.0093**	**0.764**	**0.0191**	**0.766**	**0.0326**	**0.322**	**0.0124**	**0.873**
		0.0081	*0.923*	*0.0476*	*0.035*	*0.0811*	*0.000*	*0.0354*	*0.170*
	10	**0.0086**	**0.769**	**0.0254**	**0.613**	**0.0717**	**0.000**	**0.0176**	**0.836**
		0.0082	*0.934*	*0.0562*	*0.011*	*0.1250*	*0.000*	*0.0415*	*0.135*

Table 4.5. Maximal estimated relative bias ($MAXRB$) and proportion of tests which accepted target correlation ($ACC_{0.05}$) based on $R = 1000$ replications of $N = 500$ samples of size $n = 5$ for exchangeable target correlation ρ and different count margins and parameters as in Table 4.3 (bold: C-vine sampling, italics: naive sampling).

	μ	φ	ω	$MAXRB$	$ACC_{0.05}$	$MAXRB$	$ACC_{0.05}$
Poisson	S	1	0	**0.0335**	**0.999**	*0.1014*	*0.672*
	L	1	0	**0.0241**	**0.995**	*0.0052*	*0.950*
GP	S	S	0	**0.1323**	**0.516**	*0.2456*	*0.034*
	S	L	0	**0.3822**	**0.010**	*0.5107*	*0.000*
	L	S	0	**0.0146**	**0.993**	*0.0329*	*0.913*
	L	L	0	**0.0868**	**0.914**	*0.1423*	*0.307*
ZIGP	S	S	S	**0.1377**	**0.492**	*0.2603*	*0.020*
	S	S	L	**0.1875**	**0.297**	*0.2850*	*0.007*
	S	L	S	**0.3937**	**0.005**	*0.5230*	*0.000*
	S	L	L	**0.4023**	**0.004**	*0.5682*	*0.000*
	L	S	S	**0.0570**	**0.999**	*0.1069*	*0.790*
	L	S	L	**0.0794**	**0.990**	*0.1528*	*0.460*
	L	L	S	**0.0931**	**0.924**	*0.1479*	*0.304*
	L	L	L	**0.0988**	**0.906**	*0.1514*	*0.222*
NB	S	S	0	**0.1228**	**0.615**	*0.2348*	*0.035*
	S	L	0	**0.3719**	**0.012**	*0.5146*	*0.001*
	L	S	0	**0.0150**	**0.994**	*0.0280*	*0.928*
	L	L	0	**0.0582**	**0.997**	*0.1061*	*0.544*

Influence of T:

As one would expect, the higher the dimension T, the worse the approximation gets. The reason is simply error propagation.

Influence of the distribution family:

Overdispersed settings perform worse than equidispersed ones, zero-inflation additionally increases overdispersion and hence worsens the results.

Influence of the range of parameters μ:

According to $MAXRB$ and $ACC_{0.05}$ in Table 4.5, small means produce worse approximations. Small means generate more discrete data with linear correlations that are harder to optimize.

Table 4.6. Maximal estimated relative bias ($MAXRB$) and proportion of tests which accepted target correlation ($ACC_{0.05}$) based on $R = 1000$ replications of $N = 500$ samples of size $n = 5$ for AR(1) and unstructured correlation structures and different count margins (bold: C-vine sampling, italics: naive sampling).

	AR(1)			
	Poisson	GP	ZIGP	NB
$MAXRB$	**0.0220**	**0.0218**	**0.0219**	**0.0219**
	0.0736	*0.0741*	*0.0740*	*0.0738*
$ACC_{0.05}$	**0.806**	**0.807**	**0.807**	**0.807**
	0.760	*0.760*	*0.759*	*0.760*

	unstructured			
	Poisson	GP	ZIGP	NB
$MAXRB$	**0.0244**	**0.0244**	**0.0245**	**0.0245**
	0.0932	*0.0923*	*0.0937*	*0.0928*
$ACC_{0.05}$	**0.862**	**0.862**	**0.862**	**0.861**
	0.778	*0.778*	*0.777*	*0.778*

Influence of the range of parameters φ and ω:

Small dispersion and zero-inflation parameters result in dramatically better approximations than large ones. Both large φ and ω increase heterogeneity in the data and therefore also in the empirical correlations calculated.

Also, for the AR(1) and unstructured correlation matrices in Table 4.6, the results are equally good as in the five-dimensional exchangeable settings.

4.5 Summary and Discussion

Erhardt and Czado[11] suggest an iterative method for sampling correlated count random variables. Positive definiteness of the resulting association parameters is ensured by the C-vine framework the approach is embedded in. The price of this is that some of the correlations between margins are only approximated via partial correlations. The comparison carried out in this chapter illustrates that the performance of the two approaches strongly depends on the simulation setting chosen.

Two questions are raised in this chapter. First of all, how wrong can one be when using the simplified (naive) approach? The simulation study illustrates that the desired target correlations might be clearly missed especially when the dimension, the degree of discreteness and overdispersion of the margins are high. The other question is how much better the suggested C-vine approach performs. We showed that even if it tends to be less precise in the same setting where the naive approach fails, there is a substantial improvement of accuracy.

Acknowledgments

V. Erhardt is supported by a grant from Allianz Deutschland AG. C. Czado is supported by DFG (German Science Foundation) grant CZ 86/1-3.

References

1. Aas K., Czado C., Frigessi A. and Bakken H. (2007). Pair-copula constructions of multiple dependence. *Insurance: Mathematics and Economics*, 44(2):182–198.
2. Avramidis A.N., Channouf N. and L'Ecuyer P. (2008). Efficient correlation matching for fitting discrete multivariate distributions with arbitrary marginals and normal-copula dependence. *INFORMS Journal on Computing. Articles in Advance*, 1–19.
3. Bedford T. and Cooke R.M. (2001). Monte Carlo simulation of vine dependent random variables for applications in uncertainty analysis. *Proceedings of ESREL 2001*, Turin, Italy.
4. Bedford T. and Cooke R.M. (2001). Probability density decomposition for conditionally dependent random variables modeled by vines. *Annals of Mathematics and Artificial Intelligence*, 32(1–4):245–268.
5. Bedford T. and Cooke R.M. (2002). Vines: A new graphical model for dependent random variables. *Annals of Statistics*, 30:1031–1068.
6. Cario M.C. and Nelson B.L. (1996). Autoregressive to anything: Time-series input processes for simulation. *Operations Research Letters*, 19:51–58.
7. Cario M.C. and Nelson B.L. (1997). Modeling and generating random vectors with arbitrary marginal distributions and correlation matrix. Technical Report, Department of Industrial Engineering and Management Sciences, Northwestern University, Evanston, IL.
8. Chen H. (2000). Initialization for NORTA: Generation of random vectors with specified marginals and correlations. *INFORMS Journal on Computing*, 13:312–331.
9. Consul P.C. and Jain G.C. (1970) A generalization of the Poisson distribution. *Technometrics*, 15:791–799.
10. Embrechts P., McNeil A.J. and Straumann D. (2002). Correlation and dependency in risk management: Properties and pitfalls. In *Risk Management: Value at Risk and Beyond*, M. Dempster (ed.), pp. 176–223. Cambridge University Press, Cambridge.
11. Erhardt V. and Czado C. (2008). A method for approximately sampling high-dimensional count variables with prespecified Pearson correlation. Submitted for publication.
12. Fisher R. (1921). On the 'probable error' of a coefficient of correlation deduced from a small sample. *Metron*, 1:3–32.

13. Genest C. and Neslehova J. (2007). A primer on copulas for count data. *ASTIN Bulletin*, 37:475–515.
14. Ghosh S. and Henderson S.G. (2003). Behavior of the NORTA method for correlated random vector generation as the dimension increases. *ACM Transactions on Modeling and Computer Simulation*, 13(3):276–294.
15. Joe H. (1996). Families of m-variate distributions with given margins and $m(m-1)/2$ bivariate dependence parameters. In *Distributions with Fixed Marginals and Related Topics*, L. Rüschendorf, B. Schweizer and M.D. Taylor (eds.), pp. 120–141. IMS Lecture Notes Monograph Series, 28, Institute of Mathematical Statistics, Hayward, CA.
16. Joe H. (1997). *Multivariate Models and Dependence Concepts*. Chapman & Hall, London.
17. Karlis D. and Meligkotsidou L. (2005). Multivariate Poisson regression with covariance structure. *Statistics and Computing*, 15(4):255–265.
18. Kawamura K. (1979). The structure of multivariate Poisson distribution. *Kodai Mathematical Journal*, 2:337–345.
19. Kopociński B. (1999). Multivariate negative binomial distributions generated by multivariate exponential distributions. *Journal of Applied Mathematics*, 25(4):463–472.
20. Li S. and Hammond J. (1975). Generation of pseudo-random numbers with specified univariate distributions and correlation coefficients. *IEEE Transactions on Systems, Man and Cybernetics*, 5:557–561.
21. Lurie P. and Goldberg M. (1998). An approximate method for sampling correlated random variables from partially-specified distributions. *Management Science*, 44(2):203–218.
22. Marida K.V. (1970). A translation family of bivariate distributions and Fréchet's bounds. *Sankhya*, 32:119–122.
23. Nelsen R.B. (2006). *An Introduction to Copulas*, 2nd ed., Springer Series in Statistics, New York.
24. Pearson K. (1916). On some novel properties of partial and multiple correlation coefficients in a universe of manifold characteristics. *Biometrika*, 11(3):231–238.
25. Shaffer J.P. (1995). Multiple hypothesis testing. *Annual Review of Psychology*, 46:561–584.
26. Sklar A. (1959). Fonctions de répartition à n dimensions et leurs marges. *Publications de l'Institut de Statistique de L'Université de Paris*, 8:229–231.
27. Song P.X.-K. (2007). *Correlated Data Analysis: Modeling, Analytics and Applications*. Springer-Verlag, New York.
28. Srivastava M. and Khatri C. (1979). *An Introduction to Multivariate Statistics*. Wiley, New York.
29. Tsiamyrtzis P. and Karlis D. (2004). Strategies for efficient computation of multivariate Poisson probabilities. *Communications in Statistics — Simulation and Computation*, 33(2):271–292.
30. Vernic R. (2000). A multivariate generalization of the generalized Poisson distribution. *Astin Bulletin*, 30(1):57–67.

CHAPTER 5

Micro Correlations and Tail Dependence

Roger M. Cooke*,†,§, Carolyn Kousky*,¶ and Harry Joe‡

*Resources for the Future
§Cooke@Rff.org
†Department of Mathematics, Delft University of Technology
¶Kousky@Rff.org
‡Department of Statistics, University of British Columbia
Harry.Joe@ubc.ca

An elementary though seemingly underappreciated finding shows that small global correlations are amplified by aggregation. We observe this behavior in flood damage claims in the US. We also observe that upper tail dependence seems to be amplified by aggregation in these data. We seek to understand this behavior. For sums of exponential variables which are conditionally independent given a gamma-distributed rate, we derive explicit expressions for upper tail dependence and prove that it goes to one as the number of summands goes to infinity, and that the lower tail dependence is zero. We also study sums of events under a latent variable model, where each event occurs if a uniform variable exceeds a threshold, and all uniform variables are conditionally independent given a "latent variable". We obtain a necessary and sufficient condition for strong asymptotic upper tail dependence as the number of summands goes to infinity. Curiously, the normal copula satisfies this condition, although it is not tail dependent via the usual definition. Thus, sums of events under the normal copula latent variable model have upper tail dependence increasing to 1. We also identify tail dependent-like behavior in finite sums of events with the latent variable model.

5.1	Introduction	90
5.2	Micro Correlations	90
5.3	Tail Dependence and Aggregation	92
	5.3.1 Latent variable models for tail dependence	94
	5.3.2 Sum of damages over extreme events	97
	5.3.3 L_1-symmetric measures	101
	5.3.4 Tail dependence for sums of L_1 measures	103
	5.3.5 Lower tail dependence	105

5.4 Discussion 106
Appendices 106
  A.   Proofs for the tail dependence condition involving
        $C_{1|2}(u|1)$ 106
  B.   Proof of Proposition 5.1 and an example 107
  C.   Proof that $\lambda_{U,\eta,N} \to 1$ as $N \to \infty$ 109
References 111

5.1 Introduction

Micro correlations will amplify the correlation of sums of globally correlated variables, and under certain circumstances they will also amplify tail dependence. This is of evident concern to risk managers, as it will compromise risk management based on diversification. The circumstances under which aggregation amplifies tail dependence are not well understood, and this chapter represents a first foray into the area of tail dependence amplification. We study latent variable models for sums of events, and L_1-symmetric variables. We obtain a condition that leads to upper tail dependence for two different sums of events. In the case of L_1-symmetric measures with gamma scale mixtures, we can prove that aggregation amplifies upper tail dependence.

In Section 5.2, we first discuss the issue of micro correlation and present loss data, drawn, from Kousky and Cooke,[8] where micro correlations amplify under aggregation. Section 5.3 shows results on tail dependence and aggregation, and Section 5.4 concludes with a discussion of further research.

5.2 Micro Correlations

Let $X_1, \ldots, X_N$ and $X_{N+1}, \ldots, X_{2N}$ be sets of random variables with the average variance σ^2 over the first N and second N random variables and average covariance γ within and between the two sets. The correlation of the sum of the first N and second N X's is:

$$\operatorname{corr}\left(\sum_{i=1}^{N} X_i, \sum_{i=N+1}^{2N} X_i\right) = \frac{N^2 \gamma}{N\sigma^2 + N(N-1)\gamma} = \frac{N\gamma}{\sigma^2 + (N-1)\gamma}.$$

Evidently, if $\gamma > 0$ and $\sigma < \infty$, this goes to 1 as $N \to \infty$. Since $\sigma^2 > 0$, $\frac{\sigma^2}{N-1} \geq -\gamma$ which shows that for all N sufficiently large, $\gamma \geq 0$.

We can find micro correlations in many places once we start looking for them. We illustrate with two data sets: flood insurance claims data from the

US National Flood Insurance Program (NFIP) and data on crop insurance indemnities payments from the United States Department of Agriculture's Risk Management Agency. Both data sets are aggregated by county and year for the years 1980 to 2008. The data are in constant year 2000 dollars. Over this time, there has been substantial growth in exposure to flood risk, particularly in coastal counties. To remove the effect of growing exposure, we divide the claims per county per year by personal income per county per year available from the Bureau of Economic Accounts (BEA). Thus, we study yearly flood claims per dollar income, per year per county. The crop loss claims are not exposure-adjusted, as an obvious proxy for exposure is not at hand, and exposure growth was less of a concern.

Suppose we randomly draw pairs of counties in the US and compute the correlation of their exposure-adjusted flood losses. Figure 5.1 shows the histogram of 500 such correlations. The average correlation is 0.04. A few counties have quite high correlations but the bulk is around zero. Indeed, based on the sampling distribution for the normal correlation coefficient, correlations less than 0.37 in absolute value would not be statistically distinguishable from zero at the 5% significance level. 91% of these correlations fall into that category.

Instead of looking at the correlations between two randomly chosen counties, consider summing 100 randomly chosen counties and correlating this with the sum of 100 distinct randomly chosen counties (i.e., sampling without

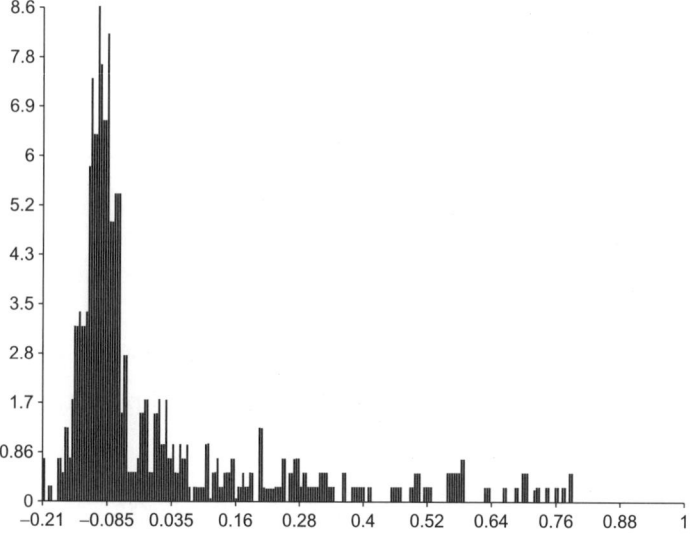

Figure 5.1. Histogram of 500 correlations of randomly paired US exposure-adjusted flood loss per county, 1980–2006. The average correlation is 0.04.

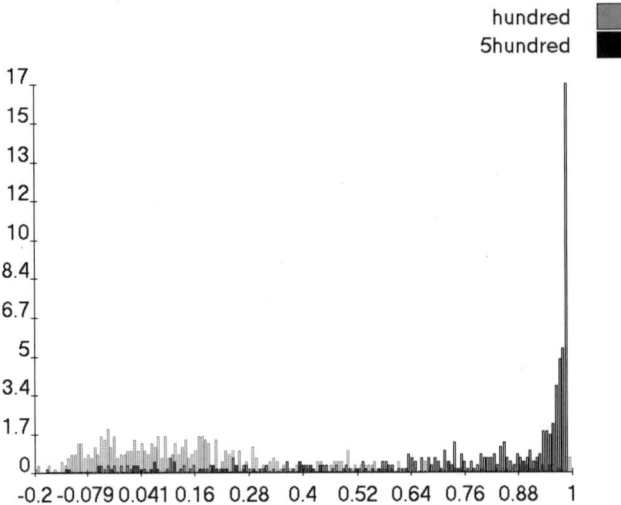

Figure 5.2. Similar to Fig. 5.1, but showing 500 correlations of random sums of 100 and 500.

replacement). If we repeat this 500 times, the centered histogram in Fig. 5.2 results; the average of 500 such correlations-of-100 is 0.23. The histogram at the upper extreme depicts 500 correlations-of-500; their average value is 0.71.

The flood damage per dollar exposure shows a lower correlation than the US crop losses in Fig. 5.3. The mean correlation is 0.13, and the mean of correlations-of-100 is 0.88.

It is interesting to compare the histograms of real loss distributions with a histogram in which each county is assigned an independent uniform variable. The histogram of 500 correlations of random pairs and correlations of random aggregations-of-500 are shown in Fig. 5.4.

5.3 Tail Dependence and Aggregation

In this section, we obtain some results on when aggregation amplifies tail dependence.

The definition of upper tail dependence is given below.

Definition 5.1 (Upper tail dependence). The upper tail dependence between random variables X and Y is

$$UTD(X,Y) = \lim_{q \to 1} \Pr(X > x_q | Y > y_q) \qquad (5.1)$$

where $x_q = F_X^{-1}(q)$ and $y_q = F_Y^{-1}(q)$.

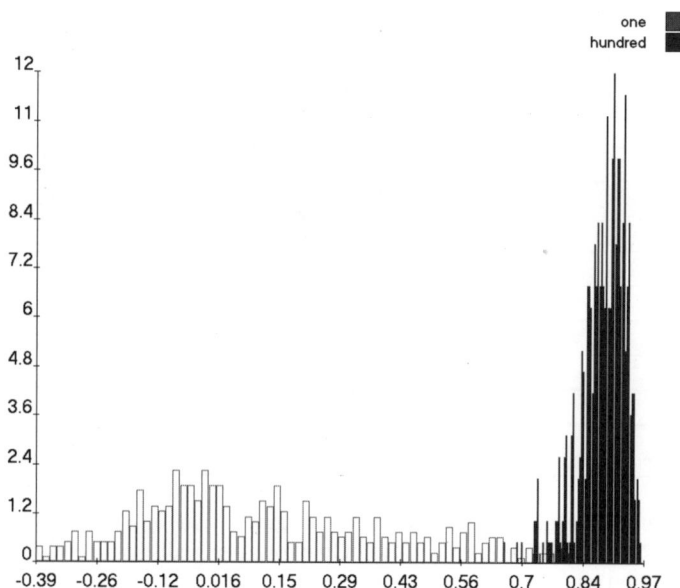

Figure 5.3. Histogram of 500 random correlations of US crop losses per county, 1980–2008, random pairs and random sums of 100.

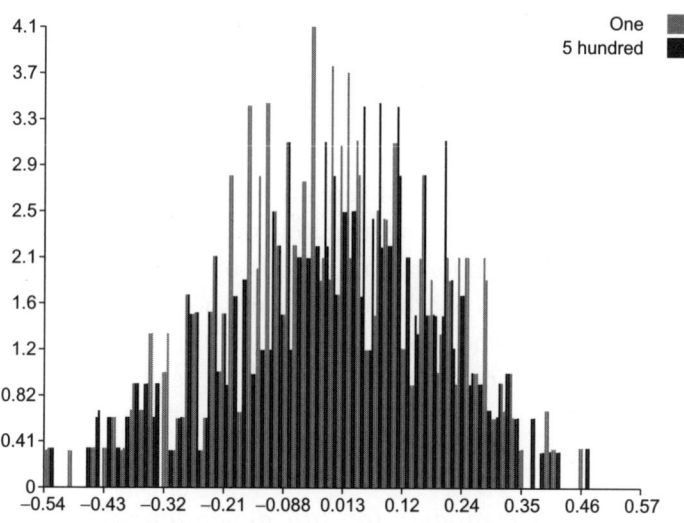

Figure 5.4. Histogram of 500 random correlations of independent uniforms assigned to each county, 1980–2008, random pairs and random sums of 500.

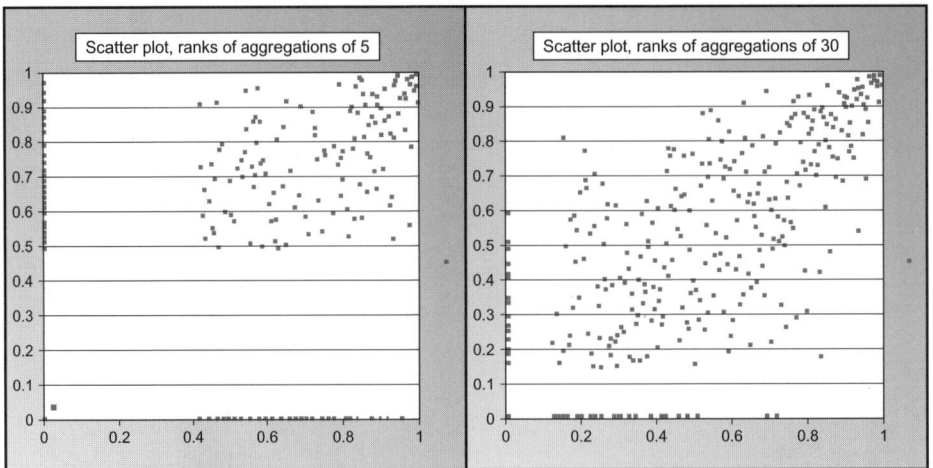

Figure 5.5. Percentile scatterplots of random aggregation of Florida county monthly flood losses. Left: two random aggregations of five counties; right: two random aggregations of 30 distinct counties.

Lower tail dependence is defined in a similar way in the lower quadrant: $LTD(X,Y) = \lim_{q \to 0} \Pr(X \leq x_q | Y \leq y_q)$. As is evident from the definition, tail dependence is a property of the copula. The normal copula has zero tail dependence for all correlation values in $(-1, 1)$; see McNeil et al.[10]

A central question is whether tail dependence is also amplified by aggregation. In loss distributions, we can see the amplification of tail dependence under aggregation. To see tail dependence, the yearly data are not sufficient. Figure 5.5 plots monthly flood loss data in the state of Florida from 1980 to 2008. We choose Florida because there are numerous counties with many non-zero losses in several months. There are two percentile scatterplots: that on the left shows two random aggregation of five counties while the plot on the right shows two random aggregations of 30 counties. Points on the axes correspond to months in which there were no losses in the corresponding aggregate variable. The plot suggests that the upper tail dependence is amplified by aggregation. We seek models to help understand why and when this happens.

5.3.1 Latent variable models for tail dependence

In simple latent variable models, a latent variable is an unobserved variable to which all observed variables are correlated, and conditional on which all observed variables are independent. Recognizing this structure as a C-vine with dependence confined to the first tree rooted at the latent variable, it is evident that this is the simplest of a wide class of models.

We first consider a finite class of events, where each event occurs when a physical variable exceeds some limit, and each physical variable is connected to a latent variable. For simplicity, let $U_1, \ldots, U_{2N}$ be uniform variables and suppose event E_i occurs if and only if $U_i > r$. Suppose further that the U_i are conditionally independent given a latent variable V, which is also uniform. To study such models, we require a copula joining U_i and V.

Specifically, $(U_i, V) \sim C(u, v)$ for all i and C is a bivariate copula. Let $C_{1|2}(u|v) = \partial C(u, v)/\partial v$ be the conditional distribution of U_i given $V = v$. We assume C has positive dependence in the sense of stochastic increasing, that is $\Pr(U_1 > u | V = v) = 1 - C_{1|2}(u|v)$ is strictly increasing in $v \in [0, 1]$ for all $0 < u < 1$. This condition is satisfied by all of the commonly used one-parameter families of copula when restricted to the region of positive dependence. For a fixed r in $(0, 1)$, let $Y_i = I(U_i > r)$ for the indicator of the extreme event E_i. Let $S_1 = S_1(N) = Y_1 + \cdots + Y_N$ and let $S_2 = S_2(N) = Y_{N+1} + \cdots + Y_{2N}$ be two aggregate numbers of extreme events in two sets. We study the (upper) tail dependence of S_1, S_2 under this simple latent variable model.

Let
$$p_r(v) = 1 - C_{1|2}(r|v), \quad q_r(v) = 1 - p_r(v), \quad 0 \leq v \leq 1. \quad (5.2)$$
For an integer k between 0 and N inclusive, and $j = 1$ or 2,
$$\Pr(S_j = k) = \int_0^1 \Pr(S_j = k | V = v) \, dv$$
$$= \int_0^1 \binom{N}{k} [p_r(v)]^k [q_r(v)]^{N-k} \, dv,$$
and
$$\Pr(S_1 = k_1, S_2 = k_2)$$
$$= \int_0^1 \Pr(S_1 = k_1 | V = v) \Pr(S_2 = k_2 | V = v) \, dv$$
$$= \int_0^1 \binom{N}{k_1} [p_r(v)]^{k_1} [q_r(v)]^{N-k_1} \binom{N}{k_2} [p_r(v)]^{k_2} [q_r(v)]^{N-k_2} \, dv.$$
For a fraction $0 < \zeta < 1$, let $\lambda_U(r, \zeta, N) = \Pr(S_2 > N\zeta | S_1 > N\zeta)$. Then
$$\lambda_U(r, \zeta, N) = \frac{\int_0^1 \sum_{k_1 \geq N\zeta, k_2 \geq N\zeta} \binom{N}{k_1} [p_r(v)]^{k_1} [q_r(v)]^{N-k_1} \binom{N}{k_2} [p_r(v)]^{k_2} [q_r(v)]^{N-k_2} \, dv}{\int_0^1 \sum_{k \geq N\zeta} \binom{N}{k} [p_r(v)]^k [q_r(v)]^{N-k} \, dv}.$$
$$(5.3)$$

The analysis of (5.3) for large N is given next. Let Z be a standard normal random variable, with cumulative distribution function Φ. Let

$$g(v) = g(v; r, \zeta) = \frac{p_r(v) - \zeta}{\sqrt{p_r(v) q_r(v)}}.$$

By the normal approximation to binomial, for large N, (5.3) can be approximated by

$$\frac{\int_0^1 \{\Pr(Z > [N\zeta - Np_r(v)]/\sqrt{Np_r(v)q_r(v)})\}^2 \, dv}{\int_0^1 \Pr(Z > [N\zeta - Np_r(v)]/\sqrt{Np_r(v)q_r(v)}) \, dv} = \frac{\int_0^1 \Phi^2[N^{1/2} g(v)] \, dv}{\int_0^1 \Phi[N^{1/2} g(v)] \, dv}. \tag{5.4}$$

From the positive dependence assumption of stochastic increasing, $p_r(v)$ in (5.2) is increasing in v. Let

$$v_0 = v_0(r, \zeta) = \sup\{v \in (0, 1) : p_r(v) \leq \zeta\}$$
$$= \sup\{v \in (0, 1) : g(v; r, \zeta) \leq 0\}.$$

Then (5.4) becomes

$$\frac{\int_0^{v_0} \Phi^2[N^{1/2} g(v)] \, dv + \int_{v_0}^1 \Phi^2[N^{1/2} g(v)] \, dv}{\int_0^{v_0} \Phi[N^{1/2} g(v)] \, dv + \int_{v_0}^1 \Phi[N^{1/2} g(v)] \, dv}. \tag{5.5}$$

If $0 \leq v_0 < 1$, then

$$\lim_{N \to \infty} \int_0^{v_0} \Phi^j[N^{1/2} g(v)] \, dv = 0, \quad \text{and}$$

$$\lim_{N \to \infty} \int_{v_0}^1 \Phi^j[N^{1/2} g(v)] \, dv = 1 - v_0, \quad j = 1, 2.$$

Therefore, $\lambda_U(r, \zeta, N)$ in (5.3) goes to 1 as $N \to \infty$ if $0 \leq v_0 < 1$, and

$$\lim_{N \to \infty} \lambda_U(r, \zeta, N) = 1 \quad \forall \, 0 < r < 1, \, 0 < \zeta < 1$$

if and only if $p_r(1) = \overline{C}_{1|2}(r|1) = 1$ for all $0 < r < 1$ or $C_{1|2}(u|1) = 0$ for all $0 < u < 1$.

If $p_r(1) = \overline{C}_{1|2}(r|1) < 1$, then $\lim_{N \to \infty} \lambda_U(r, \zeta, N) = 1$ only if ζ is small enough so that $0 < v_0 < 1$. If $v_0 = 1$ and $p_r(1) < \zeta$, then (5.5) is bounded above by $\max_{0 \leq v \leq 1} \Phi[N^{1/2} g(v)]$ and this approaches 0 as $N \to \infty$.

For numerical computations, if the limit is 1, $\lambda_U(r, \zeta, N)$ is practical only if $\Pr(S_1 > \zeta N)$ is not too small and $v_0(r, \zeta)$ is not too close to 1; this means

ζ should not be too close to 1. For fixed ζ, $\Pr(S_1 > \zeta N)$ tends to get smaller as the (upper tail) dependence of (U_i, V) gets weaker.

The condition of $\overline{C}_{1|2}(r|1) = \Pr(U > r|V = 1) = 1$ for all $0 < r < 1$ or
$$C_{1|2}(u|1) = \Pr(U \leq u|V = 1) = 0 \quad \forall 0 < u < 1 \tag{5.6}$$
is an **upper tail dependence condition**. It is the same as $[U|V = v] \underset{p}{\to} 1$, as $v \uparrow 1$.

Equation (5.6) holds for all bivariate extreme value copulae, e.g., Gumbel and Galambos. The condition for (5.6) to hold for an Archimedean copula $C_\psi(u,v) = \psi(\psi^{-1}(u) + \psi^{-1}(v))$ is $\psi'(0) = -\infty$ and this is the same condition for the usual tail dependence (Theorem 3.12 in Joe[6]). Hence, (5.6) fails to hold for the Frank copula. It also fails to hold for the Plackett copula but holds for the bivariate normal copula with positive correlation ρ. This means that (5.6) is not exactly the same as the usual tail dependence condition of $\lim_{v \uparrow 1} \overline{C}(v,v)/(1-v)$ being positive because the bivariate normal copula does satisfy this. Some proofs of the preceding cases are given in Appendix A.

Table 5.1 compares the conditional probability $\lambda_U(r, \zeta, N)$ for the Gumbel, bivariate normal and Frank copulae when $r = 0.9$, $\zeta = 0.7$, and the dependence parameters for the three copulae are chosen to get a rank correlation of 0.5.

The definition of tail dependence as limiting conditional probabilities of exceedence is not appropriate for finite sums of events. Nonetheless we can identify tail dependence-like behavior in finite sums of events. With Frank's copula, take the probability of the individual events as 0.1 and the correlation to the latent variable V as 0.9 (the parameter $\theta = 12.3$) which induces a correlation 0.36 between any two events. Figure 5.6 illustrates curious non-monotonic behavior in $P\{S_1 > i | S_2 > i\}$, for $N = 100$ and $i = 1, \ldots, 100$. This is caused by the interaction of two opposing "forces"; as i increases, $P\{S_1 > i\}$ goes down, while on the other hand, conditionalizing on $P\{S_2 > i\}$ drives the latent V up, which increases $P\{S_1 > i | S_2 > i\}$. The pattern with N fixed and ζ increasing is quite different from the pattern when ζ is fixed and N increasing.

5.3.2 Sum of damages over extreme events

Instead of the number of extreme events, consider the sum of losses or damages. The situation becomes more complex and the results depend strongly on the copula and the damage distributions. Figure 5.7 shows percentile

Table 5.1. Conditional probabilities $\Pr(S_2 > N\zeta | S_1 > N\zeta) = \lambda_U(r, \zeta, N)$ with $r = 0.9$, $\zeta = 0.7$, Spearman ρ_S = rank correlation $= 0.5$; leading to parameters $\theta = 1.54$ for the Gumbel, $\rho = 0.518$ for the bivariate normal (BVN), $\theta = 7.90$ for the Frank copulae respectively. Limit behavior depends on the comparison sign of $p_r(1) - \zeta$.

	$\lambda_U(r, \zeta, N)$		
N	Gumbel	BVN	Frank
10	0.604	0.264	0.144
20	0.687	0.411	0.067
30	0.733	0.528	0.034
40	0.763	0.620	0.019
50	0.785	0.695	0.010
60	0.801	0.755	0.006
70	0.815	0.804	0.003
80	0.826	0.845	0.002
90	0.835	0.877	0.001
100	0.843	0.903	0.001
	$p_r(v) = \overline{C}_{1|2}(r|v)$		
v	Gumbel	BVN	Frank
1	1.0	1.0	0.546
0.99999	0.994	0.861	0.546

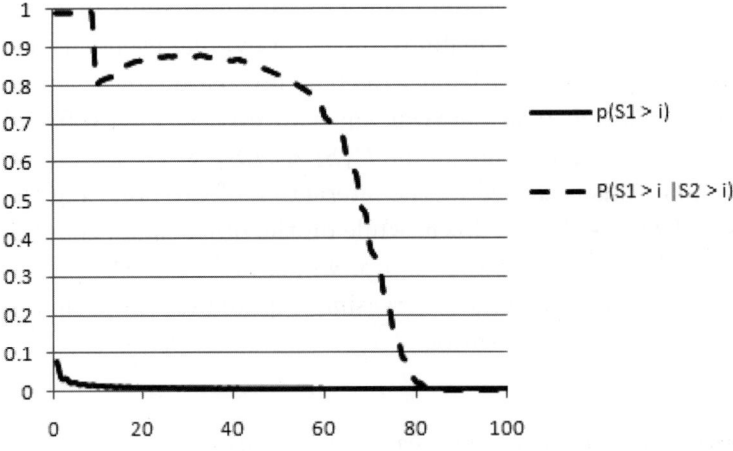

Figure 5.6. Tail dependent-like behavior of sums of events, probability of exceedence as function of i, for Frank's copula, $\theta = 12.3$, $N = 100$.

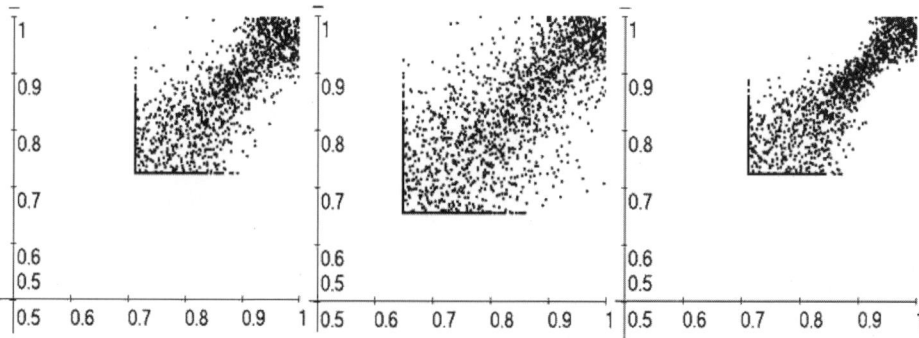

Figure 5.7. Tail dependent-like behavior of sums of events times damages. Left: Pareto 2 damages with Gumbel copula; middle: Pareto 2 damages with bivariate normal copula; right: exponential damages with Gumbel copula.

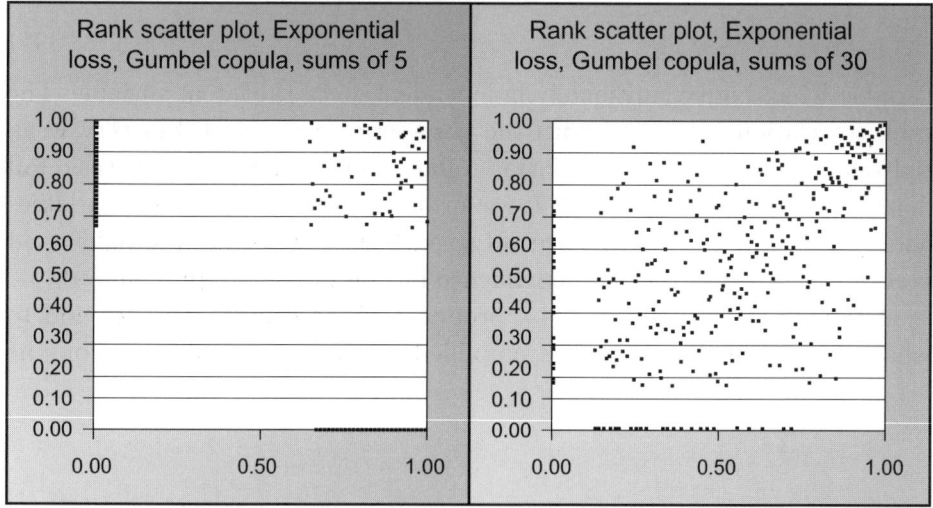

Figure 5.8. A model for Florida monthly flood damages, exponential damages linked to a latent variable with the Gumbel copula.

scatterplots of events multiplied by independent damages, and where the joining copulae are Gumbel and bivariate normal. Figure 5.8 shows exponentially distributed damages linked to a latent variable via the Gumbel copula, and parameters are chosen to resemble Fig. 5.5. This suggests that simple latent variable models may describe such loss phenomena satisfactorily.

Without considering sums of events, it is easy to construct simulations in which this amplification occurs. Figure 5.9 shows percentile plots of two normal variables X_1 and X_2 which each have rank correlation 0.1 to a latent

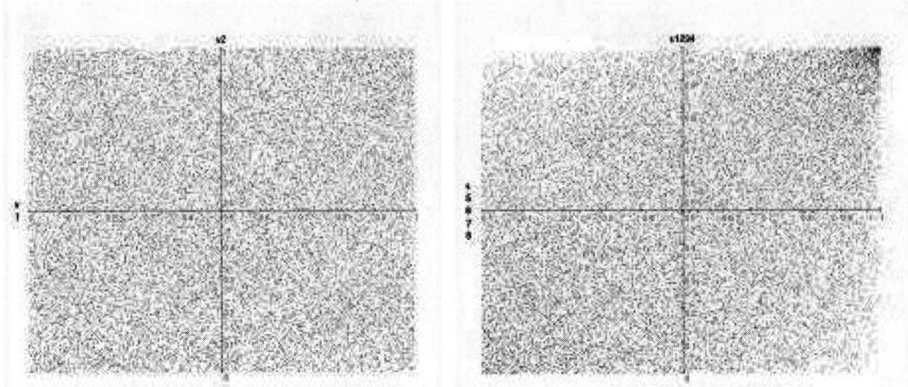

Figure 5.9. Percentile plots with tail dependence. Left: two normal variables rank-correlated 0.1 to a latent variable with Gumbel copula; right: distinct sums of 40 such variables, each similarly rank-correlated to the latent variable.

variable V, and are conditionally independent given the latent variable. The rank correlation is realized with the Gumbel copula, which has very weak tail dependence at that correlation value. This induces a very weak tail dependence between X_i and V. If we form sums of 40 such normal variables and consider the tail dependence of two such sums, we see in the right-hand plot of Fig. 5.9 that the tail dependence has become more pronounced.

Although tail dependence is a property of the copula, whether and to what degree tail dependence is amplified by aggregation depends on the

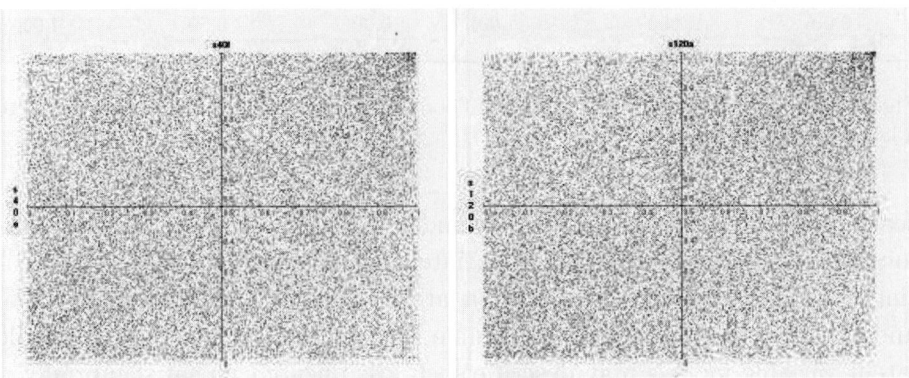

Figure 5.10. Percentile plots with tail dependence. Left: sums of 40 Pareto variables with survival function $(1/(1+x))$, each rank-correlated 0.1 to a latent variable with Gumbel copula; right: sums of 120 such variables.

marginal distributions. Figure 5.10 is similar to Fig. 5.9 except that the variables are Pareto with survival function $S(x) = (1+x)^{-1}$. The amplification of tail dependence for 120 Pareto variables is much weaker than that for 40 normal variables. This Pareto distribution does not have a finite first moment (or, of course, correlation).

In certain cases, we can prove some results for tail dependence. The following proposition, whose proof is in Appendix B, gives a lower bound for tail dependence of variables, which are tail dependent on a latent variable:

Proposition 5.1. *Suppose (U_1, V) and (U_2, V) are pairwise upper tail dependent with, respectively, coefficients $\lambda_1 > 0$ and $\lambda_2 > 0$, and (U_1, U_2) is conditionally independent given V. Also suppose that U_1 and U_2 are each stochastically increasing in V. Then (U_1, U_2) has an upper tail dependence coefficient that exceeds $\lambda_1 \lambda_2$.*

5.3.3 L_1-symmetric measures

Results relating tail dependence to aggregation are difficult to obtain, since aggregation is not simply a question of the copula, but also of the marginal distributions. One case where analytic results are possible concerns the L_p-symmetric variables with $1/p \in \mathbb{N}$.

Recall the Gamma integral:
$$\int_0^\infty y^{\eta-1} e^{-\beta y}\, dy = \frac{\Gamma(\eta)}{\beta^\eta}; \quad \beta > 0,\ \eta > 0.$$
The Gamma(η, β) density with shape η and rate β is $f(y; \eta, \beta) = \beta^\eta y^{\eta-1} e^{-\beta y}/\Gamma(\eta)$, with mean η/β and variance η/β^2.

An atomless L_p-symmetric measure on $\mathbb{R}^n$ is one whose density at $(x_1, \ldots, x_N)$ depends only on the L_p norm $(\sum |x_i|^p)^{1/p}$. Berman[2] proved that L_p-symmetric measures on $\mathbb{R}$ can be uniquely represented as conditionally independent gamma transforms with shape $1/p$. For L_1 measures, we have conditionally independent exponentials given the failure rate. $(X_1, \ldots, X_N)$ have an L_1-symmetric distribution with Gamma(η, β) mixing distribution if, for any N, the N-dimensional marginal density is given by

$$f_N(x_1, \ldots, x_N) = \int \left\{ \prod_{i=1}^N \lambda e^{-\lambda x_i} \right\} \beta^\eta \lambda^{\eta-1} e^{-\beta \lambda} d\lambda / \Gamma(\eta). \qquad (5.7)$$

Setting $N = 1$ and integrating over λ, one finds the univariate density and survivor functions:

$$f_1(x) = \frac{\eta \beta^\eta}{(\beta + x)^{\eta+1}}; \quad 1 - F_1(x) = \left(\frac{\beta}{\beta + x}\right)^\eta, \qquad (5.8)$$

which is the Pareto thick-tailed (leptokurtic) distribution with shape parameter η and scale parameter β. These multivariate distributions were first studied by Takahasi[11] and Harris.[5] Then unconditionally the joint survival function of $X_1, \ldots, X_N$ is

$$\Pr(X_1 > x_1, \ldots, X_N > x_N) = \int_0^\infty \prod_{i=1}^N e^{-\lambda x_i} \frac{\lambda^{\eta-1}\beta^\eta}{\Gamma(\eta)} e^{-\beta\lambda} d\lambda$$

$$= \frac{\beta^\eta}{[\beta + x_1 + \cdots + x_N]^\eta}. \quad (5.9)$$

This is a special case of the multivariate Burr distribution of Takahasi,[11] with type II Pareto as a special case of Burr for the univariate margins. The multivariate Pareto distribution of Mardia[9] has type I Pareto margins rather than type II Pareto. From this distribution, Cook and Johnson[4] obtained the copula (replacing Pareto survival functions) as

$$C(u_1, \ldots, u_N; \eta) = [u_1^{-1/\eta} + \cdots + u_N^{-1/\eta} - (N-1)]^{-\eta}. \quad (5.10)$$

As an aside, Kimeldorf and Sampson[7] did the same thing but only for the bivariate case; Clayton[3] has the bivariate distribution as a gamma frailty model and through a derivation from a differential equation, but does not have the multivariate case. In this parametrization, dependence increases as η decreases. The copula (5.10) has lower tail dependence and the distribution (5.9) has upper tail dependence.

Consider the sum $S = X_1 + \cdots + X_N$, where $(X_1, \ldots, X_N)$ has density (5.7). Since $S|\Lambda = \lambda \sim \text{Gamma}(n, \lambda)$,

$$f_S(r; N) = \int_0^\infty \frac{1}{\Gamma(N)} \lambda^N r^{N-1} e^{-\lambda r} \cdot \frac{1}{\Gamma(\eta)} \lambda^{\eta-1} \beta^\eta e^{-\beta\lambda} d\lambda$$

$$= \frac{r^{N-1}\Gamma(N+\eta)\beta^\eta}{\Gamma(N)\Gamma(\eta)(\beta+r)^{\eta+N}}.$$

The sums have the same tail behavior as the one-dimensional margins. From (5.8), we obtain the mean of X. The variance, covariance and product moment correlation may be obtained from (5.7) with $N = 2$, giving:

$$\mu(X) = \frac{\beta}{\eta - 1}; \quad \eta > 1$$

$$\text{Var}(X_1) = \frac{\beta^2 \eta}{(\eta-1)^2(\eta-2)}; \quad \eta > 2$$

$$\text{Cov}(X_1, X_2) = \frac{\beta^2}{(\eta-1)^2(\eta-2)}; \quad \eta > 2$$
$$\text{corr}(X_1, X_2) = \eta^{-1}; \quad \eta > 2$$
$$\text{Var}(X_1 + \cdots + X_N) = \text{Var}(X_1)\{N + N(N-1)\eta^{-1}\}; \quad \eta > 2.$$

Note that the mean exists only if $\eta > 1$, and the variance, covariance and correlation require $\eta > 2$.

5.3.4 Tail dependence for sums of L_1 measures

Computations of tail dependence for sums of L_1 measures are tractable, and the same holds for L_p measures with $1/p \in \mathbb{N}$. If X is independent of Y then $UTD(X,Y) = 0$ but not conversely. Tail dependence is invariant under a monotone transformation of X and Y. Hence, it is a property of the copula joining X and Y.

Let $(X_1, \ldots, X_{2N})$ have density (5.7) with $2N$ replacing N.

The incomplete Gamma integral with positive integer parameter m is:

$$\frac{1}{\Gamma(m)} \int_y^\infty \lambda^m z^{m-1} e^{-\lambda z} = \sum_{k=0}^{m-1} \frac{(\lambda y)^i}{i!} e^{-\lambda y}, \quad y > 0.$$

Then

$$\Pr\left(\sum_{i=1}^N X_i > r\right) = \int_0^\infty \Pr\left(\sum_{i=1}^N X_i > r \,|\, \Lambda = \lambda\right) \frac{\lambda^{\eta-1}\beta^\eta}{\Gamma(\eta)} e^{-\beta\lambda} d\lambda$$

$$= \int_0^\infty \sum_{k=0}^{N-1} \frac{(\lambda r)^k}{k!} e^{-\lambda r} \cdot \frac{\lambda^{\eta-1}\beta^\eta}{\Gamma(\eta)} e^{-\beta\lambda} d\lambda$$

$$= \left(\frac{\beta}{\beta+r}\right)^\eta \left[\sum_{k=0}^{N-1} \frac{\Gamma(\eta+k)}{k!\,\Gamma(\eta)} \frac{r^k}{(\beta+r)^k}\right]. \quad (5.11)$$

As $r \to \infty$, the bracketed term goes to

$$\left[\sum_{k=0}^{N-1} \frac{\Gamma(\eta+k)}{k!\,\Gamma(\eta)}\right].$$

Similarly,

$$\Pr\left(\sum_{i=1}^{N} X_i > r \cap \sum_{i=N+1}^{2N} X_i > r\right)$$

$$= \int_0^\infty e^{-2\lambda r} \left[\sum_{k=0}^{N-1} \frac{(\lambda r)^k}{k!}\right]^2 \cdot \frac{\lambda^{\eta-1}\beta^\eta}{\Gamma(\eta)} e^{-\beta\lambda} d\lambda$$

$$= \left(\frac{\beta}{\beta+2r}\right)^\eta \sum_{k,j=0}^{N-1} \frac{\Gamma(k+j+\eta)}{k!j!\Gamma(\eta)} \left(\frac{r}{\beta+2r}\right)^{k+j}. \qquad (5.12)$$

The tail dependence of sums of N L_1 variables is therefore the limiting ratio as $r \to \infty$ of (5.12) over (5.11):

$$\left(\frac{1}{2}\right)^\eta \frac{\sum_{k,j=0}^{N-1} 2^{-k-j} \frac{\Gamma(k+j+\eta)}{k!j!\Gamma(\eta)}}{\sum_{k=0}^{N-1} \frac{\Gamma(\eta+k)}{k!\Gamma(\eta)}}. \qquad (5.13)$$

Table 5.2 gives some values, comparing the number N of disjunct variables summed. We see that the tail dependence grows in N and decreases in the shape factor η. Also (5.13) converges to 1 as $N \to \infty$ for any $\eta > 0$ — a proof is given in Appendix C; the rate of convergence to 1 is slower for larger η.

Figure 5.11 shows rank scatterplots for sums of L_1 measures with shape $\eta = 3$. The first shows two variables, the second shows two sums of 10 variables, and the third shows two sums of 50 variables.

Table 5.2. Upper tail dependence for sums of N L_1 variables, the shape of the Gamma mixing distribution ranges from 1 to 5, 10, 15 and 20.

Shape	corr(X_1, X_2)	$N = 1$	$N = 3$	$N = 5$	$N = 10$	$N = 50$
1		0.500	0.688	0.754	0.824	0.920
2		0.250	0.453	0.549	0.664	0.842
3	0.3333	0.125	0.289	0.388	0.523	0.767
4	0.2500	0.062	0.180	0.267	0.405	0.694
5	0.2000	0.031	0.109	0.180	0.307	0.624
10	0.1000	0.001	0.007	0.019	0.061	0.338
15	0.0667	3×10^{-5}	4×10^{-4}	0.002	0.009	0.160
20	0.0500	1×10^{-6}	2×10^{-5}	1×10^{-4}	0.001	0.066

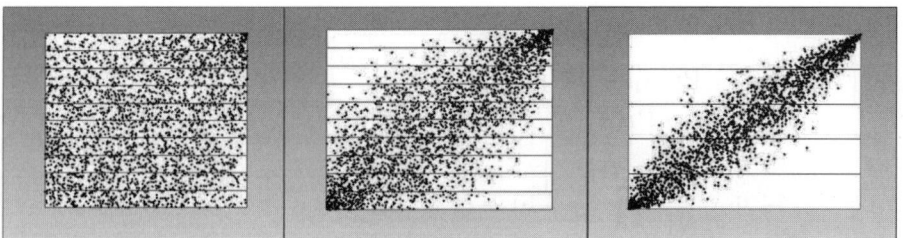

Figure 5.11. Percentile scatterplots for sums of L_1 variables, with shape of Gamma mixing distribution = 3. Left: 2 L_1 variables, rank correlation = 0.21; center: sums of 10 such variables, rank correlation = 0.77; right: sums of 50 such variables, rank correlation = 0.94.

5.3.5 Lower tail dependence

The multivariate Pareto model (5.9) does not have lower tail dependence, so it is not surprising that the aggregate losses S_1, S_2 do not have lower tail dependence. A derivation is given below, making use of the identity for the incomplete Gamma function with an integer shape parameter.

For $i = 1, 2$, let $S_i = S_i(N)$ denote the ith sum of N L_1 variables, as above. The marginal probability is

$$\Pr(S_i \leq r) = \frac{\beta^\eta}{(r+\beta)^\eta} \sum_{k=N}^{\infty} \frac{r^k}{(r+\beta)^k} \frac{\Gamma(\eta+k)}{k!\,\Gamma(\eta)}$$

and

$$\Pr(S_1 \leq r, S_2 \leq r)$$
$$= \int_0^\infty \sum_{k=N}^{\infty} \frac{(\lambda r)^k}{k!} e^{-\lambda r} \cdot \sum_{j=N}^{\infty} \frac{(\lambda r)^j}{j!} e^{-\lambda r} \cdot \frac{\lambda^{\eta-1}\beta^\eta}{\Gamma(\eta)} e^{-\beta\lambda} d\lambda$$
$$= \frac{\beta^\eta}{(2r+\beta)^\eta} \sum_{k=N}^{\infty} \sum_{j=N}^{\infty} \frac{r^{k+j}}{(2r+\beta)^{k+j}} \frac{\Gamma(\eta+k+j)}{k!\,j!\,\Gamma(\eta)}.$$

Putting $z = r/\beta$ this becomes:

$$z^{2N} \frac{\Gamma(\eta+2N)}{N!\,N!\,\Gamma(\eta)} + O(z^{2N+1}), \quad r = \beta z \to 0,$$

and

$$\Pr(S_1 \leq r) = \sum_{k=N}^{\infty} z^k (1+z)^{-\eta-k} \frac{\Gamma(\eta+k)}{k!\,\Gamma(\eta)}$$
$$= z^N \frac{\Gamma(\eta+N)}{N!\,\Gamma(\eta)} + O(z^{N+1}), \quad y = \beta z \to 0.$$

The limit of the lower tail (for a fixed N) is:

$$\lambda_L = \lim_{r \to 0} \frac{\Pr(S_1 \leq r, S_2 \leq r)}{\Pr(S_1 \leq r)} = \lim_{z \to 0} \frac{z^{2N} \frac{\Gamma(\eta+2N)}{N!N!\Gamma(\eta)} + O(z^{2N+1})}{z^N \frac{\Gamma(\eta+N)}{N!\Gamma(\eta)} + O(z^{N+1})}$$

$$= \lim_{z \to 0} \frac{z^N}{N!} \prod_{k=N}^{2N-1} (\eta+k) = 0.$$

5.4 Discussion

In this chapter, we have shown how some simple latent variable models lead to interesting results on tail dependence of aggregate losses. Further research consists of studying tail dependence on sums under more general dependence models, such as via vines. For example, $\sum_{i=1}^{N} X_i$ and $\sum_{i=N+1}^{2N} X_i$ are conjectured to have upper tail dependence of 1 as $N \to \infty$, if $X_1, \ldots, X_{2N}$ have Pareto-like upper tails and their joint distribution has upper tail dependence. In analyzing data, tail dependence-like behavior is also of interest, as this behavior may obtain for more general classes of copulae.

Appendices

A. *Proofs for the tail dependence condition involving $C_{1|2}(u|1)$*

The conditional distributions $C_{1|2}$ for the common one-parameter copula families are given on pp. 146–147 of Joe.[6]

- For the Frank copula with parameter $\theta > 0$,

$$\overline{C}_{1|2}(u|v) = [1 + e^{-\theta v} a(u)]^{-1}, \quad a(u) = (1 - e^{-\theta u})/(e^{-\theta u} - e^{-\theta}),$$

so that $\overline{C}_{1|2}(u|1) = [1 + e^{-\theta} a(u)]^{-1} = (e^\theta - e^{\theta u})/(e^\theta - 1) < 1$ for $0 < u < 1$.
- For the Plackett copula with parameter $\theta > 0$, $\overline{C}_{1|2}(u|1) = \theta(1-u)/[\theta(1-u)+u] < 1$ for $0 < u < 1$.
- For the bivariate normal copula with parameter $\rho > 0$, $\overline{C}_{1|2}(u|v) = 1 - \Phi([\Phi^{-1}(u) - \rho\Phi^{-1}(v)]/\sqrt{1-\rho^2}) \to 1 - \Phi(-\infty) = 1$ as $v \to 1$.
- For the Archimedean copula: with $C_\psi(u, v) = \psi(\psi^{-1}(u) + \psi^{-1}(v))$, where ψ is a Laplace transform,

$$C_{1|2}(u|v) = \frac{\psi'(\psi^{-1}(u) + \psi^{-1}(v))}{\psi'(\psi^{-1}(v))}$$

so that
$$\lim_{v \to 1} C_{1|2}(u|v) = \lim_{s \to 0} \frac{\psi'(\psi^{-1}(u) + s)}{\psi'(s)} = \lim_{s \to 0} \frac{\psi'(\psi^{-1}(u))}{\psi'(s)}.$$

This is 0 if $\psi'(0) = -\infty$ and is in $(0,1)$ if $-\psi'(0) < \infty$.

- For the Extreme-value copula: Let $C(u,v) = e^{-A(-\log u, -\log v)}$, where $\max\{w_1, w_2\} \le A(w_1, w_2) \le w_1 + w_2$ and A is homogeneous of order 1. Let $A_2 = \partial A/\partial w_2$ which is homogeneous of order 0. Then $C_{1|2}(u|v) = C(u,v) A_2(-\log u, -\log v) \cdot v^{-1}$ so that $C_{1|2}(u|1) = u A_2(-\log u, 0) = 0$, assuming $A(w_1, w_2) \ne w_1 + w_2$ and

$$A_2(w_1, 0) = \lim_{w_2 \to 0} \frac{\partial A(w_1, w_2)}{\partial w_2} = \frac{\partial \lim_{w_2 \to 0} A(w_1, w_2)}{\partial w_2} = \frac{\partial w_1}{\partial w_2} = 0.$$

It is easily shown directly that $A_2(w, 0) = 0$ for the Gumbel and Galambos copulae with positive dependence. For the Gumbel copula, $A(w_1, w_2) = (w_1^\theta + w_2^\theta)^{1/\theta}$ (for $\theta > 1$), and for the Galambos copula, $A(w_1, w_2) = w_1 + w_2 - (w_1^{-\theta} + w_2^{-\theta})^{-1/\theta}$ (for $\theta > 0$)

B. Proof of Proposition 5.1 and an example

Proof. Since tail dependence is invariant under monotone increasing transforms, without loss of generality, we assume that U_1, U_2, V are uniform $(0,1)$ random variables. We need to show that $\lim_{u \uparrow 1} \Pr(U_2 > u | U_1 > u) \ge \lambda_1 \lambda_2$.

Let $C_{U_1 U_2 V}(u_1, u_2, v)$ be the copula and joint distribution of U_1, U_2, V with margins $C_{U_1 V}(u_1, v), C_{U_2 V}(u_2, v)$. Let $C_{12|V}, C_{1|V}, C_{2|V}$ be the partial derivatives with respect to v, and let $\overline{C}_{12|V}, \overline{C}_{1|V}, \overline{C}_{2|V}$ be the corresponding survival functions. Note that for $0 < u < 1$,

$$\Pr(U_2 > u | U_1 > u) \ge \Pr(U_2 > u, V > u | U_1 > u)$$
$$= \frac{\Pr(U_2 > u, V > u, U_1 > u)}{1 - u} = (1-u)^{-1} \int_u^1 \overline{C}_{12|V}(u, u|v) \, dv$$
$$= (1-u)^{-1} \int_u^1 \overline{C}_{1|V}(u|v) \overline{C}_{2|V}(u|v) \, dv, \qquad (5.14)$$

where the last equality comes from conditional independence. The right-hand side of (5.14) is the same as

$$\mathbb{E}\left[\overline{C}_{1|V}(u|Z) \overline{C}_{2|V}(u|Z)\right], \qquad (5.15)$$

where Z is uniform on $[u, 1]$. With the stochastically increasing assumption, $\overline{C}_{1|V}(u|v)$ and $\overline{C}_{2|V}(u|v)$ are increasing in $v \in [u, 1)$. By positive dependence from Fréchet upper bound or co-monotonicity, the covariance of two increasing functions of a random variable is non-negative (if it exists), and hence (5.15) exceeds

$$\mathbb{E}[\overline{C}_{1|V}(u|Z)] \cdot \mathbb{E}[\overline{C}_{2|V}(u|Z)]$$
$$= (1-u)^{-1} \int_u^1 \overline{C}_{1|V}(u|v)\, dv \cdot (1-u)^{-1} \int_u^1 \overline{C}_{2|V}(u|v)\, dv$$
$$= \Pr(U_1 > u | V > u) \cdot \Pr(U_2 > u | V > u). \qquad (5.16)$$

Take the limit of (5.14) and (5.16) to get:

$$\lim_{u \uparrow 1} \Pr(U_2 > u | U_1 > u)$$
$$\geq \lim_{u \uparrow 1} \Pr(U_1 > u | V > u) \cdot \lim_{u \uparrow 1} \Pr(U_2 > u | V > u) = \lambda_1 \lambda_2 > 0. \qquad \square$$

Remark 5.1. Note that the stochastic increasing condition can be weakened to "$\Pr(U_i > u | V = v)$ is increasing in $v \in [u, 1)$ for all u near 1". Hence, it is a weak condition that would be expected to hold if there is tail dependence. The stochastic increasing condition, as given in Proposition 5.1, usually holds in models with conditional independence given a latent variable, as shown in the example below.

Example 5.1. For the multivariate Pareto distribution (5.9) that derives from a Gamma mixture of exponentials, let (X_1, X_2) be such that $X_i | \Lambda = a$ are conditional exponential with mean a^{-1}, and $\Lambda \sim \text{Gamma}(\eta, \beta)$. Then with $U_1 = X_1$, $U_2 = X_2$, $V = \Lambda^{-1}$, U_1, U_2 are each stochastically increasing in V. From the copula (5.10), the bivariate upper tail dependence parameter of (X_1, X_2) is $2^{-\eta}$. We next obtain the common tail dependence parameter λ_1 for (X_i, V) for $i = 1, 2$ and show the inequality from the proposition. Because of scale invariance, we assume $\beta = 1$ for the following calculations. Let $G(z; \eta) = [\Gamma(\eta)]^{-1} \int_0^z y^{\eta-1} e^{-y}\, dy$ be the cumulative distribution function of the Gamma$(\eta, 1)$ random variable Λ. Then

$$\Pr(X_1 > x | \Lambda^{-1} > v) = \Pr(X_1 > x, \Lambda < v^{-1}) / \Pr(\Lambda^{-1} > v), \qquad (5.17)$$
$$\Pr(X_1 > x, \Lambda < v^{-1})$$
$$= \Gamma^{-1}(\eta) \int_0^{v^{-1}} e^{-ax} a^{\eta-1} e^{-a}\, da = (1+x)^{-\eta} G(v^{-1}(1+x); \eta). \qquad (5.18)$$

X_1 has cumulative distribution function $F(x) = 1 - (1+x)^{-\eta}$ $(x > 0)$ and inverse cumulative distribution function $F^{-1}(p) = (1-p)^{-1/\eta} - 1$ $(0 < p < 1)$. For z near 0, $G(z; \eta) \approx z^\eta / \Gamma(\eta + 1)$. For $0 < u < 1$ that is close to 1, let $x = F^{-1}(u) = (1-u)^{-1/\eta} - 1$ and $v(u)$ be the u quantile of Λ^{-1}, so that $[v(u)]^{-1}$ is the lower $1-u$ quantile of Λ or $[v(u)]^{-1} \approx [(1-u)\Gamma(\eta+1)]^{1/\eta}$. Substitute into (5.17) and (5.18) to get:

$$\lim_{u \uparrow 1} \Pr(X_1 > F^{-1}(u) | \Lambda^{-1} > v(u))$$

$$= \lim_{u \uparrow 1} \frac{(1-u) \, G(\Gamma^{1/\eta}(\eta+1)(1-u)^{1/\eta}(1-u)^{-1/\eta}; \eta)}{1-u} = G(\Gamma^{1/\eta}(\eta+1); \eta).$$

To match Proposition 5.1, $\lambda_1 = \lambda_2 = G(\Gamma^{1/\eta}(\eta + 1); \eta)$ and it can be shown numerically that

$$\lim_{u \uparrow 1} \Pr(X_2 > F^{-1}(u) | X_1 > F^{-1}(u)) = 2^{-\eta} \geq [G(\Gamma^{1/\eta}(\eta+1); \eta)]^2.$$

C. Proof that $\lambda_{U,\eta,N} \to 1$ as $N \to \infty$

Rewrite (5.13) as:

$$\lambda_{U,\eta,N} = 2^{-\eta} \frac{\sum_{k=0}^{N-1} \sum_{j=0}^{N-1} \frac{\Gamma(\eta+k+j)}{\Gamma(\eta) \, 2^{k+j} k! j!}}{\sum_{k=0}^{N-1} \frac{\Gamma(\eta+k)}{\Gamma(\eta) \, k!}}. \tag{5.19}$$

The numerator on the right-hand side of (5.19) can be written as

$$\sum_{\ell=0}^{2N-2} \frac{\Gamma(\eta+\ell)}{\Gamma(\eta) \, \ell!} A_{\ell,N}, \tag{5.20}$$

where

$$A_{\ell,N} = \sum_{0 \leq k,j \leq N-1: k+j = \ell} \frac{\ell!}{2^{k+j} k! j!}.$$

For $0 \leq \ell \leq N-1$, then $A_{\ell,N} = 1$ from a binomial sum, and for $N \leq \ell \leq 2N-2$,

$$A_{\ell,N} = \sum_{i=\ell-N+1}^{N-1} \binom{\ell}{i} 2^{-\ell}.$$

It is shown in Lemma 5.2 below that $A_{\ell,N} \to 1$ as $N \to \infty$ for (approximately) fixed ℓ/N.

Next, (5.20) can be written as (with $k = \ell - N$ in second summation):

$$\sum_{k=0}^{N-1} \frac{\Gamma(\eta+k)}{\Gamma(\eta)\,k!} + \sum_{k=0}^{N-2} \frac{\Gamma(\eta+k+N)}{\Gamma(\eta)\,(k+N)!} A_{k+N,N} = D + \sum_{k=0}^{N-2} \frac{\Gamma(\eta+k+N)}{\Gamma(\eta)\,(k+N)!} A_{k+N,N}$$

where D is the denominator in (5.19). The proof is complete by showing that as $N \to \infty$,

$$D^{-1} \sum_{k=0}^{N-2} \frac{\Gamma(\eta+k+N)}{\Gamma(\eta)\,(k+N)!} A_{k+N,N} \to 2^\eta - 1$$

because then (5.19) goes to $2^{-\eta}[1 + (2^\eta - 1)] = 1$. This follows from the two lemmas below, together with the Lebesgue Dominated Convergence Theorem.

Lemma 5.1. *Let*

$$d_{\eta,k} = \frac{\Gamma(\eta+k)}{\Gamma(\eta)\,k!}, \quad k = 1, 2, \ldots.$$

As $N \to \infty$,

$$\frac{\sum_{k=0}^{N-2} \frac{\Gamma(\eta+k+N)}{\Gamma(\eta)\,(k+N)!}}{\sum_{k=0}^{N-1} \frac{\Gamma(\eta+k)}{\Gamma(\eta)\,k!}} = \frac{\sum_{k=0}^{N-2} d_{\eta,k+N}}{\sum_{k=0}^{N-1} d_{\eta,k}} \to 2^\eta - 1.$$

Proof. This is split into cases.

- $\eta = 1$: $d_{\eta,i} = 1$ for all i so the ratio is $1 = 2^1 - 1$.
- $\eta = 2$: $d_{\eta,k} = (k+1)$, $d_{\eta,k+N} = (k+N+1)$. Hence

$$\frac{\sum_{k=0}^{N-2} d_{\eta,k+N}}{\sum_{k=0}^{N-1} d_{\eta,k}} = \frac{\sum_{k=0}^{N-2}(k+N+1)}{\sum_{k=0}^{N-1}(k+1)} = \frac{3N(N-1)/2}{N(N+1)/2} \to 3 = 2^2 - 1.$$

- $\eta = 3$: $d_{\eta,k} = (k+2)(k+1)/2!$, $d_{\eta,k+N} = (k+N+2)(k+N+1)/2!$. Hence for large N,

$$\frac{\sum_{k=0}^{N-2} d_{\eta,k+N}}{\sum_{k=0}^{N-1} d_{\eta,k}} = \frac{\sum_{k=0}^{N-2}(k+N+2)(k+N+1)}{\sum_{k=0}^{N-1}(k+2)(k+1)}$$

$$\approx \frac{\int_0^N (x+N)^2\,dx}{\int_0^N x^2\,dx} = \frac{(2^3-1)N^3/3}{N^3/3} = 2^3 - 1.$$

- General $\eta > 0$: Since $\Gamma(\eta+i)/i!$ behaves like $i^{\eta-1}$ for large i (by applying Stirling's formula), then for large N,

$$\frac{\sum_{k=0}^{N-2} d_{\eta,k+N}}{\sum_{k=0}^{N-1} d_{\eta,k}} \approx \frac{\int_0^N (x+N)^{\eta-1} dx}{\int_0^N x^{\eta-1} dx} = \frac{(2^\eta - 1)N^\eta/\eta}{N^\eta/\eta} = 2^\eta - 1. \quad \square$$

Lemma 5.2. $A_{\ell_N, N} \to 1$ as $N \to \infty$ with $\ell_N/N \to a \in [1,2)$.

Proof. $A_{\ell,N} = \Pr(\ell - N + 1 \leq Y \leq N - 1)$, where $N \leq \ell \leq 2N - 2$ and $Y \sim \text{Binomial}(\ell, \frac{1}{2})$. By the normal approximation for large N and ℓ, this is approximately

$$\Pr\left(\frac{\ell - N + \frac{1}{2} - \frac{1}{2}\ell}{\frac{1}{2}\sqrt{\ell}} \leq Z \leq \frac{N - \frac{1}{2} - \frac{1}{2}\ell}{\frac{1}{2}\sqrt{\ell}}\right)$$

$$= \Phi\left(\frac{2N - 1 - \ell}{\sqrt{\ell}}\right) - \Phi\left(\frac{\ell - 2N + 1}{\sqrt{\ell}}\right)$$

where $Z \sim N(0,1)$ and Φ is the standard normal cumulative distribution function. Let $\ell = \ell_N = [aN]$ where $1 \leq a < 2$. Then, as $N \to \infty$,

$$\Phi\left(\frac{(2-a)N}{\sqrt{aN}}\right) - \Phi\left(\frac{-(2-a)N}{\sqrt{aN}}\right) \to 1. \quad \square$$

References

1. Abramowitz M. and Stegun I.A. (1972). *Handbook of Mathematical Functions*, Dover, New York.
2. Berman S. (1980). Stationarity, isotropy and sphericity in l_p. *Probability Theory and Related Fields*, 54(1):21–23.
3. Clayton D.G. (1978). A model for association in bivariate life tables and its application in epidemiological studies of familial tendency in chronic disease incidence. *Biometrika*, 65:141–151.
4. Cook R.D. and Johnson M.E. (1981). A family of distributions for modelling non-elliptically symmetric multivariate data. *Journal of the Royal Statistical Society: Series B*, 43:210–218.
5. Harris C. (1968). The Pareto distribution as a queue service discipline. *Operations Research*, 16:307–316.
6. Joe H. (1997). *Multivariate Models and Dependence Concepts*. Chapman & Hall, London.
7. Kimeldorf G. and Sampson A.R. (1975). Uniform representations of bivariate distributions. *Communications in Statistics*, 4:617–627.
8. Kousky C. and Cooke R.M. (2009). Climate change and risk management: Challenges for insurance, adaptation and loss estimation. Discussion Paper RFF-DP-09-03, Resources for the Future, Washington DC.

9. Mardia K.V. (1962). Multivariate Pareto distributions. *Annals of Mathematical Statistics*, 33:1008–1015.
10. McNeil A.J., Frey R. and Embrechts P. (2005). *Quantitative Risk Management: Concepts, Techniques and Tools*. Princeton University Press, New Jersey.
11. Takahasi K. (1965). Note on the multivariate Burr's distribution. *Annals of the Institute of Statistical Mathematics*, 17:257–260.

CHAPTER 6

The Copula Information Criterion and Its Implications for the Maximum Pseudo-Likelihood Estimator

Steffen Grønneberg

Department of Mathematics
University of Oslo
P.O. Box 1053 Blindern
N-0316 Oslo, Norway
steffeng@math.uio.no

This chapter surveys the asymptotic theory of estimation of a copula from a frequentistic perspective and presents the problems involved in frequentistic model selection among several candidate copulae when using the maximum pseudo-likelihood estimator (MPLE). Frequentistic copula model selection has recently been addressed through the development of the copula information criterion (CIC) — a model selection formula which extends the maximum likelihood-based Akaike information criterion (AIC) to the MPLE. We present the developments leading to the CIC with a focus on its implications, while deferring proofs of underlying limit theorems to the original CIC paper.

The CIC is in fact two different formulae, one for misspecified copula models and another for correctly specified copula models, paralleling the Takeuchi information criterion and the Akaike information criterion respectively.

These formulae show that there does not exist (in a certain technical sense) an AIC formula for MPL estimation when the parametric copula has extreme behavior near the edge of the unit cube. This means that one cannot make first-order bias-correction terms of a desired part of the attained Kullback–Leibler divergence between the MPL-estimated copula and the data-generating copula in a class of copulae which has received much attention in econometrics. This may be seen as a demarcation of which types of copulae that should be estimated with the MPLE. Interestingly, the main motivating factor for using the MPLE is also the reason for the non-existence of a general MPLE-based AIC formula. A further conclusion is that the CIC provides a counterexample to the often acclaimed intrinsic connection between the AIC and Occam's Razor.

6.1 Introduction 114
6.2 The Developments Leading to the CIC 116
    6.2.1 The fully parametric MLE 117
    6.2.2 Kullback–Leibler divergence and model selection . . 119
    6.2.3 The MPLE, the empirical copula and invariance considerations 122
    6.2.4 What about semiparametric efficiency? 124
    6.2.5 Large-sample theory for the MPLE 125
6.3 Model Selection with the MPLE 126
    6.3.1 Non-existence of bias-correction terms and implications for the MPLE 130
    6.3.2 Philosophical implications of the CIC 132
6.4 Illustrations 132
6.5 Concluding Remarks 136
References 137

6.1 Introduction

Suppose n-dimensional stochastic vectors $X_1, X_2, \ldots, X_N$ are observed, which are independent of each other, and all coming from the same, unknown data-generating distribution

$$F^\circ(x) = C^\circ(F_1^\circ(x_1), \ldots, F_n^\circ(x_n)). \tag{6.1}$$

We assume that F° is continuous and we wish to model the copula C° through one or perhaps several parametric classes. In the praxis of parametric copula modeling, there are four basic problems which are naturally met in any investigation. First, if our model is

$$f_\theta(x) = c_\theta(F_1^\circ(x_1), \ldots, F_n^\circ(x_n)) \prod_{i=1}^n f_i^\circ(x_i)$$

where the marginals F_i° are completely unknown, how should θ be estimated? Second, how should the parametric form of c_θ be chosen? Third, how should one select among several candidate models on the basis of observed data? And fourth, is the final model (or models) adequate?

The first problem has various solutions, among which the maximum pseudo-likelihood estimator (MPLE) discussed in Genest[5] is the most popular. The second problem is implicit in all multivariate model building, and much of this book is devoted solely to providing flexible solutions to this problem. The fourth problem is usually dealt with through goodness-of-fit tests which are based on the MPLE, and there exist several investigations in the area (See Genest[6]).

The development of the CIC started when it was noticed that the third issue had been ignored or dealt with in an incorrect manner. Several published papers, and many practitioners, have used the "AIC formula"

$$\text{AIC}^\bullet = 2\ell_{N,\max} - 2\,\text{length}(\theta) \tag{6.2}$$

as a model selection criterion, with $\ell_{n,\max} = \ell_n(\hat\theta)$ being the maximum pseudo-likelihood, from the traditional Akaike information criterion

$$\text{AIC} = 2\ell^{\#}_{N,\max} - 2\,\text{length}(\theta),$$

where $\ell^{\#}_{N,\max}$ is the usual maximum likelihood for a fully parametric model. One computes this AIC$^\bullet$ score for each candidate model and in the end chooses the model with the highest score.

This ignores the fact that the pseudo-likelihood is not a proper likelihood, and unfortunately it does not lead to a correct formula. Grønneberg[7] derived a proper generalization of the AIC for the MPLE and named it the copula information criterion (CIC). The formula is given by

$$\text{CIC} = 2\ell_{N,\max} - 2(\hat p^* + \hat q^* + \hat r^*) \tag{6.3}$$

with expressions for $\hat p^* + \hat q^* + \hat r^*$ different from (and more complicated than) merely length(θ). These quantities even vary non-trivially with the model parameter — in clear contrast with length(θ) which is invariant to the actual value of θ.

But the story does not end here, as the CIC formula derived in Ref. 7 does not exist for a large class of copula families such as copulae with extreme tail dependence. This lack of existence is, however, not a deficiency of the arguments used in Ref. 7, but is an inherent limitation of the asymptotic behavior of the MPLE. This makes model selection with the MPLE a more complex problem than the fully parametric case, and the CIC formula can only attack model selection problems concerning copulae which are sufficiently well-behaved along the edges of the unit cube. The implications of this is discussed in the conclusion of the chapter.

To understand these developments and the difficulties involved in the model selection problem for copula estimation with the MPLE, one needs to understand some fundamental issues concerning the MLE, the AIC and the MPLE. The present chapter is, in addition to the introduction and concluding remarks, divided into three parts. The first part is Section 6.2, which presents the MLE, the AIC and the MPLE from a perspective which naturally leads to the CIC formula. The second part of our story is Section 6.3, which derives the two CIC formulae. Finally, we include a brief simulation example in Section 6.4. Although we will omit the technical asymptotic developments needed to make the arguments rigorous, we will discuss the needed mathematical structures to such a degree that the above-mentioned exploding bias-correction terms can be presented without simplification.

Let us first introduce some general notation that we use throughout the chapter. Let $F_1^\circ, F_2^\circ, \ldots, F_n^\circ$ be the marginal distributions of F°, and let

$$F_\perp^\circ(x) := (F_1^\circ(x_1), F_2^\circ(x_2), \ldots, F_n^\circ(x_n))$$

be the vector of marginal distributions. We will denote all sizes related to the true data-generating distribution F° by circle superscripts, and all empirical estimates through replacing the circle with a hat, so that for example $\hat{F}_N$ can be seen right away to estimate F°. The assumed continuity of F° implies the existence of a unique copula C° defined implicitly through

$$F^\circ(x) = C^\circ(F_\perp^\circ(x)) \tag{6.4}$$

or equivalently through the more explicit

$$C^\circ(v) = F^\circ(F_\perp^{\circ\,-1}v(u)) \tag{6.5}$$

where

$$F_\perp^{\circ\,-1}v(u) = (F_1^{\circ\,-1}v(u_1), F_2^{\circ\,-1}v(u_2), \ldots, F_n^{\circ\,-1}v(u_n))$$

is the vector of inverse marginal distributions.

6.2 The Developments Leading to the CIC

The MPLE and the AIC both generalize the MLE, but in completely different ways. The AIC generalizes the MLE to multimodel estimation, while the MPLE generalizes the MLE to situations where the marginals are unknown. The CIC generalize both the MPLE and the AIC in that it implements the AIC-generalization of the MLE to the MPL estimator. In order to present this generalization, we thus need to present the fundamentals of the MLE, the AIC and the MPLE.

The MPLE sets out to estimate a copula parameter θ in a parametric model

$$f_\theta(x) = c_\theta(F_\perp^\circ(x)) \prod_{i=1}^n f_i^\circ(x_i)$$

where the marginal distributions $F_\perp^\circ$ are completely unspecified. Its precise form is defined through the following two considerations.

(1) It asymptotically minimizes the Kullback–Leibler divergence between the true data-generating copula c° and a parametric copula c_θ. This generalizes the standard MLE.
(2) The estimation of the θ that minimizes the Kullback–Leibler divergence between c° and c_θ is invariant to a large class of symmetries. An empirical estimate $\hat\theta$ should be invariant to the same symmetries.

Although the motivation for using the ML estimator to estimate a parametric model that is correctly specified is well-known, its connection to the minimization of Kullback–Leibler divergence in the general case is not. This perspective naturally leads to the model selection strategy of Akaike, and Sections 6.2.1 and 6.2.2 treat these two themes. The above-mentioned invariance considerations are even less well-known (it seems not to have been made explicit in any previous expositions), and we use Section 6.2.3 to discuss it and to define the MPLE precisely. Finally, Section 6.2.4 discusses the fact that the MPLE is not semiparametrically efficient, and argues that the concept of semiparametric efficiency is a very different way of constructing estimators, and is in natural opposition to symmetry considerations. The central argument is that the MPLE is not a semiparametric estimator per se, but focuses on estimating the copula parameter θ° which is least false with respect to the Kullback–Leibler divergence while respecting the related symmetry considerations. In doing so, it does provide nonparametric estimates of the vector of marginal distributions $F_\perp^\circ$, but this infinite-dimensional part of the MPLE is merely a by-product of symmetry considerations.

6.2.1 The fully parametric MLE

Let us quickly review how the MLE is justified when we refuse to make the assumption of having the true data-generating distribution f° contained in the parametric model to be fitted. For more details with a model selection perspective in mind, see Ref. 4. Suppose (for the moment) that we wish to fit a *fully parametric* density

$$f_{\theta,\gamma}(x) = c_\theta(F_{1,\gamma(1)}(x_1), \ldots, F_{n,\gamma(d)}(x_n)) \prod_{i=1}^{n} f^\circ_{i,\gamma(i)}(x_i)$$

to observed data $X_1, \ldots, X_N \sim F^\circ$. The MLE paradigm tries to estimate

$$(\theta^\circ_{\mathrm{ML}}, \gamma^\circ_{\mathrm{ML}}) = \underset{\theta,\gamma}{\mathrm{argmax}} \int \log f_{\theta,\gamma} \, dF^\circ \qquad (6.6)$$

from empirical data through replacing the unknown F° with the known multivariate empirical distribution $\hat{F}_n$ defined by

$$\hat{F}_N(x) := \frac{1}{N}\sum_{i=1}^{N}\prod_{j=1}^{n} I\{X_{j,i} \leq x_j\} = \frac{1}{N}\sum_{i=1}^{N} I\{X_i \leq x\}.$$

Recall that $\int \log f_{\theta,\gamma}\, dF^\circ$ is a so-called multivariate Lebesgue–Stieltjes integral, and is just another way of writing $\mathbb{E}\log f_{\theta,\gamma}(X)$. We will use this notation throughout the chapter, as it leads to a very simple and rather general principle that often gives consistent empirical estimators for many quantities of interest[a] through replacing "the circle with a hat" in F° and $\hat{F}_N$. The Lebesgue–Stieltjes integral has certain continuity properties, so that under quite general conditions "uniform (strong) consistency" of $\hat{F}_N$, meaning that

$$\lim_{N\to\infty}\sup_{x\in\mathbb{R}^n}|\hat{F}_N(x) - F^\circ(x)| = 0 \text{ almost surely},$$

implies that for each θ we have

$$\lim_{N\to\infty}\int \log f_{\theta,\gamma}\, d\hat{F}_N = \int \log f_{\theta,\gamma}\, dF^\circ \text{ almost surely}. \qquad (6.7)$$

This is close to showing that the plug-in step of "putting a hat on" F° works in the sense that $(\hat{\theta}, \hat{\gamma}) \xrightarrow[N\to\infty]{\text{a.s}} (\theta^\circ, \gamma^\circ)$. For $\hat{F}_N$, we have

$$\int \log f_{\theta,\gamma}\, d\hat{F}_N = \frac{1}{N}\sum_{i=1}^{N} \log f_{\theta,\gamma}(X_i),$$

so Eq. (6.7) is just another way of stating the strong law of large numbers. But this perspective will give us a simple way of making the consistency of the MPLE plausible. For the standard MLE, the "plug-in" step takes us from

$$(\theta^\circ_{\text{ML}}, \gamma^\circ_{\text{ML}}) = \operatorname*{argmax}_{\theta,\gamma} \int \log f_{\theta,\gamma}\, dF^\circ$$

to the empirical estimate

$$(\hat{\theta}_{\text{ML}}, \hat{\gamma}_{\text{ML}}) = \operatorname*{argmax}_{\theta,\gamma} \int \log f_{\theta,\gamma}\, d\hat{F}_N,$$

which is also the standard definition of the MLE.

---

[a]This functional perspective comes from the theory of stochastic processes, but simplifies and clarifies many asymptotic developments for random variables as well. It has many implications, and leads to a very intuitive and transparent point of view for the asymptotic properties of almost all common statistical estimators. See Ref. 18 for examples and mathematical developments.

The ML estimator was originally motivated by assuming that $f^\circ = f_{\theta_{\text{ML}}^\circ, \gamma_{\text{ML}}^\circ}$ and then proceeding to find the estimator which asymptotically has the least variance for the true parameter. In spite of this motivation, the MLE can be calculated even when f° is not assumed to be expressible through $f_{\theta,\gamma}$ and the above consistency result is valid no matter what the true density f° is. Hence, the maximum likelihood estimator will consistently maximize $\int \log f_{\theta,\gamma} \, \mathrm{d}F^\circ$. We now show that the parameter configuration which maximizes $\int \log f_{\theta,\gamma} \, \mathrm{d}F^\circ$ is a "least false" parameter in the following sense.

The relative entropy ("Kullback–Leibler divergence") between f° and $f_{\theta,\gamma}$ is

$$\mathrm{KL}(f^\circ, f_{\theta,\gamma}) = \int f^\circ \log \frac{f^\circ}{f_{\theta,\gamma}} \, \mathrm{d}x = \int f^\circ \log f^\circ \, \mathrm{d}x - \int f^\circ \log f_{\theta,\gamma} \, \mathrm{d}x$$

where the second term is recognized from Eq. (6.6). As the first term in the above display does not vary with (θ, γ), we have

$$\operatorname*{argmin}_{\theta,\gamma} \mathrm{KL}(f^\circ, f_{\theta,\gamma}) = \operatorname*{argmax}_{\theta,\gamma} \int \log f_{\theta,\gamma} \, \mathrm{d}F^\circ = (\theta_{\text{ML}}^\circ, \gamma_{\text{ML}}^\circ),$$

so that finding the maximum likelihood estimate will asymptotically reach the parameter $(\theta^\circ, \gamma^\circ)$ which minimizes the Kullback–Leibler divergence between f° and $f_{\theta,\gamma}$. We call $(\theta^\circ, \gamma^\circ)$ the least false parameter (with respect to the Kullback–Leibler divergence).

The Kullback–Leibler divergence $\mathrm{KL}(f, g)$ is zero if and only if $f = g$ almost surely with respect to the Lebesgue measure, which means that we can use the Kullback–Leibler divergence to distinguish between two densities. This property is the absolute minimal assumption needed to provide motivation to minimize $\mathrm{KL}(f^\circ, f_{\theta,\gamma})$ with respect to the parameter sets. There are also deeper motivations for using precisely the Kullback–Leibler divergence, and not just any other function which is zero if and only if $f = g$ almost surely, as it is connected with the mathematical concept of information and entropy. See Ref. 4 for a general discussion.

6.2.2 *Kullback–Leibler divergence and model selection*

Maximizing the likelihood function asymptotically reaches the parameter configuration that minimizes the Kullback–Leibler divergence between f° and $f_{\theta,\gamma}$. In the presence of several competing parametric models

$$f_{1,\alpha(1)}, \ldots, f_{K,\alpha(K)},$$

it is natural to define the best model as the model which minimizes the Kullback–Leibler divergence to the truth. Let

$$\alpha(k)^\circ = \operatorname*{argmin}_{\alpha(k)} \operatorname{KL}(f^\circ, f_{k,\alpha(k)})$$

denote the least false parameter configuration when constrained to the kth parametric class, so that the parametric model with the index

$$k^\circ = \operatorname*{argmin}_{1 \le k \le K} \operatorname{KL}(f^\circ, f_{k,\alpha(k)^\circ})$$

is the best (in the Kullback–Leibler sense) model *among the ones we are presently considering*, i.e., the global minimizer of Kullback–Leibler divergence in the space of all parameter configurations possible among all considered models. As k° only depends on the data-generating distribution F° through a multivariate Lebesgue–Stieltjes integral, the plug-in principle lets us define the empirical estimators of k° by using

$$\tilde{k}_N = \operatorname*{argmax}_{1 \le k \le K} \int \log f_{\hat{\alpha}(k)} \, \mathrm{d}\hat{F}_N$$

where

$$\hat{\alpha}_N(k) = \operatorname*{argmax}_{\alpha(k)} \int \log f_{k,\alpha(k)} \, \mathrm{d}\hat{F}_N.$$

This is the *main conceptual step in developing the Akaike information criterion*, and the precise AIC formula is simply a refinement of this observation. Although $\tilde{k}_N$ is a consistent estimator, it has non-negligible bias for small[b] N. The above definition of $\tilde{k}_N$ simply defines the estimated best model as the one with the highest log-likelihood at the maximum likelihood estimate, and the standard AIC formula derives first-order bias corrections *in a rather specific way*. A Taylor expansion together with the well-known asymptotic likelihood theory show that

$$\int \log f_{k,\hat{\alpha}(k)} \, \mathrm{d}\hat{F}_N - \int \log f_{k,\hat{\alpha}(k)} \, \mathrm{d}F^\circ = \bar{Z}_N + \frac{1}{N} p_N(k) + o_p(N^{-1})$$

in which $\mathbb{E}\bar{Z}_N = 0$ while $p_N(k)$ converges in distribution to $p(k)$ with expectation $p^*(k)$. Asymptotic likelihood theory provides the distribution of $p^*(k)$, and so we can estimate it. This leads to a first-order bias-correction term of

$$\int \log f_{k,\hat{\alpha}(k)} \, \mathrm{d}\hat{F}_N,$$

---

[b]First-order bias-correction terms are insignificant for very large N, and so if N is sufficiently large, the estimator $\tilde{k}_N$ yields a sensible model selection strategy.

in which it is crucial to notice that this expression is defined in terms of $\hat{\alpha}(k)$, the empirical estimate which is potentially being used, and not $\alpha^\circ(k)$, the least false parameter configuration which is unknown. If we work under the assumption that f° is in the parametric class under consideration, we get the rather amazing conclusion that $p^*(k) = \text{length}(\alpha(k))$, giving the famous AIC strategy

$$\hat{k}_N^{\text{AIC}} = \underset{1 \leq k \leq K}{\operatorname{argmax}} \left[\int \log f_{k,\hat{\alpha}(k)} \, \mathrm{d}\hat{F}_N - \frac{1}{N} \text{length}(\hat{\alpha}(k)) \right]$$

requiring *no empirical estimation of the bias-correction term*. For this strategy to be conceptually and formally consistent, we need to assume nested models. If this assumption cannot be justified, one can use the Takeuchi information criterion, which uses plug-in estimators of $p^*(k)$, and hence is of higher variability. See Ref. 4 for a more detailed discussion. We will define the development of first-order bias-correction terms as *the AIC programme*, and it is this we will carry out to conclude with the copula information criterion. We stress the importance of the $o_p(N^{-1})$ term and note that it is the N^{-1} which defines to what resolution we need to provide bias corrections if we are to implement the above "AIC programme".

Notice that the AIC formula derives bias corrections of $\int \log f_{k,\alpha(k)} \, \mathrm{d}\hat{F}_N$ instead of $\tilde{k}_N$, which is what we are actually going to use. In this connection, we note that more sophisticated non-asymptotic approximations are developed in Ref. 12 through powerful concentration inequalities. These developments are much more mathematically complex than the methodology of the standard AIC formula and are not generalized to the MPLE setting. We will be content with working with the above programme.

Another feature of the AIC formula is that we work with the expectation of $p(k)$, the weak limit of $p_N(k)$. This is perhaps first and foremost motivated through mathematical convenience as there is no general expression for $\mathbb{E}p_N(k)$. However, a more subtle point is that $\mathbb{E}p_N(k)$ can be infinite for even simple models such as the binomial model (Chapter 2 of Ref. 4). The AIC formula solves this potential explosion (that is, the non-existence of expectations) through going to the limit, and there everything works out nicely. For the CIC case, which transfers the above derivations to parameter estimates based on the MPLE and not the MLE, we get an additional bias-correction term r_n which has the unfortunate feature that going to the limit does not avoid the possibility of a non-existing expectation. Several common copulae models have an exploding exploding $\mathbb{E}r_N$, leading to non-existing bias-correction terms *with respect to the above defined AIC programme*.

6.2.3 The MPLE, the empirical copula and invariance considerations

We would like to fit a parametric copula c_θ without specifying the marginal distributions. So we work under the assumption that observed data have a parametric distribution given by

$$f_\theta(x) = c_\theta(F_1^\circ(x_1), \ldots, F_n^\circ(x_n)) \prod_{j=1}^n f_j^\circ(x_j).$$

If the parametric form of the copula includes the correct copula c°, we wish to find the true parameter value. Otherwise, we wish to find the θ which minimizes the Kullback–Leibler divergence between f_θ and the true density

$$f^\circ(x) = c^\circ(F_1^\circ(x_1), \ldots, F_n^\circ(x_n)) \prod_{j=1}^n f_k^\circ(x_j).$$

That is, the loss function we wish to minimize is $d(\theta) = \mathrm{KL}(f^\circ, f_\theta)$, where the minimum will be zero if and only if the model is correctly specified. Notice that we do not focus on estimating the marginals f_i°, but only on finding the least false copula inside the parametric class under consideration.

In many cases, the nonspecification of the marginals comes from lack of *a priori* knowledge of parametric forms for the marginals. If this is the case, the above estimation problem has important symmetry properties, which motivates the use of the MPLE from equivariance considerations of classical point estimation theory, as described, e.g., in Ref. 10. First, the copula of any stochastic vector is left invariant to any (not necessarily linear) change in scale for the data. More precisely, assume that a stochastic vector X has distribution function $C^\circ(F_\perp^\circ)$. The copula C° of X is then invariant to the whole class of functions

$$\mathcal{S} := \{H : \mathbb{R}^n \mapsto \mathbb{R}^n : H(x_1, \ldots, x_n) = (H_1(x_1), H_2(x_2), \ldots, H_n(x_n)),$$
$$\text{and each } H_i \text{ is monotonously increasing}\} \qquad (6.8)$$

in the sense that for an $H \in \mathcal{S}$, the random vector $H(X)$ also has the copula C°. To see this, notice that the marginal distributions of $H(X)$ are given by $F_{H_i(X_i)}(v) = P\{H_i(X_i) \leq v\} = P\{X_i \leq H_i^{-1}(v)\}$, and so $F_{H(X),\perp}(x) = F_\perp(H^{-1}(v))$. Thus,

$$F_{H(X),\perp}(x)(H(X)) = F_\perp \circ H^{-1} \circ H(X) = F_\perp(X) \sim C^\circ,$$

which demonstrates the invariance. As the copula C° is completely unaffected under $\mathcal{S}$-transformations, this invariance will be shared by any parametric copula family c_θ. This should also be intuitively clear, as the copula represents the dependency structure of X, and each H in $\mathcal{S}$ merely changes the scale of each coordinate. This change in scale does transform the (intuitive notion of) dependency among the elements of X.

The loss function $d(\theta) = \mathrm{KL}(f^\circ, f_\theta)$ is also invariant to the class $\mathcal{S}$, as it in fact does not depend on the marginals $F_\perp^\circ$. To see this, notice that

$$\mathrm{KL}(f^\circ, f_\theta) = \int \log \frac{f^\circ}{f_\theta} \, \mathrm{d}F^\circ$$

$$= \int \log c^\circ(F_1^\circ(x_1), \ldots, F_n^\circ(x_n)) \, \mathrm{d}F^\circ + \sum_{j=1}^{n} \int \log f_j^\circ(x_k) \, \mathrm{d}F^\circ$$

$$- \int \log c_\theta(F_1^\circ(x_1), \ldots, F_n^\circ(x_n)) \, \mathrm{d}F^\circ - \sum_{j=1}^{n} \int \log f_j^\circ(x_j) \, \mathrm{d}F^\circ$$

$$= \int \log \frac{c^\circ(F_1^\circ(x_1), \ldots, F_n^\circ(x_n))}{c_\theta(F_1^\circ(x_1), \ldots, F_n^\circ(x_n))} \, \mathrm{d}F^\circ \qquad (6.9)$$

$$= \int \log \frac{c^\circ(v_1, \ldots, v_n)}{c_\theta(v_1, \ldots, v_n)} \, \mathrm{d}C^\circ(v) \qquad (6.10)$$

$$= \mathrm{KL}(c^\circ, c_\theta),$$

where the transition from Eqs. (6.9) to (6.10) applies the change of variables formula for multivariate Lebesgue–Stieltjes integrals.

This validates the principle of equivariance (see Ref. 10), meaning that any estimator of $\hat{\theta}$ should be invariant to transformations of $\mathcal{S}$. It is well-known from the problem of testing independence that multivariate rank statistics are "maximally invariant" (see Ref. 11 for precise definitions) with respect to the transformations in $\mathcal{S}$, and so our estimator needs to be a functional of multivariate rank statistics.

Univariate ranks are equivalently represented through the marginal empirical distribution function. Analogously, multivariate ranks are equivalently represented through the empirical copula

$$\hat{C}_N(v) = \frac{1}{N} \sum_{i=1}^{N} \prod_{j=1}^{n} I\{\hat{F}_{N,j}(X_{i,j}) \leq v_j\}$$

$$= \frac{1}{N+1} \sum_{i=1}^{N} I\{\hat{F}_{N,\perp}(X_i) \leq v\},$$

so that any functional of the multivariate ranks is a functional of the empirical copula. Here $\hat{F}_{N,\perp}$ is the vector of marginal empirical distributions multiplied by $\frac{N}{(N+1)}$ to keep the observations away from the edge of the unit

cube. That is,
$$\hat{F}_{N,\perp}(x) = \left(\hat{F}_{N,1}(x_1), \hat{F}_{N,2}(x_2), \ldots, \hat{F}_{N,n}(x_n)\right)$$
where
$$\hat{F}_{N,j}(x_j) = \frac{1}{N+1}\sum \prod_{j=1}^{n} I\{X_{ij} \leq x_j\}$$

When observing that the least false copula parameter $\theta°$ can be written as
$$\theta° = \underset{\theta}{\operatorname{argmin}}\, \mathrm{KL}(f°, f_\theta) = \underset{\theta}{\operatorname{argmax}} \int \log c_\theta \, \mathrm{d}C°,$$
and when one knows that the empirical copula is a uniformly strongly consistent estimator of the data-generating copula in the sense that
$$\sup_{v}|\hat{C}_N(v) - C°(v)| \xrightarrow[N\to\infty]{\text{a.s.}} 0, \qquad (6.11)$$
a very natural estimator of $\theta°$ is the MPLE given by
$$\hat{\theta} = \underset{\theta}{\operatorname{argmax}} \int \log c_\theta \, \mathrm{d}\hat{C}_N = \underset{\theta}{\operatorname{argmax}} \frac{1}{N} \sum_{i=1}^{N} \log c_\theta(\hat{F}_{N,\perp}(X_i)).$$

6.2.4 What about semiparametric efficiency?

It is well-known that the MPLE is not universally semiparametrically efficient in the sense of, e.g., Ref. 1. In the context of model selection of semiparametric copula models, lack of semiparametric efficiency is not a serious deficiency. The semiparametric efficiency concept is defined for models that include the true data-generating distribution, which is certainly not the case in any investigation where non-nested model selection is needed.

Although there does exist a semiparametric copula estimation routine which is universally semiparametrically efficient (given in Ref. 3), it does not respect the symmetry considerations leading to the MPLE. While the Chen[3] method is well-motivated only when the parametric copula model includes the data-generating copula, the symmetry considerations motivating the MPLE are valid no matter which copula is the data-generating one. Although it would be desirable for the MPLE to be semiparametrically efficient, this is not the problem the MPLE sets out to solve. There should be no surprise if estimators derived from equivariance considerations, and that happen to be interpretable also as semiparametric estimators, are not semiparametrically efficient, as these two concepts most often represent opposing interests.[c]

6.2.5 Large-sample theory for the MPLE

In Section 6.2.2, we saw that the large-sample theory of the MLE was needed to derive bias corrections that motivated the AIC formula. This section will state the large-sample results which form a basis for the CIC. The results are justified in Refs. 2, 5 and 16, and we state them without further justification.

Recall the definition of $\hat{F}_{N,\perp}$ and define

$$\ell_N(\theta) = \sum_{i=1}^{N} \log c_\theta(\hat{F}_{N,\perp}(X_i))$$

as the "pseudo-likelihood" function, and let

$$\hat{A}_N(\theta) = \frac{1}{N}\ell_N(\theta) = \int \log c_\theta \, d\hat{C}_N$$

be the normalized pseudo-likelihood function so that

$$\hat{\theta} = \operatorname*{argmax}_\theta \ell_N(\theta) = \operatorname*{argmax}_\theta \hat{A}_N(\theta).$$

And while $\ell_N(\theta) \to \infty$, we have normalized $\hat{A}_N$ so that

$$\hat{A}_N(\theta) \xrightarrow[N\to\infty]{\text{a.s.}} \int \log c_\theta \, dC^\circ =: A(\theta).$$

Classical Taylor expansion-based proofs of normality for M-estimators (estimators which optimize a criterion function) require the asymptotic distribution of the score function

$$U_N := \frac{\partial \hat{A}_N(\theta_0)}{\partial \theta}.$$

As $U_N = \int \phi(v, \theta_0) \, d\hat{C}_N$, where $\phi(\cdot, \theta) = \partial/\partial\theta \log c(\cdot, \theta)$, the score function is a multivariate rank statistic, whose asymptotic behavior is derived in

---

[c] However, Ref. 8 gives sufficient conditions for estimators derived through invariance considerations to be semiparametrically efficient.

Refs. 14 and 15. We get

$$\sqrt{N}\, U_N \xrightarrow[N\to\infty]{W} U \sim N_p(0, \Sigma),$$

where Σ is somewhat inflated compared to the standard maximum likelihood setting. We have

$$\Sigma = \mathcal{I} + \mathrm{Cov}\left\{ \sum_{j=1}^{n} \int_{[0,1]^n} \frac{\partial \phi(v, \theta_0)}{\partial v_j} (I\{\xi_j \leq v_j\} - v_j)\, \mathrm{d}C^\circ(v) \right\}$$

in which $\mathcal{I}$ is the information matrix, $\mathcal{I} = \mathbb{E}\phi(\xi, \theta_0)\phi(\xi, \theta_0)^t$, and $\xi = (\xi_1, \xi_2, \ldots, \xi_n)$ is a random vector distributed according to C°. Note that the above covariance is taken with respect to ξ.

Regularity conditions then secure

$$\sqrt{N}(\hat{\theta} - \theta_0) \xrightarrow[N\to\infty]{W} J^{-1} U \sim N_p(0, J^{-1}\Sigma J^{-1}), \qquad (6.12)$$

where

$$J = -A''(\theta_0) = -\int_{[0,1]^n} \frac{\partial^2 \log c_{\theta_0}(v)}{\partial \theta \partial \theta^t}\, \mathrm{d}C^\circ.$$

If $c^\circ = c_{\theta_0}$, the well-known information matrix equality $J = \mathcal{I}$ is valid. This means that the limit covariance of Eq. (6.12) is simplified to

$$J^{-1} + J^{-1} \mathrm{Cov}\left\{ \sum_{j=1}^{n} \int_{[0,1]^n} \frac{\partial \phi(v, \theta_0)}{\partial v_j} (I\{\xi_j \leq v_j\} - v_j)\, \mathrm{d}C^\circ(v) \right\} J^{-1}.$$

6.3 Model Selection with the MPLE

We are now ready to implement the AIC programme for the MPLE paralleling the developments of Section 6.2.2. All proofs and technical subtleties are omitted, for which the reader can refer to Grønneberg.[7]

Suppose we have K copula models $c_{1,\theta(1)}, \ldots, c_{K,\theta(K)}$ and wish to choose which to use on the basis of empirical data. We assume that the MPLE is to be used in the estimation of the copula parameters. This means we define the best parameter configuration for each of the models to be the $\theta^\circ(k)$ which minimizes the Kullback–Leibler divergence between c° and $c_{k,\theta(k)}$. From this perspective, there is only one natural way to extend the AIC principle to our current setting, and that is to define the best copula model to be the one with index

$$k° := \underset{1 \leq k \leq K}{\operatorname{argmin}} \operatorname{KL}(c°, c_{k,\theta°(k)}).$$

As for the AIC case, we can naively use

$$\tilde{k}_N := \underset{1 \leq k \leq K}{\operatorname{argmax}} \int \log c_{\hat{\theta}(k)} \, \mathrm{d}\hat{C}_N, \qquad (6.13)$$

which is consistent but with poor small-sample behavior. We can make small-sample corrections to the estimate $\tilde{k}_N$ analogous to the AIC formula. The definition of $k°$ as the best parametric copula model is the decisive step in the development to the CIC. The remaining steps are entirely analogous to Section 6.2.2, and although their validity requires some mathematical sophistication, the conceptual side of the CIC is now fully developed.

As in the development of the AIC formula, we can use a Taylor expansion together with the limit theorems of Section 6.2.5 to conclude that

$$\hat{A}_N(\hat{\theta}) - A(\hat{\theta}) = \bar{Z}_N + N^{-1} p_N + \hat{A}_N(\theta°) - A(\theta°) + o_P(N^{-1})$$

where $\mathbb{E}\bar{Z}_N = 0$ and p_N is of a known form and converges to a Gaussian distribution.

But in contrast to the developments of the standard AIC in Section 6.2.2, this expansion is not sufficient to conclude with a model selection formula. To see this, notice that in the standard ML case with known marginals, the $\hat{A}_N(\theta°) - A(\theta°)$ would be included in the mean zero variable $\bar{Z}_N$, as we *would* have

$$\mathbb{E}\hat{A}_N(\theta°) = \mathbb{E} \int \log c_{\theta°}(v) \tilde{C}_N = \mathbb{E} \frac{1}{N} \sum_{i=1}^{N} \log c_{\theta°}(F_\perp°(X_i))$$

$$= \int \log c_{\theta°}(F_\perp°(x)) \, \mathrm{d}F° = \int \log c_{\theta°}(v) \, \mathrm{d}C° = A(\theta°) \qquad (6.14)$$

in which $\tilde{C}_N$ is the empirical distribution based on observations $F_\perp°(X_1)$, ..., $F_\perp°(X_N)$. As we are interested in bias-correction terms and accordingly only focus on the mean value behavior, we could in the classical ML case ignore both $\bar{Z}_N$ and $\hat{A}_N(\theta°) - A(\theta°)$. We then only had to investigate the behavior of p_N, and find an estimator $\hat{p}^*$ for $p^* = \mathbb{E}p$ where $p_N \xrightarrow[N \to \infty]{W} p$ to get the classical AIC formula.

In the MPLE case, we encounter the complication that

$$\mathbb{E}\hat{A}_N(\theta°) = \mathbb{E} \int \log c_{\theta°}(v) \tilde{C}_N = \frac{1}{N} \sum_{i=1}^{N} \mathbb{E} \log c_{\theta°}(F_{N,\perp}(X_i)) \neq A(\theta°),$$

in which we have the stochastic and far from trivial function $F_{N,\perp}(X_i)$ inside of $c_{\theta°}$ — in contrast to the $F_\perp°(X_i)$ we had in Eq. (6.14). Remember that the AIC gives bias corrections up to the $o_P(N^{-1})$ precision level. As we define this to be the AIC programme, we have to take the behavior of $\hat{F}_{N,\perp}$ into consideration to provide a genuine extension of the standard AIC. A Taylor expansion of $\log c_{\theta^\circ}(\cdot)$ around $F_\perp^\circ(X_i) - \hat{F}_{N,\perp}(X_i)$ replaces the problematic $\hat{F}_{N,\perp}$ with $F_\perp^\circ$ — which we had in the standard ML case — and also quantifies the magnitude of error we are committing. This error is of the desired order $o_P(N^{-1})$. We get that

$$\hat{A}_N(\theta^\circ) = N^{-1} \sum_{i=1}^{N} \left[\log c(F_\perp^\circ(X_i), \theta^\circ) + \zeta'(F_\perp^\circ(X_i), \theta^\circ)^t(\hat{V}_i - F_\perp^\circ(X_i)) \right.$$
$$\left. + \frac{1}{2}(\hat{V}_i - F_\perp^\circ(X_i))^t \zeta''(F_\perp^\circ(X_i), \theta^\circ)(\hat{V}_i - F_\perp^\circ(X_i)) \right]$$
$$+ O_P(N^{-1}) \qquad (6.15)$$

where

$$\zeta'(v, \theta) = \frac{\partial \log c(v, \theta)}{\partial v} \quad \text{and} \quad \zeta''(v, \theta) = \frac{\partial^2 \log c(v, \theta)}{\partial v \partial v^t}$$

are the vector of derivatives and matrix of double derivatives of the log copula density, respectively.

The first summation term of Eq. (6.15) has expectation $A(\theta^\circ)$, as in the ML case, but we also end up with two additional terms to deal with.

Through the use of empirical process theory, Grønneberg[7] concludes that

$$\hat{A}_N(\hat{\theta}) - A(\hat{\theta}) = \tilde{Z}_N + N^{-1}(p_N + q_N + r_N) + o_P(N^{-1})$$

in which $\tilde{Z}_N = 0$, and $p_N = O_P(1)$, $q_N = O_P(\sqrt{N})$ and $r_N = O_p(1)$. Further,

$$q_N^* = q_N \to \int_{[0,1]^n} \zeta'(v; \theta_0)^t (\mathbf{1} - v) C^\circ(v)$$
$$r_N^* = r_N \to r^* = \mathbf{1}^t \Upsilon \mathbf{1}$$

where $\Upsilon = (\Upsilon_{a,b})_{1 \leq a,b \leq n}$ is the symmetric matrix with

$$\Upsilon_{a,a} = \int_{[0,1]^n} \zeta''_{a,a}(u; \theta_0) u_a(1 - u_a) C^\circ,$$

$$\Upsilon_{a,b} = \int_{[0,1]^n} \zeta''_{a,b}(u; \theta_0) [C_{a,b}(u_a, u_b) - u_a v_b] C^\circ$$

and r_N is finite only if Υ is. Here $C_{a,b}$ is the cumulative copula of $(X_{1,a}, X_{1,b})$.

Empirical estimates of these correction terms can readily be made. We deal with correctly specified and misspecified models separately. We construct an "AIC like" CIC, valid under the assumptions of a correctly specified parametric copula model, and also a "TIC-like" CIC which estimates the bias-correction terms consistently even without the assumption of a correctly specified parametric copula model.

In the "AIC-like" CIC formula, simplifications can be made, and we get a formula which is visually very similar to the classical AIC. We get
$$\widehat{\text{CIC}}_{\text{AIC}} = 2\ell_{N,\max} - 2(\hat{p}^* + \hat{r}^*).$$
The estimator $\hat{p}^*$ is given by
$$\hat{p}^* = \text{length}(\theta) + (\hat{\mathcal{I}}^{[-1]}\hat{W}),$$
where $\hat{\mathcal{I}}^{-1}$ and $\hat{W}$ is the empirical estimates formed through using $c_{\hat{\theta}}$ as plug-in estimates of c° in the defining formulae of $\mathcal{I}$ and W, where $\hat{\mathcal{I}}^{-1}$ is a generalized inverse of $\hat{\mathcal{I}}$. The estimator $\hat{r}^*$ is given by $\hat{r}^* = \mathbf{1}^t \hat{\Upsilon} \mathbf{1}$, defined in terms of the plug-in estimators

$$\hat{\Upsilon}_{a,a} = \int_{[0,1]^n} c(v;\hat{\theta}) \zeta''_{a,a}(v;\hat{\theta}) v_a(1-v_a) v,$$

$$\hat{\Upsilon}_{a,b} = \int_{[0,1]^n} c(v;\hat{\theta}) \zeta''_{a,b}(v;\hat{\theta}) \left[C_{a,b}(v_a, v_b; \hat{\theta}) - v_a v_b \right] v$$

where $C_{a,b}(v_a, v_b; \theta)$ is the cumulative copula of (Y_a, Y_b) where $(Y_1, Y_2, \ldots, Y_d) \sim C_\theta$.

The formula for $\hat{p}^*$ is almost the same as $\hat{p}^* = \text{length}(\theta)$ in the AIC formula, but with an extra term $\left(\hat{\mathcal{I}}^{-1}\hat{W} \right)$ which is *always positive*. However, $\hat{r}^*$ *can be both positive and negative* — depending on the estimated dependency structure of the parametric copula.

One of the main advantages of the original AIC formula compared to the TIC is that the bias-correction term is only length(θ), which does not have to be estimated on the basis of observed data. The "AIC-like" CIC does not have this advantage and we need to estimate high-order cumulants to apply it. An interpretation of the terms in the "AIC-like" CIC formula is that $\text{Tr}(\hat{\mathcal{I}}^{[-1]}\hat{W})$ takes into consideration the inflated (compared to the standard ML) covariance matrix of the asymptotic limit of the score function, while $\hat{r}^*$ stabilizes the effects of using nonparametric marginal estimates $\hat{F}_{N,\perp}$ instead of the correct $F_\perp^\circ$.

If we do not assume a correctly specified model, we get the more complicated and more general "TIC-like" CIC formula
$$\widehat{\text{CIC}}_{\text{TIC}} = 2\ell_{N,\max} - 2(\hat{p}^* + \hat{q}^* + \hat{r}^*),$$
which is always valid. We use
$$\hat{p}^* = (\hat{J}^- \hat{\Sigma}), \quad \hat{q}^* = \int_{[0,1]^n} \zeta'(v;\hat{\theta})^t (\mathbf{1}-v) \hat{C}(v),$$
$$\hat{r}^* = \mathbf{1}^t \hat{\Upsilon} \mathbf{1}$$

where now

$$\hat{\Upsilon}_{a,a} = \int_{[0,1]^n} \zeta''_{a,a}(v;\hat{\theta}) v_a (1-v_a) C_N,$$

$$\hat{\Upsilon}_{a,b} = \int_{[0,1]^n} \zeta''_{a,b}(v;\hat{\theta}) \left[\hat{C}_{N,a,b}(v_a, v_b) - v_a v_b\right] C_N$$

where $C_{N,a,b}$ is the empirical copula based on $(X_{1,a}, X_{1,b}), (X_{2,a}, X_{2,b}), \ldots, (X_{N,a}, X_{N,b})$. We use the standard empirical estimates of $\hat{J}^-$ and $\hat{\Sigma}$ given in, e.g. Ref. 2, where $\hat{J}^-$ is a generalized inverse of $\hat{J}$.

6.3.1 Non-existence of bias-correction terms and implications for the MPLE

Many practitioners of copulae are mainly interested in the copulae which have extreme tail dependence (see Ref. 9). However, the bias-correction terms q^* and r^* are defined through the differentials of $\log c_\theta(v)$ with respect to v. These will continuously grow when extreme behavior near the edge of the unit cube is introduced, until they explode and do not have a finite expectation. Let us agree to call parametric copula models with non-existent q^* and r^* "edge-extreme". The implication of these exploding terms is that empirical estimates of q^* and r^* do not exist, as it simply does not make sense to estimate anything non-existent. Hence, there cannot be any generally applicable model selection formula in the sense of providing a first-order bias correction to the model-relevant part of the attained Kullback–Leibler divergence between the MPL estimated model and c°. This poses a limitation for the use of the MPLE, which is shared by all two-stage copula estimators that estimate the marginals non-parametrically, say, with $\tilde{F}_{N,\perp}$, and the copula through minimizing a pseudo-likelihood

$$\sum_{i=1}^{N} \log c_\theta(\tilde{F}_{N,\perp}(X_i)).$$

To see this, notice the following.

The q^* and r^* terms can be traced back to Section 6.3 where we observed that

$$\mathbb{E}\hat{A}_N(\theta^\circ) \neq A(\theta^\circ). \tag{6.16}$$

But this is actually the case for all two-stage estimators,[d] such as the IFM discussed in Ref. 9. In the IFM case, we have parametric marginal estimates. Going through the same procedures as Section 6.3 shows that

$$\hat{A}_N(\theta°) = \frac{1}{N} \sum_{i=1}^{N} \log c_\theta(F_{\hat{\gamma},\perp}(X_i))$$

where $F_{\hat{\gamma},\perp}$ is the vector of estimated marginal cumulative distributions found through standard ML estimates. If $F_\perp° = F_{\gamma°,\perp}$, so that the parametric class of marginal models is correctly specified, a Taylor expansion of

$$\log c_\theta(v)|_{v=F_{\perp,\hat{\gamma}}(X_i)},$$

not in the full v, but for $\gamma \mapsto F_{\perp,\gamma}$ around $\hat{\gamma} - \gamma°$, yields terms paralleling q^* and r^* of the CIC that always exist under classical regularity conditions for all copulae. So the problem does not come from Eq. (6.16), but from the need to perform a Taylor-expansion around v in terms such as

$$\log c_\theta(v)|_{v=\tilde{F}_{N,\perp}(X_i)}. \tag{6.17}$$

Unless empirical estimators of $F_\perp°$ can be found such that $N \sup |\tilde{F}_{N,\perp} - F_\perp°| = O_P(1)$, this cannot be avoided at the precision level we have defined as the "AIC programme". And one would even then have to demand regularity conditions on the $C°$ integrability of functions of ζ' and ζ''. This would still be confining with respect to the types of parametric copulae that could have been estimated while still having AIC-like model selection formulae.

Finally, we note that a solution which might seem promising is to utilize univariate extreme value theory (EVT) to estimate the tails of the marginals. EVT gives general conditions for when the tails of univariate distributions can be approximated by generalized Pareto distributions, and there is a well-developed machinery for finding empirical estimates for the parameters involved. As this would reduce the estimation of the functional form of the tails of the distributions to a low-dimensional problem, it would seem that a possible solution to the above problems is to define $\hat{F}_{N,\perp}$ coordinate-wise as the standard univariate empirical distribution functions below thresholds, while using n estimated generalized Pareto distributions above these

---

[d]This seems to be a new observation, whose consequences have not been properly dealt with. The inequality (6.16) invalidates the AIC formula for all multi-stage estimation routines, and through following the derivation of the CIC it is not difficult to provide modifications of (or quantify consequences of using) the standard AIC formula in these settings.

thresholds. Such an approach for estimating the univariate distributions is discussed in Ref. 13, but the plug-in step of using such an $\hat{F}_{N,\perp}$ seems to be new. However, there are two problems with such an approach. First, such EVT-estimates require the specification of a point over the threshold which is defined either algorithmically or manually. In practice this hinders a mathematical theory of estimation based on asymptotics. Second, simulations show that standard *automated* routines for specifying the points over the threshold and estimating the parameters of the generalized Pareto distributions introduce so much new noise in the estimation process that the resulting copula parameter estimates are mostly inferior to the MPL estimates. These two issues show that such an EVT-based solution is not fruitful.

6.3.2 *Philosophical implications of the CIC*

This very brief section discusses what implications the CIC formula has for the interpretation of the standard AIC formula.

The AIC formula is often seen heuristically as expressing a formalization of Occam's Razor. This interpretation is often presented as some kind of general principle, intrinsic to the arguments underlying the AIC formula.

Although the $\hat{p}^*$ in the CIC formula retains the interpretation of being a "penalty for complexity", the full CIC formula has additional terms which can be *both positive and negative*, and the "penalization term" can all in all be negative. Examples of two such cases are found in Section 6.4. Hence the bias-correction term of the CIC no longer has the straightforward interpretation of "penalizing for complexity", and can no longer be directly interpreted as a formalized Occam's Razor.

As the CIC is motivated through the same steps as the AIC, we see that the "penalization for complexity" interpretation of the AIC — although valid in the AIC case — is not a general principle which always follows from the underlying ideas of the AIC. The CIC seems to be the first information-based model selection criterion that provides such a counterexample, hence the importance of this observation.

6.4 Illustrations

We include a brief illustration of the computational aspects of using the CIC, while confirming its validity numerically. Consider the Frank and the Plackett copulae (families B3 and B2 in Ref. 9, respectively) and denote their cumulative distribution functions by $C_{F,\delta}$ and $C_{P,\delta}$. Figures 6.1(a)–6.1(d)

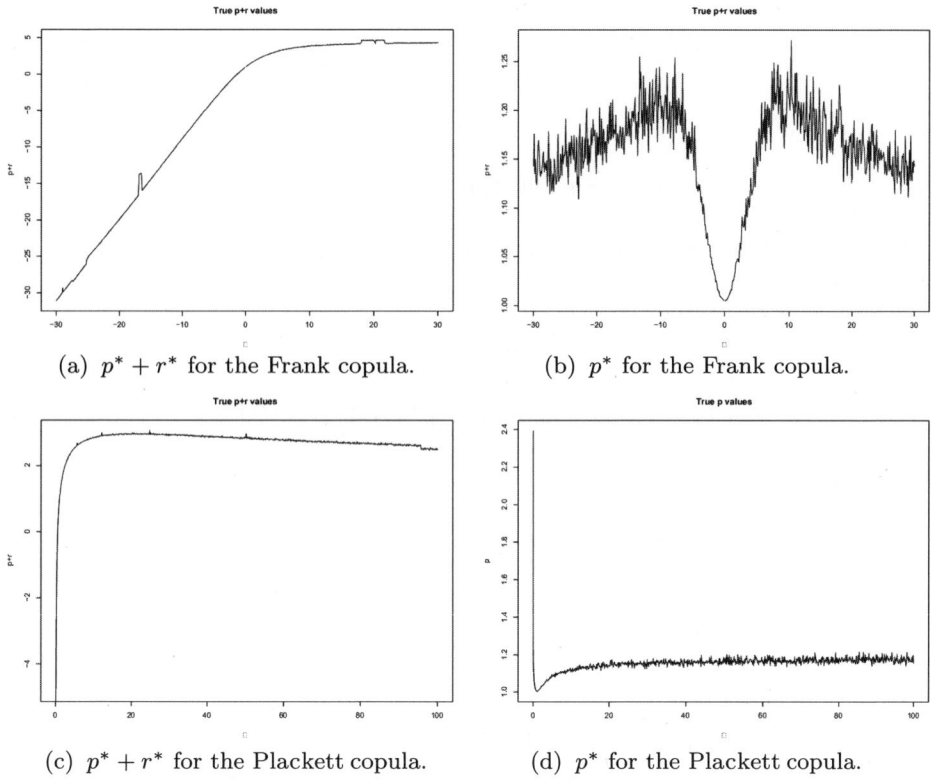

Figure 6.1. Plots of true CIC values under the assumption of a correctly specified parametric model for the Frank and Plackett copulae with varying dependence parameters.

show the CIC values for the two models with varying δ. It is clear that the r^* term dominates the CIC value, and that it reflects the degree of positive or negative dependence in the data, which any sensible set of models should agree on. The random noise in the approximated p^* values is due to variation inherent in Monte Carlo integration. Notice that for large degrees of negative dependence, both copulae give CIC formulae that are negative.

Assume $X \sim \mathcal{N}(0,1)$ and $Y \sim \mathcal{N}(0,1)$ while the copula of (X,Y) is a copula mixture of the form $\lambda C_{F,\delta} + (1-\lambda) C_{P,\delta}$ with $\lambda = 80\%$. We want to use the known (near) unbiasedness of the AIC in the fully parametric case to illustrate that the CIC works as it should. We can do this by the following.

If we restrict attention to parametric models with normal marginals and either a Frank or a Plackett copula, we have

$$f_i(x,y;\delta) = c_i(\Phi^{-1}(x), \Phi^{-1}(y); \delta)\phi(x)\phi(y)$$

using the information that both marginals are known to be standard normal and where $i \in \{F, P\}$. The true copula is known to be a mixture of the two. Denote this density by c°, and let f° be the full data-generating mechanism of (X, Y). We have

$$f^\circ(x,y) = c^\circ(\Phi^{-1}(x), \Phi^{-1}(y))\phi(x)\phi(y).$$

This means that the Kullback–Leibler divergence between f° and $f_{i,\delta}$ is

$$\mathrm{KL}(f^\circ, f_{i,\delta}) = \log \frac{f^\circ(X,Y)}{f_{i,\delta}(X,Y)}$$
$$= \log \frac{c^\circ(\Phi^{-1}(X), \Phi^{-1}(Y))}{c_i(\Phi^{-1}(X), \Phi^{-1}(Y); \delta)} = \mathrm{KL}(c^\circ, c_{i,\delta}),$$

implying

$$\Delta\mathrm{KL}(f^\circ) := \mathrm{KL}(f^\circ, f_{F,\delta_F}) - \mathrm{KL}(f^\circ, f_{P,\delta_P})$$
$$= \mathrm{KL}(c^\circ, c_{F,\delta_F}) - \mathrm{KL}(c^\circ, c_{P,\delta_P}). \qquad (6.18)$$

Consider the following three formulae:

1. The standard AIC formula $2\ell^{\#}_{N,\max} - 2\,\mathrm{length}(\delta)$ where $\ell^{\#}_{N,\max}$ is the observed maximum likelihood of the full likelihood of (X, Y) under the assumption that $X \sim \mathcal{N}(\mu_1, \sigma_1^2)$ and $Y \sim \mathcal{N}(\mu_2, \sigma_2^2)$ and with either a Frank or a Plackett copula specifying their simultaneous distribution. Denote the observed AIC scores simply by AIC_F for the Frank-copula case and AIC_P for the Plackett-copula case and let $\Delta\mathrm{AIC} = \mathrm{AIC}_F - \mathrm{AIC}_P$.
2. The AIC-like formula $2\ell_{N,\max} - 2\,\mathrm{length}(\theta)$, where $\ell_{N,\max}$ is the observed maximum pseudo-likelihood for the copula model. Denote the observed (but unjustified) AIC scores by $\mathrm{AIC}^\bullet_F$ and $\mathrm{AIC}^\bullet_P$ and let $\Delta\mathrm{AIC}^\bullet = \mathrm{AIC}^\bullet_F - \mathrm{AIC}^\bullet_P$.
3. The CIC formula $2\ell_{N,\max} - 2(p^* + r^*)$ calculated under the assumption of a correctly specified model. Denote the observed CIC scores by CIC_F and CIC_P and let $\Delta\mathrm{CIC} = \mathrm{CIC}_F - \mathrm{CIC}_P$.

Equation (6.18) shows that if the $\mathrm{AIC}^\bullet$ formula is correct, $\Delta\mathrm{AIC}^\bullet$ should be approximately equal to $\Delta\mathrm{AIC}$, but if the CIC formula is correct, $\Delta\mathrm{CIC}$ should be approximately equal to $\Delta\mathrm{AIC}$. A simulated sample of (X, Y) with the mixture copula is illustrated in Figs. 6.2(a)–6.2(d) with $N = 2000$. It is not obvious which model is the best, as the fit of the MPLE models seems to vary in different parts of the sample space. However, assume that we

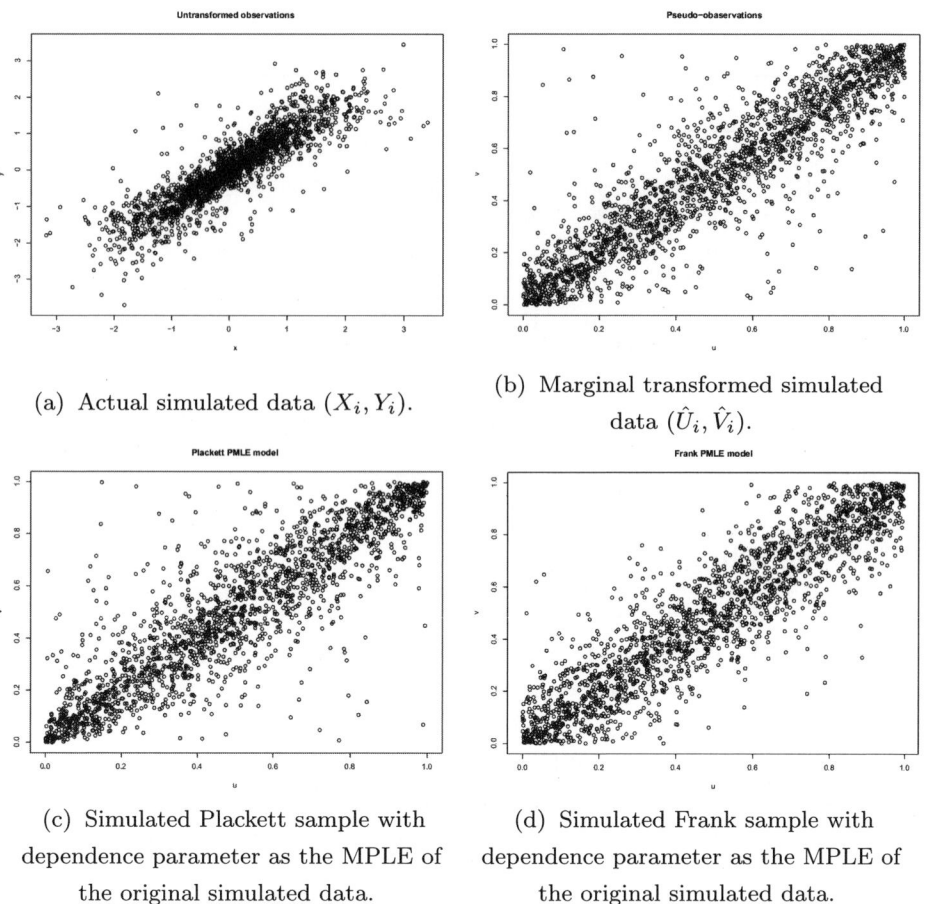

(a) Actual simulated data (X_i, Y_i).

(b) Marginal transformed simulated data $(\hat{U}_i, \hat{V}_i)$.

(c) Simulated Plackett sample with dependence parameter as the MPLE of the original simulated data.

(d) Simulated Frank sample with dependence parameter as the MPLE of the original simulated data.

Figure 6.2. Plots of simulated data.

want to know which model has the least Kullback–Leibler divergence to the true model. Notice that we use the AIC-like formulae and not the TIC-like formulae, which is an approximation typical in model selection practice as the TIC-like formulae have a much higher variability than the AIC-like formulae.

We ran 500 simulations as above — each with 2000 sample points — and for each simulation calculated the AIC, AIC$^\bullet$ and CIC values. Table 6.1 shows that the CIC formulae on average agrees with the fully parametric AIC value, while the mean of the incorrectly motivated AIC$^\bullet$ misses the mean of AIC almost exactly by the average of $-2\Delta(p^* + r^*)$, the correction term which separates AIC$^\bullet$ and CIC.

Table 6.1. Summary statistics for the simulation of 500 data sets each consisting of 2000 samples.

	Min.	1st Qu.	Median	Mean	3rd Qu.	Max.
ΔAIC	−108.80	−26.73	−6.13	−5.28	16.87	84.95
ΔCIC	−122.90	−28.80	−4.65	−5.00	18.14	93.15
ΔAIC$^{\bullet}$	−120.30	−26.23	−2.07	−2.43	20.72	95.72
ΔAIC − ΔCIC	−27.52	−7.42	−0.64	−0.28	6.51	39.26
ΔAIC − ΔAIC$^{\bullet}$	−30.10	−9.99	−3.22	−2.85	3.94	36.69
MPLE δ_F	12.80	13.50	13.77	13.77	14.03	15.04
MPLE δ_P	43.06	47.05	48.74	48.71	50.12	56.13
$p_P^* + r_P^*$	2.78	2.83	2.84	2.84	2.85	2.96
$p_F^* + r_F^*$	4.00	4.04	4.06	4.06	4.08	4.13
$2\Delta(p^* + r^*)$	−2.65	−2.50	−2.44	−2.44	−2.39	−2.23

6.5 Concluding Remarks

Standard semiparametric estimation theory, as summarized in Ref. 1, postulates that the true data-generating distribution is included in the space of all models spanned by the semiparametric model. The infinite-dimensional part of semiparametric models often spans such a large space that it is realistic to make this assumption. But for most practical uses of semiparametric copula models, this is not realistic and motivates the investigation of semiparametric model selection techniques in the style of the AIC.

Standard semiparametric estimation theory is based on the assumption that the rationale for using a semiparametric model (in contrast with using a fully nonparametric model) is that the investigator possesses *a priori* knowledge of the correct finite-dimensional part of the data-generating distribution. This is often not the case in copula estimation.

The basis for the CIC investigation of Ref. 7 was to assess the consequences of using the "AIC formula" of Eq. (6.2). The main conclusions were

- The "penalization" for dimensionality of the copula model is only part of the story, and the correct sum of all bias-correction terms can be negative.
- No proper generalization of the AIC formula exists for "edge extreme" copulae when parameters are estimated with the MPLE. The class of edge extreme copulae includes most copula models in common use.

Both of these points have practical implications for copula users. The first point has an obvious implication: do not use the AIC$^{\bullet}$ formula of Eq. (6.2) — its rationale is unjustified. The second point has more subtle implications. It indicates that the estimation of parametric edge extreme copulae is fundamentally more complex without the knowledge of finite-dimensional parametric marginals. Edge extreme copulae are often used to provide multivariate extreme value estimates such as Value At Risk calculations for the sum of dependent vectors for high quantiles. If this is the aim of the study at hand, the MPLE seems not to be the best choice.

A possible solution to the second point is to ignore the bias-correction term which gets us into trouble, and work directly with $\tilde{k}_N$ of Eq. (6.13). If N is sufficiently large, first-order bias corrections are insignificant (see footnote b), making this a sensible model selection routine in some circumstances. This is implicitly done in Ref. 2 (although they did not notice that the "AIC formula" of Eq. (6.2) is unjustified for the MPLE), and they provide statistical tests to assess the conclusion of the resulting model selection strategy.

Another way to address the second point is to look for alternative estimators of the copula parameter. It seems that the only well-known alternative to the MPLE is the sieve-based estimator proposed in Ref. 3, motivated through semiparametric efficiency considerations. But the concept of semiparametric efficiency is defined only when the model in question is correctly specified. This is clearly not the case for any investigation in which the (non-nested) model selection problem appears.

A third possible approach to the second point is to develop an analog to the impressive machinery of Ref. 12 for the current situation. This seems currently out of reach, and would lead to a theory based on fundamentally different principles than the comparatively simple AIC formula.

If none of the candidate copula models are edge extreme, the CIC formula provides a general model selection strategy, but if at least one copula under consideration is edge extreme, there are currently no fully satisfactory solutions to the model selection problem. Finally, we note that model selection by cross-validation and boot-strap procedures are reasonable methods also for the MPLE. However, their theoretical properties are not yet well-understood.

Acknowledgments

This work is funded by Statistics for Innovation $(sfi)^2$.

References

1. Bickel P., Klaassen A., Ritov Y. and Wellner J. (1993). *Efficient and Adaptive Inference in Semi-Parametric Models*. Johns Hopkins University Press, Baltimore.

2. Chen X. and Fan Y. (2005). Pseudo-likelihood ratio tests for semiparametric multivariate copula model selection. *Canadian Journal of Statistics*, 33(3):389–414.
3. Chen X., Fan Y. and Tsyrennikov V. (2006). Efficient estimation of semiparametric copula models. *Journal of the American Statistical Association*, 101:1228–1240.
4. Claeskens G. and Hjort N. (2008). *Model Selection and Model Averaging*, Cambridge University Press, New York.
5. Genest C., Ghoudi K. and Rivest L.-P. (1995). A semiparametric estimation procedure of dependence parameters in multivariate families of distributions. *Biometrika*, 82(3):543–552.
6. Genest C., Quessy J.-F. and Remillard B. (2006). Goodness-of-fit procedures for copula models based on the probability integral transform. *Scandinavian Journal of Statistics*, 33:337–366.
7. Grønneberg S. and Hjort N. (2008). The copula information criterion. Technical Report 7, Department of Mathematics, University of Oslo.
8. Hallin M. and Werker B. (2003). Semi-parametric efficiency, distribution-freeness and invariance. *Bernoulli*, 9(1):137–165.
9. Joe H. (1997). *Multivariate Models and Dependence Concepts*. Chapman & Hall, London.
10. Lehmann E. and Casella G. (1998). *Theory of Point Estimation*. Springer, New York.
11. Lehmann E. and Romano J. (2005). *Testing Statistical Hypotheses*. Springer, New York.
12. Massart P. (2007). *Concentration Inequalities and Model Selection*. Springer, New York.
13. McNeil A. and Saladin T. (1997). The peaks over thresholds method for estimating high quantiles of loss distributions. *Proceedings of 28th International ASTIN Colloquim*, Vol. 28.
14. Ruymgaart F.H. (1974). Asymptotic normality of nonparametric tests for independence. *Annals of Statistics*, 2(5):892–910.
15. Ruymgaart F.H., Shorack G.R. and van Zwet W.R. (1972). Asymptotic normality of nonparametric tests for independence. *Annals of Mathematical Statistics*, 43(4):1122–1135.
16. Tsukahara H. (2005). Semiparametric estimation in copula models. *Canadian Journal of Statistics*, 33(3):357–375.
17. Vapnik V.N. (2000). *The Nature of Statistical Learning Theory*. Springer, New York.
18. van der Vaart A.W. (1998). *Asymptotic Statistics*. Cambridge University Press, Cambridge.

CHAPTER 7

Dependence Comparisons of Vine Copulae with Four or More Variables

Harry Joe

Department of Statistics
University of British Columbia
Vancover BC, Canada, V6T 1Z2

It is known that there is one equivalence class of regular vines for $n = 3$ and two equivalence classes for $n = 4$. Through an enumeration, we show that there are six equivalence classes for $n = 5$ with boundary classes of C-vines and D-vines, so that the dimension of $n = 5$ is manageable in terms of studying the intermediate regular vines. For vine copulae, which obtain from associating a bivariate copula with each pair in the vine, we develop an approach that leads to algorithms for C-vines, D-vines and intermediate vines. Some similarities between different vine copulae can be seen from how they are simulated. For Gaussian and non-Gaussian vines, we compare the marginal bivariate dependence of different vine copulae when the level ℓ bivariate linking copulae are all set at C_ℓ, $\ell = 1, \ldots, n-1$.

7.1	Introduction	139
7.2	Equivalence Classes of Regular Vines	140
7.3	Simulation from Vine Copulae	145
7.4	Comparing Dependence of Vine Copulae	154
7.5	Gaussian Vines and Generalized Toeplitz Matrices	156
7.6	More Comparisons of Dependence for Different Vines	159
7.7	Discussion and Further Research	162
	References	163

7.1 Introduction

The aim of this chapter is to compare the similarities and differences in dependence for different regular vines on n variables, with $n \geq 4$. Understanding how different vines compare can be useful in deciding on appropriate vine copula models for high-dimensional multivariate data. This might be important, for example, in adding extra variables to an existing vine distribution with m variables. There are different ways of linking the new variables to the existing vine.

The dimension $n = 5$ is the first interesting case to get intermediate vines that are between the boundary cases of D-vines and C-vines (canonical vines). See Bedford and Cooke[2,3] and Kurowicka and Cooke[13] for the definitions of vines and different ways of viewing them. To help enumerate the regular vines on n variables, Section 7.2 gives a definition of an equivalence class of regular vines in terms of the conditioned and conditioning indices of the variables in the conditional distributions. One new result is the enumeration of the six equivalence classes for dimension $n = 5$. For $n > 5$ the number of equivalence classes increases exponentially, so $n = 5$ is the dimension for which extra insight can be obtained into the intermediate vines.

In financial and other applications, it is important to be able to simulate from the copula in order to assess probabilities relative to risks. In Section 7.3, algorithms for vine copulae are developed in a unified approach for simulating from D-vines and C-vines in dimension n, and also from any intermediate vine in dimension n. For the intermediate vines in dimension 5, we show that some similarity results for different vines are most easily seen from the simulation algorithms.

In Sections 7.4, 7.5 and 7.6, comparisons are made for different vines based on analytic and simulation results, when the level ℓ bivariate linking copulae are all set at C_ℓ, $\ell = 1, \ldots, n-1$. For $n = 5$, some extra results can be obtained for Gaussian vines which have a partial correlation α_ℓ for the level ℓ of the vine; the Gaussian vines with this constraint sometimes lead to generalized Toeplitz matrices.

Section 7.7 concludes with a summary and some discussion of future research.

7.2 Equivalence Classes of Regular Vines

In this section, we define equivalence classes of regular vines, give examples to show how different vines are distinguished, and enumerate and discuss the equivalence classes for $n = 5$. Keeping track of the order of the nodes for trees at different levels of the vine is a quick way to differentiate some of the regular vines.

To define equivalence classes of (regular) vines and list them for dimensions $n = 3, 4, 5$, we use shorthand notation for conditional distributions. For $\ell \geq 2$, the notation $\{i_1, i_2\} \mid \{j_1, \ldots, j_{\ell-1}\}$ refers to the conditional distribution of variables with indices i_1, i_2, given the variables with indices $j_1, \ldots, j_{\ell-1}$. If $\ell = 1$, then the above refers to the marginal distribution of the variables with indices i_1, i_2.

In the above notation, the two boundary classes of D-vines and C-vines are written as follows. Some of the braces { and } and commas are omitted below for brevity.

- D-vine: $12; 23; \ldots; n-1, n; 13|2; 24|3; \ldots; n-2, n|n-1; \ldots, 1, n|2 \ldots n-1$.
  That is, $\{\{i_1, i_2\}|\{i_1+1, \ldots, i_2-1\} : 1 \leq i_1 < i_2 \leq n\}$ to cover $\binom{n}{2}$ pairs.

- C-vine: $12; 13; \ldots; 1, n; 23|1; 24|1; \ldots; 2, n|1; \ldots, n-1, n|1 \ldots n-2$.
  That is, $\{\{i_1, i_2\}|\{1, \cdots, i_1-1\} : 1 \leq i_1 < i_2 \leq n\}$ to cover $\binom{n}{2}$ pairs.

For a conditional distribution of variables indexed by i_1, i_2, we refer to level ℓ if cardinality of conditioning set $D(i_1, i_2, \ell) = \{j_1, \ldots, j_{\ell-1}\}$ is $\ell - 1$ for $\ell = 1, \ldots, n-1$.

Vines in the same equivalence classes can be matched after a permutation of indices. Our notation for a permutation π of $\{1, \ldots, n\}$, is $\pi(1 \cdots n) = k_1 \cdots k_n$ or $\pi = \begin{pmatrix} 1 & 2 & \cdots & n \\ k_1 & k_2 & \cdots & k_n \end{pmatrix}$ and components of π are summarized as $\pi(i) = k_i$. Also for a subset D, we use the notation $\pi(D) = \{\pi(j) : j \in D\}$.

Definition 7.1 (Equivalence classes of vines). Consider a regular vine written in the form of a set of $\binom{n}{2}$ conditional distributions:

$$\{\{i_1, i_2\}|D(i_1, i_2, \ell) : 1 \leq i_1 < i_2 \leq n, \text{ with } n - \ell \text{ pairs at level} \\ \ell, \ell = 1, \ldots, n-1\}. \quad (7.1)$$

Let π be a permutation. Then the regular vines $\{\{i_1, i_2\} \mid D(i_1, i_2, \ell)\}$ and $\{\{\pi(i_1), \pi(i_2)\} \mid \pi(D(i_1, i_2, \ell))\}$ are said to be in the *same equivalence class*.

Alternatively, two regular vines summarized as $\{\{i_1, i_2\} \mid D(i_1, i_2, \ell)\}$ and $\{\{i_1', i_2'\} \mid D(i_1', i_2', \ell)\}$ are in the *same equivalence class* if there is a permutation π such that $\{\{\pi(i_1), \pi(i_2)\} \mid \pi(D(i_1, i_2, \ell))\}$ and $\{\{i_1', i_2'\} \mid D(i_1', i_2', \ell)\}$ are the same.

Note that in the set notation, the order within the subset of conditioned variables or within the subset of conditioning variables does not matter. Also, the order of the listing of the distributions does not matter.

The above notation means that a pair (i_1, i_2) with $i_1 \neq i_2$ appears as the conditioned variables in exactly one level. There are $n - \ell$ pairs for which the number of conditioning variables is $\ell - 1$ for $\ell = 1, \ldots, n - 1$. There are additional constraints linking the sets $D(i_1, i_2, \ell)$ at different levels in (7.1); see Section 4.4 of Kurowicka and Cooke.[13] For example, if $\{i_1, i_2\} \mid D(i_1, i_2, \ell)$ is at level ℓ for $\ell > 1$, then there exists $j, j' \in D(i_1, i_2, \ell)$ (which could be the same or different) such that two distributions at level $\ell - 1$ are $\{i_1, j\} \mid D(i_1, i_2, \ell) \backslash \{j\}$ and $\{i_2, j'\} \mid D(i_1, i_2, \ell) \backslash \{j'\}$.

We illustrate the above definition and notation to show that there is one equivalence class of vines for $n = 3$ and two equivalence classes of vines for $n = 4$.

For $n = 3$, we can assume that 12 is a pair at level 1, then the second pair is either 23 (D-vine) or 13 (C-vine). They are listed below.

- D-vine: 12; 23; 13|2 in standard form;
- C-vine: 12; 13; 23|1 in standard form.

For the C-vine, permute $\pi(123) = 213$ to get $\{21; 23; 13|2\}$, which is the same as the D-vine. Note that for this vine in the graph representation, the level 1 tree has one node of order 2 and two nodes of order 1.

For $n = 4$, for the level 1 tree, two nodes can have order 2 and the other two have order 1 (like the D-vine), or one node can have order 3 and the other three have order 1 (like the C-vine). The discussions of level 2 are given below, leading to two equivalence classes: the C-vine and D-vine.

- D-vine: 12; 23; 34; 13|2; 24|3; 14|23. Note that there is no alternative conditioning after level 1 since the margins 12 and 34 cannot be paired to have a common conditioning index.
- C-vine: 12; 13; 14; 23|1; 24|1; 34|12 in standard form;
- An alternative conditioning starting from $\{12; 13; 14\}$ at level 1 is 12; 13; 14; 23|1; 34|1; 24|13. Permute $\pi(1234) = 1324$ to get 13; 12; 14; 32|1; 24|1; 34|12, so this is now a C-vine in standard form.
- A second alternative conditioning starting from $\{12; 13; 14\}$ at level 1 is 12; 13; 14; 24|1; 34|1; 23|14. Permute $\pi(1234) = 1432$ to get 14; 13; 12; 42|1; 32|1; 43|12, and this is also a C-vine in standard form.

Before enumerating the equivalence classes of regular vines for $n = 5$, we indicate a way to construct a five-dimensional regular vine that need not be a C-vine or D-vine.

$$\begin{array}{cccc} & & & 15 \\ & & 14 & \\ & 13 & & 25 \\ 12 & & 24 & \\ & 23 & & 35 \\ & & 34 & \\ & & & 45 \end{array}$$

For the first level or tree of the vine, pick a pair from each column; if two pairs are chosen from the same column, then the graph representation of the resulting pairs has a cycle. Choosing always the top pair leads to the C-vine and choosing always the bottom pair leads to the D-vine. Other choices can also lead to C-vines or D-vines (in non-standard form), for example 12; 13; 34; 25 at level 1 leads to a D-vine.

An example that does not lead to the C-vine or D-vine after permutation is given next.

- Start with 12; 23; 24; 35 and form Tree 1. Nodes 12 and 23 have *edge* 13|2, nodes 12 and 24 have *edge* 14|2, nodes 12 and 35 lead to ∅ for no edge, nodes 23 and 24 have *edge* 34|2, nodes 23 and 35 have *edge* 25|3, nodes 24 and 35 lead to ∅.
- For Tree 2, choose three edges and avoid a cycle, say, 13|2; 14|2; 25|3 (with 34|2 not included).
  Nodes 13|2 and 14|2 have *edge* 34|12; nodes 13|2 and 25|3 have *edge* 15|23; nodes 14|2 and 25|3 lead to a non-regular case because there are not two indices in the intersection.
- For Tree 3, choose nodes 34|12 and 15|23 with *edge* 45|123.

Next, we give a brief derivation to show that there are six equivalence classes of regular vines for $n = 5$. The enumeration starts by noting that there are three possibilities for the level 1 tree: (a) three nodes with order 2 and the other two with order 1 (like the D-vine), or (b) one node with order 3, one node with order 2 and the other three with order 1, or (c) one node with order 4 and the other four with order 1 (like the first tree of the C-vine).

(a) D-vine 12; 23; 34; 45; 13|2; 24|3; 35|4; 14|23; 25|34; 15|234 (first level order of nodes 1,2,2,2,1, and there are unique trees for levels 2,3,4 like for $n = 4$).
(b) 12; 13; 14; 25 (first level order of nodes 3,2,1,1,1). *Second level possibilities are 23|1, 24|1, 34|1, 15|2, of which two of the first three can be chosen (as they form a cycle).*

Variables 3 and 4 are symmetric so there are two distinct cases at level 2:

(i) 23|1; 24|1; 15|2, leading to *third level possibilities 34|12, 35|12, 45|12 — two cases for next step*:

(i') 34|12; 35|12; 45|123 or (i'') 35|12; 45|12; 34|125, or,

(ii) 23|1; 34|1; 15|2, leading to the remainder 24|13; 35|12; 45|123.

(c) 12; 13; 14; 15 (first level order of nodes 4,1,1,1,1). *Second level possibilities are 23|1, 24|1, 34|1, 25|1, 35|1, 45|1 — can choose three of six that do not form cycle.*

(i) 23|1; 24|1; 25|1 (C-vine-like) leads to *third level possibilities 34|12, 35|12, 45|12*, and remaining step is a C-vine for any choices, say 34|12; 35|12; 45|123, or

(ii) 23|1; 24|1; 35|1 (which is D-vine 4-2-3-5), so following are 34|12; 25|13; 45|123.

For the remainder of this chapter, we use labels for the six (equivalence classes of) vines: D from (a); B1, B2, B3 from (b); C, B0 from (c). The B is shorthand for "Between" because B0,B1,B2,B3 are between the C-vine and D-vine. It will turn out below that the B2 vine is halfway between in one sense.

For ease of dependence comparisons in subsequent sections, we permute the indices so that 4 and 5 are labels of the conditioned variables for the final conditional distribution 45|123. The sequences are given in the bottom part of Table 7.1. Note that B0 and B2 have more symmetry than B1 and B3; there is symmetry with $\pi(12345) = 13254$ in B0, and symmetry with $\pi(12345) = 12354$ in B2. The extra symmetry affects whether there are generalized Toeplitz matrices associated with the vine; see Section 7.5.

It is informative to write down the matrices with level by order of nodes (at variable 1,2,3,4,5); see Table 7.1. For level order $\ell \geq 2$, in $\{i_1, i_2\}|D(i_1, i_2, \ell)$, there is a count for the order of the node for the conditioned variables i_1 and i_2 only. Two vines, where the matrix of level by order of nodes are the same after column permutation, must be in different equivalence classes. However, (b)(i') and (b)(i'') have "isomorphic" matrices but different equivalence classes.

For $n = 6$, an enumeration (by hand, similar to the preceding for $n = 5$) led to 40 equivalence classes, and this was confirmed using the algorithm in Chapter 10.

From the enumeration method of regular vines in Morales-Nápoles et al.,[17] a bound is $2^{(n-2)(n-3)/2}$; this bound is 8 for $n = 5$ and 64 for $n = 6$.

Table 7.1. Matrices with level by order of nodes (at variable 1,2,3,4,5).

Case (a): D-vine		Case (b)(i′): B1-vine		Case (b)(i″): B2-vine	
level	variables 12345	level	variables 12345	level	variables 12345
1	22211	1	32111	1	32111
2	12111	2	12111	2	12111
3	10111	3	00211	3	00211
4	00011	4	00011	4	00011

Case (b)(ii): B3-vine		Case (c)(i): C-vine		Case (c)(ii): B0-vine	
level	variables 12345	level	variables 12345	level	variables 12345
1	32111	1	41111	1	41111
2	11211	2	03111	2	02211
3	01111	3	00211	3	01111
4	00011	4	00011	4	00011

D: 15; 12; 23; 34; 25|1; 13|2; 24|3; 35|12; 14|23; 45|123
B1: 12; 13; 14; 25; 23|1; 24|1; 15|2; 34|12; 35|12; 45|123
B2: 12; 15; 14; 23; 25|1; 24|1; 13|2; 35|12; 34|12; 45|123
B3: 12; 13; 14; 25; 23|1; 34|1; 15|2; 24|13; 35|12; 45|123
C: 12; 13; 14; 15; 23|1; 24|1; 25|1; 34|12; 35|12; 45|123
B0: 12; 13; 14; 15; 23|1; 24|1; 35|1; 34|12; 25|13; 45|123

7.3 Simulation from Vine Copulae

The main goal of this section is to provide algorithms for simulation from vine copulae. We provide pseudo-codes for C-vines and D-vines, in standard form, of any dimension n. These improve a little on the algorithms in Aas et al.,[1] mainly in the naming of intermediate variables. We also extend the pseudo-codes to any regular vine, but assume a certain indexing of the vine. These pseudo-codes have more specific details than the descriptions of sampling a regular vine given in Section 6.4.2 of Kurowicka and Cooke.[13]

Besides simulation, the pseudo-codes can be used to show how similar different vines are. Extra details will be listed for the B0, B1, B2, B3-vines for $n = 5$ in order to show their pairwise similarity. The proof that some of the different pairs of five-dimensional vine copulae can have a common four-dimensional margin comes from the algorithms.

For an n-dimensional vine copula, there is a bivariate copula family associated with each bivariate or conditional bivariate margin of form

$\{i_1, i_2\} \mid D(i_1, i_2, \ell) = \{j_1, \ldots, j_{\ell-1}\}$. The notation is $C_{i_1,i_2|j_1,\ldots,j_{\ell-1}}$, and for $\ell \geq 2$, this is assumed to be independent of the values of the conditioning variables, that is, conditional copulae constant over $u_{j_1}, \ldots, u_{j_{\ell-1}}$. As an illustration of the notation, the ten copulae for the six five-dimensional vines in the previous section are:

C: $C_{12}, C_{13}, C_{14}, C_{15}, C_{23|1}, C_{24|1}, C_{25|1}, C_{34|12}, C_{35|12}, C_{45|123}$.

B0: $C_{12}, C_{13}, C_{14}, C_{15}, C_{23|1}, C_{24|1}, C_{35|1}, C_{34|12}, C_{25|13}, C_{45|123}$.

B1: $C_{12}, C_{13}, C_{14}, C_{25}, C_{23|1}, C_{24|1}, C_{15|2}, C_{34|12}, C_{35|12}, C_{45|123}$.

B2: $C_{12}, C_{15}, C_{14}, C_{23}, C_{25|1}, C_{24|1}, C_{13|2}, C_{34|12}, C_{35|12}, C_{45|123}$.

B3: $C_{12}, C_{13}, C_{14}, C_{25}, C_{23|1}, C_{34|1}, C_{15|2}, C_{24|13}, C_{35|12}, C_{45|123}$.

D: $C_{15}, C_{12}, C_{23}, C_{34}, C_{25|1}, C_{13|2}, C_{24|3}, C_{35|12}, C_{14|23}, C_{45|123}$.

In the simulation algorithms below, we have the following additional notation. For a bivariate copula C_{ij}, $C_{i|j} = \partial C_{ij}/\partial u_j$ and $C_{j|i} = \partial C_{ij}/\partial u_i$ are conditional distributions. For a conditional bivariate copula $C_{i_1,i_2|j_1,\ldots,j_{\ell-1}}$, the conditional distributions are denoted as: $C_{i_1|i_2;j_1,\ldots,j_{\ell-1}} = \partial C_{i_1,i_2|j_1,\ldots,j_{\ell-1}}/\partial u_{i_2}$ and $C_{i_2|i_1;j_1,\ldots,j_{\ell-1}} = \partial C_{i_1,i_2|j_1,\ldots,j_{\ell-1}}/\partial u_{i_1}$. For the algorithms, the outputs are n dependent uniform random deviates $U_1, \ldots, U_n$. Apply quantile functions $F_1^{-1}(U_1), \ldots, F_n^{-1}(U_n)$ to get dependent random deviates with the distributions $F_1, \ldots, F_n$.

The algorithm for the C-vine is given in the first part of Table 7.2. It is simplest in that the "backward" steps are not needed. For the D-vine in standard form, the algorithm is given in the second part of Table 7.2; backward steps are needed, and matrices (a_{ij}) and (b_{ij}) are used rather than $(v_{ij})_{1 \leq i \leq n, 1 \leq j \leq 2n}$ in Aas et al.[1] The pseudo-code for the D-vine extends to any regular vine (see the third part of Table 7.2) that has been indexed so that if one deletes all $\{i_1, i_2\}|D(i_1, i_2, \ell)$ that include variables $m+1, \ldots, n$ ($m = 2, 3, \ldots, n-1$), then the remainder is a regular vine in variables $1, \ldots, m$. All of the five-dimensional vines above are indexed in this way; that is, all regular vines can have the variables permuted to satisfy this condition. The condition in the third part of Table 7.2 is similar to and slightly more general than that given on p. 173 of Kurowicka and Cooke.[13]

The main idea for the forward steps (a_{ij} in Table 7.2) will be shown in a couple of specific examples, first for the four-dimensional C-vine, and then a four-dimensional D-vine with permuted indices.

Here, we need to use F for the copula of the constructed vine, in order to avoid notation confusion with the conditional copulae; $F_{1j} = C_{1j}$ for the

Table 7.2. Pseudo-codes for the C-vine, D-vine and regular vines in dimension n.

C-vine

Generate $w_1, \ldots, w_n$ to be $U(0,1)$ random deviates.
$u_1 \leftarrow w_1$, $u_2 \leftarrow C_{2|1}^{-1}(w_2|w_1)$.
for $i = 3, \ldots, n$:
  $t \leftarrow w_i$, for $j = i-1, i-2, \ldots, 1$: $t \leftarrow C_{i|j:1,\ldots,j-1}^{-1}(t|w_j)$;
  $u_i \leftarrow t$.

D-vine

Generate $w_1, \ldots, w_n$ to be $U(0,1)$. Arrays a, b are $n \times n$, but only upper diagonals used.
$u_1 \leftarrow w_1$, $a_{11} \leftarrow w_1, b_{11} \leftarrow w_1$.
for $i = 2, \ldots, n$:
  $a_{i1} \leftarrow w_i$, for $j = 2, \ldots, i$: $a_{ij} \leftarrow C_{i|j-1:j\cdots i-1}^{-1}(a_{i,j-1}|b_{i-1,j-1})$;
  $u_i \leftarrow a_{ii}$, $b_{ii} \leftarrow a_{ii}$;
  for $j = i-1, \ldots, 1$: $b_{ij} \leftarrow C_{j|i:j+1\cdots i-1}(b_{i-1,j}|a_{i,j+1})$.

Regular vine with indexing as follows (this is a necessary but not sufficient condition for a vine):
12; $k_{31}3$; $k_{32}3|k_{31}$; $\ldots$; $k_{i1}i$; $k_{i2}i|k_{i1}$; $\ldots$; $k_{i,i-1}i|k_{i1}, \ldots, k_{i,i-2}$; $\ldots$;
$k_{n1}n$; $k_{n2}n|k_{n1}$; $\ldots$; $k_{n,n-1}n|k_{n1}, \ldots, k_{n,n-2}$,
where $(k_{i1}, \ldots, k_{i,i-1})$ is a permutation of $(1, \ldots, i-1)$ for $i = 2, \ldots, n$.
(For D-vine, $k_{ij} = i - j$; for C-vine $k_{ij} = j$.)

Generate $w_1, \ldots, w_n$ to be $U(0,1)$. Arrays a, b are $n \times n$, but only upper diagonals used.
$u_1 \leftarrow w_1$, $a_{11} \leftarrow w_1, b_{11} \leftarrow w_1$.
for $i = 2, \ldots, n$:
  $M \leftarrow \{k_{i1}, \ldots, k_{i,i-2}\}$, $v_{i,k_{i,i-1},M} = v_{i,\{k_{i,i-1}\cup M\}} \leftarrow w_i$, $a_{i1} \leftarrow v_{i,k_{i,i-1},M}$,
  for $j = 2, \ldots, i$:
    $M \leftarrow \{k_{i1} \cdots k_{i,i-j}\}$
    $v_{i,M} \leftarrow C_{i|k_{i,i+1-j}:M}^{-1}(v_{i,k_{i,i+1-j},M}|v_{k_{i,i+1-j},M}) = C_{i|k_{i,i+1-j}:M}^{-1}(a_{i,j-1}|v_{k_{i,i+1-j},M})$
    $[v_{k_{i,i+1-j},\{k_{i1}\cdots k_{i,i-j}\}}$ from a previous backward step];
    $a_{ij} \leftarrow v_{i,M}$
  end j loop
  $u_i \leftarrow a_{ii}$, $b_{ii} \leftarrow a_{ii}$;
  for $j = i-1, \ldots, 1$: (some backward steps)
    $M \leftarrow \{k_{i1} \cdots k_{i,i-1-j}\}$;
    $v_{k_{i,i-j},i,M} \leftarrow C_{k_{i,i-j}|i:M}(v_{k_{i,i-j},M}|v_{i,M}) = C_{k_{i,i-j}|i:M}(v_{k_{i,i-j},M}|a_{i,j+1})$
    $[v_{k_{i,i-j},\{k_{i1}\cdots k_{i,i-1-j}\}}$ from a previous step];
    $b_{ij} \leftarrow v_{k_{i,i-j},i,M}$;
  end j loop

---

first level of the vine, and then using the notation mentioned in Chapter 1, the copula for the four-dimensional C-vine is

$$F_{1234}(\boldsymbol{u}) = \int_0^{u_1} \int_0^{u_2} C_{34|12}(F_{3|21}(u_3|v_2, v_1), F_{4|21}(u_4|v_2, v_1))$$
$$f_{12}(v_1, v_2) \, dv_2 dv_1,$$

where
$$F_{12j}(u_1, u_2, u_j) = \int_0^{u_1} C_{2j|1}(F_{2|1}(u_2|v), F_{j|1}(u_j|v)) dv, \quad j = 3, 4.$$
By differentiation,
$$F_{3|12}(u_3|u_1, u_2) = \frac{\partial^2 F_{123}/\partial u_1 \partial u_2}{\partial^2 F_{12}/\partial u_1 \partial u_2} = C_{3|2:1}(F_{3|1}(u_3|u_1)|F_{2|1}(u_2|u_1)),$$
which implies for $0 < q < 1$ that
$$F_{3|12}^{-1}(q|u_1, u_2) = F_{3|1}^{-1}(C_{3|2:1}^{-1}[q|F_{2|1}(u_2|u_1)]|u_1)$$
$$= C_{3|1}^{-1}(C_{3|2:1}^{-1}[q|C_{2|1}(u_2|u_1)]|u_1), \qquad (7.2)$$
and there is a similar expression for $F_{4|12}^{-1}(q|u_1, u_2)$. Similarly,
$$F_{4|123}(u_4|u_1, u_2, u_3) = \frac{\partial^3 F_{1234}/\partial u_1 \partial u_2 \partial u_3}{\partial^3 F_{123}/\partial u_1 \partial u_2 \partial u_3}$$
$$= C_{4|3:12}(F_{4|12}(u_4|u_1, u_2)|F_{3|12}(u_3|u_1, u_2))$$
and, for $0 < q < 1$,
$$F_{4|123}^{-1}(q|u_1, u_2, u_3) = F_{4|12}^{-1}[C_{4|3:12}^{-1}(q|F_{3|12}(u_3|u_1, u_2))|u_1, u_2]$$
$$= C_{4|1}^{-1}(C_{4|2:1}^{-1}[u_{43}|C_{2|1}(u_2|u_1)]|u_1),$$
$$u_{43} = C_{4|3:12}^{-1}(q|F_{3|12}(u_3|u_1, u_2)).$$
So if $W_1, \ldots, W_4$ are independent $U(0,1)$ random variables, then $(U_1, \ldots, U_4) \sim F_{1234}$ if $U_1 = W_1$, $U_2 = C_{2|1}^{-1}(W_2|W_1)$,
$$U_3 = C_{3|1}^{-1}(C_{3|2:1}^{-1}(W_3|C_{2|1}(U_2|U_1))|U_1) = C_{3|1}^{-1}(C_{3|2:1}^{-1}(W_3|W_2)|U_1), \quad (7.3)$$
and
$$U_4 = C_{4|1}^{-1}(C_{4|2:1}^{-1}[U_{43}|C_{2|1}(U_2|U_1)]|U_1) = C_{4|1}^{-1}(C_{4|2:1}^{-1}[U_{43}|W_2]|W_1),$$
with $U_{43} = C_{4|3:12}^{-1}(W_4|F_{3|12}(U_3|U_1, U_2)) = C_{4|3:12}^{-1}(W_4|W_3)$, via (7.3) and (7.2), or
$$U_4 = C_{4|1}^{-1}(C_{4|2:1}^{-1}[C_{4|3:12}^{-1}(W_4|W_3)|W_2]|W_1).$$
This pattern explains the pseudo-code for the C-vine in Table 7.2.

For non C-vines, we introduce the following notation to explain the steps. With the notation introduced earlier, $C_{i|j:M}$ is the conditional distribution of the linking copula $C_{ij|M}$ in the vine, where i, j are integers and M is a subset (possibly empty). The bottom part of Table 7.2 makes use of the following:
$$v_{i,j,M} = v_{i,\{j\} \cup M} = C_{i|j:M}(v_{i,M}|v_{j,M}),$$
$$v_{i,M} = C_{i|j:M}^{-1}(v_{i,j,M}|v_{j,M}). \qquad (7.4)$$

Note that some of these v's appear in the pseudo-code without identifying which a_{ij} and b_{ij} they correspond to; this match can be done generically only for the C-vine and D-vine.

To understand the more general pseudo-code, let us look at the same steps for the four-dimensional D-vine summarized as 12; 13; 34; 23|1; 14|3; 24|13. F_{123} is the same as for the above C-vine, and then with similar steps,

$$F_{4|123}(u_4|u_1, u_2, u_3) = C_{4|2;1,3}(F_{4|13}(u_4|u_1, u_3)|F_{2|13}(u_2|u_1, u_3))$$

where

$$F_{4|13}(u_4|u_1, u_3) = C_{4|1;3}(C_{4|3}(u_4|u_3)|C_{1|3}(u_1|u_3)),$$
$$F_{2|13}(u_2|u_1, u_3) = C_{2|3;1}(C_{2|1}(u_2|u_1)|C_{3|1}(u_3|u_1)).$$

Let $U_1, U_2, U_3, W_1, W_2, W_3$ be the same as above, and then

$$U_4 = C_{4|3}^{-1}(C_{4|1;3}^{-1}[U_{42}|C_{1|3}(U_1|U_3)]|U_3) \quad \text{with}$$
$$U_{42} = C_{4|2:13}^{-1}(W_4|F_{2|13}(U_2|U_1, U_3)),$$
$$F_{2|13}(U_2|U_1, U_3) = C_{2|3;1}(W_2|C_{3|1}(U_3|U_1)). \tag{7.5}$$

This is more complicated than for the C-vine because other functions of the U_i's or W_i's are needed. To see the algorithmic approach, the pseudo-code in the bottom part of Table 7.2, with upper case U_i, W_i, becomes:

- a_{11}, b_{11}: $U_1 = W_1 = v_1$
- a_{21}: $v_{2,1} = W_2$
- a_{22}: $U_2 = v_2 = C_{2|1}^{-1}(v_{2,1}|v_1) = C_{2|1}^{-1}(W_2|W_1)$
- b_{21}: $v_{1,2} = C_{1|2}(v_1|v_2) = C_{1|2}(W_1|U_2)$
- a_{31}: $v_{3,2,1} = W_3 = v_{3,12}$
- a_{32}: $v_{3,1} = C_{3|2:1}^{-1}(v_{3,2,1}|v_{2,1}) = C_{3|2:1}^{-1}(W_3|W_2)$
- a_{33}: $U_3 = v_3 = C_{3|1}^{-1}(v_{3,1}|v_1) = C_{3|1}^{-1}(v_{3,1}|W_1)$
- b_{32}: $v_{1,3} = C_{1|3}(v_1|v_3) = C_{1|3}(W_1|U_3)$
- b_{31}: $v_{2,3,1} = v_{2,13} = C_{2|3:1}(v_{2,1}|v_{3,1}) = C_{2|3:1}(W_2|v_{3,1})$
- a_{41}: $v_{4,2,13} = W_4 = v_{4,213}$
- a_{42}: $v_{4,1,3} = v_{4,13} = C_{4|2:13}^{-1}(v_{4,2,13}|v_{2,13}) = C_{4|2:13}^{-1}(W_4|v_{2,13})$
- a_{43}: $v_{4,3} = C_{4|1:3}^{-1}(v_{4,1,3}|v_{1,3})$
- a_{44}: $U_4 = v_4 = C_{4|3}^{-1}(v_{4,3}|v_3) = C_{4|3}^{-1}(v_{4,3}|U_3).$

This matches with (7.5) because for step b_{31},

$$C_{3|1}(U_3|U_1) = C_{3|1}(C_{3|1}^{-1}(v_{3,1}|W_1)|U_1) = v_{3,1}.$$

Table 7.3. Five-dimensional D-vine (standard order) for assigning a_{ij} and b_{ij}.

Let $w_1, w_2, \ldots$ be a sequence of uniform $(0,1)$ rvs.
For each step i, do all forward substeps before backward.

i	Forward	Backward
1	a_{11}: $v_1 = w_1$	b_{11}: $u_1 = v_1$
2	a_{21}: $v_{2,1} = w_2$	b_{22}: $u_2 = v_2$
	a_{22}: $v_2 = C_{2\|1}^{-1}(v_{2,1}\|v_1)$	b_{21}: $v_{1,2} = C_{1\|2}(v_1\|v_2)$
3	a_{31}: $v_{3,1,2} = w_3$	b_{33}: $u_3 = v_3$
	a_{32}: $v_{3,2} = C_{3\|1:2}^{-1}(v_{3,1,2}\|v_{1,2})$	b_{32}: $v_{2,3} = C_{2\|3}(v_2\|v_3)$
	a_{33}: $v_3 = C_{3\|2}^{-1}(v_{3,2}\|v_2)$	b_{31}: $v_{1,3,2} = v_{1,23} = C_{1\|3:2}(v_{1,2}\|v_{3,2})$
4	a_{41}: $v_{4,1,23} = w_4$	b_{44}: $u_4 = v_4$
	a_{42}: $v_{4,2,3} = v_{4,23} = C_{4\|1:23}^{-1}(v_{4,1,23}\|v_{1,23})$	b_{43}: $v_{3,4} = C_{3\|4}(v_3\|v_4)$
	a_{43}: $v_{4,3} = C_{4\|2:3}^{-1}(v_{4,2,3}\|v_{2,3})$	b_{42}: $v_{2,4,3} = v_{2,34} = C_{2\|4:3}(v_{2,3}\|v_{4,3})$
	a_{44}: $v_4 = C_{4\|3}^{-1}(v_{4,3}\|v_3)$	b_{41}: $v_{1,4,23} = v_{1,234} = C_{1\|4:23}(v_{1,23}\|v_{4,23})$
5	a_{51}: $v_{5,1,234} = w_5$	b_{55}: $u_5 = v_5$
	a_{52}: $v_{5,2,34} = v_{5,234}$ $= C_{5\|1:234}^{-1}(v_{5,1,234}\|v_{1,234})$	b_{54}: $v_{4,5} = C_{4\|5}(v_4\|v_5)$
	a_{53}: $v_{5,3,4} = v_{5,34} = C_{5\|2:34}^{-1}(v_{5,2,34}\|v_{2,34})$	b_{53}: $v_{3,5,4} = v_{3,45} = C_{3\|5:4}(v_{3,4}\|v_{5,4})$
	a_{54}: $v_{5,4} = C_{5\|3:4}^{-1}(v_{5,3,4}\|v_{3,4})$	b_{52}: $v_{2,5,34} = v_{2,345} = C_{2\|5:34}(v_{2,34}\|v_{5,34})$
	a_{55}: $v_5 = C_{5\|4}^{-1}(v_{5,4}\|v_4)$	b_{51}: $v_{1,5,234} = v_{1,2345}$ $= C_{1\|5:234}(v_{1,234}\|v_{5,234})$

Next, the detailed steps for the five-dimensional D-vine are given in Table 7.3 before we show some steps for the other vines that make use of (7.4). Let $w_1, w_2, \ldots$ be a sequence of uniform $(0,1)$ rvs.

More details for the other five-dimensional vines are summarized into a few tables. The C-vine, B0-vine and B1-vine have the same first four steps; see Table 7.4. Table 7.5 has the fifth step for the B0-vine and B1-vine. Table 7.6 has the third and fifth steps for the B2-vine. Table 7.7 has the fourth step for the B2-vine. Table 7.8 shows how the D-vine can be simulated with different permutations of the variables.

For the B-vines in the forms given above, at most only two of the b_{ij}'s in the reverse direction are needed:

- b_{21}: $v_{1,2} = C_{1|2}(v_1|v_2)$ needed for the B1-vine, B2-vine and B3-vine; this step is coded as $C_{1|2}(a_{11}|a_{22})$
- b_{31}: $v_{2,3,1} = v_{2,13} = C_{2|3:1}(v_{2,1}|v_{3,1})$ needed for the B0-vine and B3-vine; this step is coded as $C_{2|3:1}(a_{21}|a_{32})$.

Table 7.4. Step $i = 1$ to 4 for C-vine, B0-vine and B1-vine.

i	Forward substeps
1	a_{11}: $u_1 = v_1 = w_1$
2	a_{21}: $v_{2,1} = w_2$
	a_{22}: $u_2 = v_2 = C_{2\|1}^{-1}(v_{2,1}\|v_1)$
3	a_{31}: $v_{3,2,1} = w_3 = v_{3,12}$
	a_{32}: $v_{3,1} = C_{3\|2:1}^{-1}(v_{3,2,1}\|v_{2,1})$
	a_{33}: $u_3 = v_3 = C_{3\|1}^{-1}(v_{3,1}\|v_1)$
4	a_{41}: $v_{4,3,12} = w_4 = v_{4,123}$
	a_{42}: $v_{4,2,1} = v_{4,12} = C_{4\|3:12}^{-1}(v_{4,3,12}\|v_{3,12})$
	a_{43}: $v_{4,1} = C_{4\|2:1}^{-1}(v_{4,2,1}\|v_{2,1})$
	a_{44}: $u_4 = v_4 = C_{4\|1}^{-1}(v_{4,1}\|v_1)$

Table 7.5. Step $i = 5$ of B0-vine and B1-vine.

	B0: 12; 13; 14; 15; 23|1; 24|1; 35|1; 34|12; 25|13; 45|123	B1: 12; 13; 14; 25; 23|1; 24|1; 15|2; 34|12; 35|12; 45|123
	B0, b_{31}: $v_{2,3,1} = v_{2,13} = C_{2\|3:1}(v_{2,1}\|v_{3,1})$	B1, b_{21}: $v_{1,2} = C_{1\|2}(v_1\|v_2)$
a_{51}	$v_{5,4,123} = w_5$	$v_{5,4,123} = w_5$
a_{52}	$v_{5,2,13} = v_{5,123} = C_{5\|4:123}^{-1}(v_{5,4,123}\|v_{4,123})$	$v_{5,3,12} = v_{5,123} = C_{5\|4:123}^{-1}(v_{5,4,123}\|v_{4,123})$
a_{53}	$v_{5,3,1} = v_{5,13} = C_{5\|2:13}^{-1}(v_{5,2,13}\|v_{2,13})$ [b_{31}]	$v_{5,1,2} = v_{5,12} = C_{5\|3:12}^{-1}(v_{5,3,12}\|v_{3,12})$
a_{54}	$v_{5,1} = C_{5\|3:1}^{-1}(v_{5,3,1}\|v_{3,1})$	$v_{5,2} = C_{5\|1:2}^{-1}(v_{5,1,2}\|v_{1,2})$ [needs b_{21}]
a_{55}	$u_5 = v_5 = C_{5\|1}^{-1}(v_{5,1}\|v_1)$	$u_5 = v_5 = C_{5\|2}^{-1}(v_{5,2}\|v_2)$

The fourth and fifth steps of the B2-vine are the same as for the C-vine. The third and fifth steps of the B2-vine are given in Table 7.6. The B2-vine has the same 1245 four-dimensional margins as for the C-vine by noting that the fourth and fifth steps of the B2-vine do not depend on $v_{3,2}$ and v_3 (these steps have $v_{3,1,2} = v_{3,12}$ which is the same for the two vines).

Table 7.7 has the fourth step of the B3-vine. The fifth step is the same as for the B1-vine, and the first three steps are the same as for the C-vine.

For the simulation comparisons in the next section, it is useful to simulate the D-vine with orders 51234 and 52134. The D-vine in the order 51234 has the same 1235 margin as the B2-vine; the algorithm is the same as that for the B2-vine except step 4, which is given in Table 7.6. For the D-vine in

Table 7.6. Steps $i = 3$ and $i = 5$ of B2-vine.

i	B2: 12; 15; 14; 23; 25|1; 24|1; 13|2; 35|12; 34|12; 45|123 B2, b_{21}: $v_{1,2} = C_{1\|2}(v_1\|v_2)$
3	a_{31}: $v_{3,1,2} = v_{3,12} = w_3$
	a_{32}: $v_{3,2} = C^{-1}_{3\|1:2}(v_{3,1,2}\|v_{1,2})$ [needs b_{21}]
	a_{33}: $u_3 = v_3 = C^{-1}_{3\|2}(v_{3,2}\|v_2)$
5	a_{51}: $v_{5,4,123} = w_5$
	a_{52}: $v_{5,3,12} = v_{5,123} = C^{-1}_{5\|4:123}(v_{5,4,123}\|v_{4,123})$
	a_{53}: $v_{5,2,1} = v_{5,12} = C^{-1}_{5\|3:12}(v_{5,3,12}\|v_{3,12})$
	a_{54}: $v_{5,1} = C^{-1}_{5\|2:1}(v_{5,2,1}\|v_{2,1})$ [differs from B1]
	a_{55}: $u_5 = v_5 = C^{-1}_{5\|1}(v_{5,1}\|v_1)$ [differs from B1]

Table 7.7. Step $i = 4$ of B3-vine, fifth step same as B1-vine.

B3: 12; 13; 14; 25; 23|1; 34|1; 15|2;
24|13; 35|12; 45|123
B0, b_{31}: $v_{2,13} = C_{2|3:1}(v_{2,1}|v_{3,1})$

a_{41}: $v_{4,2,13} = w_4 = v_{4,123}$

a_{42}: $v_{4,3,1} = v_{4,13} = C^{-1}_{4|2:13}(v_{4,2,13}|v_{2,13})$ [needs b_{31}]

a_{43}: $v_{4,1} = C^{-1}_{4|3:1}(v_{4,3,1}|v_{3,1})$

a_{44}: $u_4 = v_4 = C^{-1}_{4|1}(v_{4,1}|v_1)$

the order 52134, the algorithm is the same as for the B1-vine and B3-vine except step 4.

All of the above algorithms were implemented in the C programming language for use in the comparisons in Section 7.6.

We conclude this section with Table 7.9, which shows the similarity between the different vines in terms of the maximum lower-dimensional margin that can be matched. This is partly deduced from comparing the simulation algorithms above. For example, from the list of the sequences of conditional distributions, it is clear that the C-vine and B1-vine have the same 1234 four-dimensional margins. The overlapping four variables of the B2-vine with the C-vine and D-vine (form D1) are mentioned above.

Table 7.9 assumes $\{i_1, i_2\} \mid \{j_1, \ldots, j_{\ell-1}\}$ means the same thing if it appears in different vines; the notation for the vines are repeated in the

Table 7.8. Third backward and fourth forward step of D-vine in two different orders.

Step	D1 form: order 51234
3	b_{33}: $u_3 = v_3$ b_{32}: $v_{2,3} = C_{2|3}(v_2|v_3)$ [or $C_{2|3}(a_{22}|a_{33})$]; b_{31}: $v_{1,23} = C_{1|3:2}(v_{1,2}|v_{3,2})$ [or $C_{1|3:2}(b_{21}|a_{32})$]
4	a_{41}: $v_{4,1,23} = w_4 = v_{4,123}$ [same as B2-vine] a_{42}: $v_{4,2,3} = v_{4,23} = C^{-1}_{4|1:23}(v_{4,1,23}|v_{1,23})$ [needs b_{31}] [not used in step5] a_{43}: $v_{4,3} = C^{-1}_{4|2:3}(v_{4,2,3}|v_{2,3})$ [needs b_{32}] [not used in step5] a_{44}: $u_4 = v_4 = C^{-1}_{4|3}(v_{4,3}|v_3)$ [not used in step5]

Step	D2 form: order 52134
3	b_{33}: $u_3 = v_3$ b_{32}: $v_{1,3} = C_{1|3}(v_1|v_3)$ [or $C_{1|3}(a_{11}|a_{33})$] b_{31}: $v_{2,13} = C_{2|3:1}(v_{2,1}|v_{3,1})$ [or $C_{2|3}(a_{21}|a_{32})$]
4	a_{41}: $v_{4,2,13} = w_4 = v_{4,123}$ a_{42}: $v_{4,1,3} = v_{4,13} = C^{-1}_{4|2:13}(v_{4,2,13}|v_{2,13})$ [needs b_{31}] a_{43}: $v_{4,3} = C^{-1}_{4|1:3}(v_{4,1,3}|v_{1,3})$ [needs b_{32}] a_{44}: $u_4 = v_4 = C^{-1}_{4|3}(v_{4,3}|v_3)$

Table 7.9. Similarity table of the five-dimensional vines.

—	C	B0	B1	B2	B3	D
C	—	1234	1234	1245	123	123/D2
B0	1234	—	1234	124	123	123/D2
B1	1234	1234	—	124	1235	1235/D2
B2	1245	124	124	—	123/B2'	1235/D1
B3	123	123	1235	123	—	1235/D2
D	123	123	1235	1235	1235	—

C: 12; 13; 14; 15; 23|1; 24|1; 25|1; 34|12; 35|12; 45|123
B0: 12; 13; 14; 15; 23|1; 24|1; 35|1; 34|12; 25|13; 45|123
B1: 12; 13; 14; 25; 23|1; 24|1; 15|2; 34|12; 35|12; 45|123
B2: 12; 15; 14; 23; 25|1; 24|1; 13|2; 35|12; 34|12; 45|123

B2' means with 3,5 interchanged above.

B3: 12; 13; 14; 25; 23|1; 34|1; 15|2; 24|13; 35|12; 45|123
D1: 51; 12; 23; 34; 25|1; 13|2; 24|3; 35|12; 14|23; 45|123
D2: 52; 21; 13; 34; 15|2; 23|1; 14|3; 35|12; 24|13; 45|123

table. The last column shows that the most overlap with the D-vine depends on the permutation of the D1 or D2 forms. For the B2-vine and B3-vine, there is more similarity when two variables in B2 are permuted.

7.4 Comparing Dependence of Vine Copulae

The comparisons made in this section and the next two sections are motivated by some numerical results of the range of tail dependence parameters for four-dimensional vines in Joe et al.[11] We want to understand conditions where one vine can have more marginal bivariate dependence than another when the set of bivariate copulae used to specify the vine copulae are the same.

- D-vine: 12; 23; 34; 13|2; 24|3; 14|23: permute $\pi(1234) = 3214$ to get 32; 21; 14; 31|2; 24|1; 34|21.
- C-vine: 12; 13; 14; 23|1; 24|1; 34|12 in standard form.

The third level is 34|12 in both cases, and

- for the C-vine, the (3,4) bivariate margin can be the Fréchet upper bound (co-monotonic). It is co-monotonic if $C_{23} = C_{24}$, $C_{13|2} = C_{14|2}$ (in which case $C_{213} = C_{214}$) and $C_{34|12}$ is co-monotonic.
- for the (permuted) D-vine, the (3,4) bivariate margin cannot be co-monotonic even if $C_{34|12}$ is co-monotonic.

For the C-vine and permuted D-vine, the stronger marginal dependence of the $(3, 4)$ bivariate margin occurs in some conditions but not always.

To understand the conditions we look at Gaussian vines where all of the bivariate copulae are Gaussian. In this case, the distribution is specified from a sequence of correlations and partial correlations.

Consider the Gaussian C-vine and D-vine with the parameters in Table 7.10.

From Ref. 12, the determinants of the two resulting correlation matrices are both $(1-\rho_a^2)(1-\rho_b^2)(1-\rho_c^2)(1-\alpha_d^2)(1-\alpha_e^2)(1-\alpha_f^2)$. So in one sense, the Gaussian C-vine and D-vine have the same amount of overall dependence. However, their correlation ρ_{34} will be different. From the above specifications it can be shown that ρ_{24} is the same for the two distributions and that ρ_{23} for the C-vine is the same as ρ_{13} for the D-vine. By applying the rules for partial correlations, the following are obtained.

Table 7.10. Parameters for 4-dimensional Gaussian vines.

C-vine	D-vine	correlation
12	12	ρ_a
14	14	ρ_b
13	23	ρ_c
24|1	24|1	α_d
23|1	13|2	α_e
34|12	34|12	α_f

- Both: $\rho_{24} = \alpha_d \sqrt{(1-\rho_a^2)(1-\rho_b^2)} + \rho_a \rho_b$.
- $C-vine: \rho_{34;1} = \alpha_f \sqrt{(1-\alpha_e^2)(1-\alpha_d^2)} + \alpha_e \alpha_d$
- $C-vine: \rho_{34} = \rho_{34;1}\sqrt{(1-\rho_c^2)(1-\rho_b^2)} + \rho_c \rho_b$
- $D-vine: \rho_{13} = \alpha_e \sqrt{(1-\rho_a^2)(1-\rho_c^2)} + \rho_a \rho_c$
- $D-vine: \rho_{23;1} = (\rho_c - \rho_a \rho_{13})/\sqrt{(1-\rho_a^2)(1-\rho_{13}^2)}$
- $D-vine: \rho_{34;1} = \alpha_f \sqrt{(1-\rho_{23;1}^2)(1-\alpha_d^2)} + \rho_{23;1}\alpha_d$
- $D-vine: \rho_{34} = \rho_{34;1}\sqrt{(1-\rho_{13}^2)(1-\rho_b^2)} + \rho_{13}\rho_b$

There are some special cases where ρ_{34} is larger for the C-vine:

(i) Markov model with partial correlations of zero for levels 2 and above, and positive correlation for level 1: $\alpha_d = \alpha_e = \alpha_f = 0$. In this case, ρ_{34} simplifies to $\rho_b \rho_c$ for the C-vine and $\rho_a \rho_b \rho_c$ for the D-vine.

(ii) Constant correlation or partial correlation for each level, so that there are three parameters: $\alpha_1 = \rho_a = \rho_b = \rho_c$, $\alpha_2 = \alpha_d = \alpha_e$ and $\alpha_3 = \alpha_f$.

Case (i) above extends to higher dimensions. For the Gaussian D-vine in dimension n, with correlations $\rho_1, \ldots, \rho_{n-1}$ at level 1 and zero partial correlations at other levels, the marginal correlation for the highest level tree is $\rho_{1n} = \prod_{j=1}^{n-1} \rho_j$. For the Gaussian C-vine in dimension n, with correlations $\rho_1, \ldots, \rho_{n-1}$ at level 1 and zero partial correlations at other levels, the marginal correlation for the highest level tree is $\rho_{n-1,n} = \rho_{n-2}\rho_{n-1}$, so that this is larger if all of the level 1 correlations are positive.

The generalization of case (ii) is studied in the next section. It has been proved for dimensions $n = 3, 4, 5$ and it has been shown to hold in numerical simulations for $n > 5$.

7.5 Gaussian Vines and Generalized Toeplitz Matrices

Gaussian vines are important special cases that motivated the development of vine copulae. Multivariate Gaussian distributions have the property that conditional copulae do not depend on the values of the conditioning variables; the conditional means change but not the conditional covariance matrices. Multivariate t-distributions also have the property that conditional copulae do not depend on the values of the conditioning variables, but the degree of freedom of the conditional copula increases with the amount of conditioning, and the conditional variances and means depend on the values of the conditioning variables.

We compare n-dimensional Gaussian vines, when the partial correlation is a constant α_ℓ for level ℓ of the vine, for $\ell = 1, \ldots, n-1$. From Kurowicka and Cooke[12] or Theorem 4.5 in Kurowicka and Cooke,[13] these vines all have the same amount of dependence based on the determinant of the resulting correlation matrix:

$$\det(R) = \prod_{\ell=1}^{n-1} \alpha_\ell^{n-\ell}.$$

Consider correlation matrices based on partial correlation α_ℓ at level ℓ: not all equivalence classes of vines lead to *generalized Toeplitz matrices* (with $n-1$ distinct correlations). The corresponding correlations ρ_ℓ can be obtained from the mappings of correlations to partial correlations. For any vine, $\rho_1 = \alpha_1$ and $\rho_2 = \rho_1^2 + \alpha_2(1-\rho_1^2)$. Let ρ_3 and ρ_3' be level 3 correlations for the C-vine and the D-vine respectively.

The equations for $n = 4$ are given below. For the C-vine,

$$\rho_{34;1} = \alpha_2^2 + \alpha_3(1-\alpha_2^2),$$
$$\rho_3 = \rho_{34} = \alpha_1^2 + \rho_{34;1}(1-\alpha_1^2) = \alpha_1^2 + [\alpha_2^2 + \alpha_3(1-\alpha_2^2)](1-\alpha_1^2).$$

For the D-vine,

$$\alpha_3 = \frac{\rho_3' - \frac{\rho_1(2\rho_2-\rho_2^2-\rho_1^2)}{1-\rho_1^2}}{1 - \frac{\rho_1^2+\rho_2^2-2\rho_1^2\rho_2}{1-\rho_1^2}} = \frac{\rho_3'(1-\rho_1^2) - \rho_1(2\rho_2-\rho_2^2-\rho_1^2)}{1-2\rho_1^2-\rho_2^2+2\rho_1^2\rho_2}$$

so that

$$\rho_3' = \frac{\alpha_3(1-2\rho_1^2-\rho_2^2+2\rho_1^2\rho_2) + \rho_1(2\rho_2-\rho_2^2-\rho_1^2)}{1-\rho_1^2}.$$

With algebraic substitutions (using symbolic manipulation software),

$$\rho_3 = [\alpha_1^2 + \alpha_2^2 - \alpha_1^2\alpha_2^2](1-\alpha_3) + \alpha_3,$$
$$\rho_3' = \alpha_1^3(1-\alpha_2)^2 + \alpha_3(1-\alpha_2^2)(1-\alpha_1^2) + \alpha_1(2\alpha_2-\alpha_2^2).$$

The difference is
$$\rho_3 - \rho_3' = (1-\alpha_1)(\alpha_1\alpha_2 + \alpha_2 - \alpha_1)^2 \geq 0.$$
Hence the C-vine has stronger marginal bivariate dependence for the level 3 correlation.

For $n = 5$ and the six vines mentioned in the preceding sections, the level 3 correlation is one of ρ_3, ρ_3', but both can appear in some cases. For the fourth correlations, we use this notation: ρ_4 for C, ρ_4' for D, $\rho_{4B0}, \rho_{4B1}, \rho_{4B2}, \rho_{4B3}$ for B0, B1, B2, B3 respectively. The five-dimensional correlation matrices for the C, B0, B1, B2, B3 and D1 (the 51234 order of the D-vine) vines are given next.

$$\begin{pmatrix} 1 & \rho_1 & \rho_1 & \rho_1 & \rho_1 \\ \rho_1 & 1 & \rho_2 & \rho_2 & \rho_2 \\ \rho_1 & \rho_2 & 1 & \rho_3 & \rho_3 \\ \rho_1 & \rho_2 & \rho_3 & 1 & \rho_4 \\ \rho_1 & \rho_2 & \rho_3 & \rho_4 & 1 \end{pmatrix} \begin{pmatrix} 1 & \rho_1 & \rho_1 & \rho_1 & \rho_1 \\ \rho_1 & 1 & \rho_2 & \rho_2 & \rho_3 \\ \rho_1 & \rho_2 & 1 & \rho_3 & \rho_2 \\ \rho_1 & \rho_2 & \rho_3 & 1 & \rho_{4B0} \\ \rho_1 & \rho_3 & \rho_2 & \rho_{4B0} & 1 \end{pmatrix} \begin{pmatrix} 1 & \rho_1 & \rho_1 & \rho_1 & \rho_2 \\ \rho_1 & 1 & \rho_2 & \rho_2 & \rho_1 \\ \rho_1 & \rho_2 & 1 & \rho_3 & \rho_3' \\ \rho_1 & \rho_2 & \rho_3 & 1 & \rho_{4B1} \\ \rho_2 & \rho_1 & \rho_3' & \rho_{4B1} & 1 \end{pmatrix}$$

$$\begin{pmatrix} 1 & \rho_1 & \rho_2 & \rho_1 & \rho_1 \\ \rho_1 & 1 & \rho_1 & \rho_2 & \rho_2 \\ \rho_2 & \rho_1 & 1 & \rho_3' & \rho_3' \\ \rho_1 & \rho_2 & \rho_3' & 1 & \rho_4 \\ \rho_1 & \rho_2 & \rho_3' & \rho_4 & 1 \end{pmatrix} \begin{pmatrix} 1 & \rho_1 & \rho_1 & \rho_1 & \rho_2 \\ \rho_1 & 1 & \rho_2 & \rho_3 & \rho_1 \\ \rho_1 & \rho_2 & 1 & \rho_2 & \rho_3' \\ \rho_1 & \rho_3 & \rho_2 & 1 & \rho_{4B3} \\ \rho_2 & \rho_1 & \rho_3' & \rho_{4B3} & 1 \end{pmatrix} \begin{pmatrix} 1 & \rho_1 & \rho_2 & \rho_3' & \rho_1 \\ \rho_1 & 1 & \rho_1 & \rho_2 & \rho_2 \\ \rho_2 & \rho_1 & 1 & \rho_1 & \rho_3' \\ \rho_3' & \rho_2 & \rho_1 & 1 & \rho_4' \\ \rho_1 & \rho_2 & \rho_3' & \rho_4' & 1 \end{pmatrix}$$

To determine if a third-order correlation is ρ_3 or ρ_3', the appropriate four-dimension sub-vine must be looked at to determine if it is a C-vine or D-vine. For example, for the B1-vine, (a) by deleting the fifth row and column of the above correlation matrix (equivalently deleting all conditional distributions with 5), a C-vine on variables 1,2,3,4 is obtained; (b) by deleting the fourth row and column of the above correlation matrix (equivalently deleting all conditional distributions with 4), a D-vine on variables 1,2,3,5 with order 3125 is obtained. For the B1-vine and B3-vine, note that a constant α_3 does not imply that there is a constant third-order correlation.

The equation for the fourth-order correlation is obtained as follows. Let $\Sigma = \begin{pmatrix} \Sigma_{11} & \Sigma_{12} \\ \Sigma_{21} & \Sigma_{22} \end{pmatrix}$ be one of the above matrices, where Σ_{11} is 3×3 and Σ_{22} is 2×2. From the form of the conditional covariance matrix of a multivariate normal distribution, let y_{11}, y_{12} be defined from

$$\Sigma_{22} - \Sigma_{21}\Sigma_{11}^{-1}\Sigma_{21} = \begin{pmatrix} 1 & \rho \\ \rho & 1 \end{pmatrix} - \begin{pmatrix} y_{11} & y_{12} \\ y_{12} & y_{11} \end{pmatrix}.$$

Then
$$\rho = y_{12} + \alpha_4(1 - y_{11}), \quad \rho \text{ one of } \rho_4, \rho_{4B0}\rho_{4B1}, \rho_{4B2}, \rho_{4B3}, \rho'_4.$$

With the help of symbolic manipulation software, the following inequalities are obtained.

- $\rho_4 = [\alpha_1^2 + \alpha_2^2 + \alpha_3^2 - \alpha_1^2\alpha_2^2 - \alpha_1^2\alpha_3^2 - \alpha_2^2\alpha_3^2 + \alpha_1^2\alpha_2^2\alpha_3^2](1 - \alpha_4) + \alpha_4,$
- $\rho_{4B2} = \rho_4,$
- $\rho_{4B0} = \rho_4 - (1 - \alpha_1^2)(1 - \alpha_2)(\alpha_2\alpha_3 + \alpha_3 - \alpha_2)^2 \leq \rho_4,$
- $\rho_{4B1} = \rho_4 - (1 - \alpha_1)(\alpha_1\alpha_2 + \alpha_2 - \alpha_1)^2 \leq \rho_4,$
- $\rho'_4 = \rho_4 - (1 - \alpha_1^2)(1 - \alpha_2)(\alpha_2\alpha_3 + \alpha_3 - \alpha_1 + \alpha_1\alpha_2)^2 \leq \rho_4,$
- $\rho_{4B3} = \rho_4 - (1 - \alpha_1)\zeta \leq \rho_4,$

$$\zeta = h_0 + h_1\alpha_2 + h_2\alpha_2^2 + h_3\alpha_2^3,$$
$$h_3 = -h_1 = (1 - \alpha_3)(1 + \alpha_1)(\alpha_1 + \alpha_3),$$
$$h_0 = \alpha_1^2 + \alpha_3^2 - \alpha_1\alpha_3 + \alpha_1\alpha_3^2 - \alpha_1^2\alpha_3,$$
$$h_2 = (1 - \alpha_3)(1 + \alpha_1)(1 + \alpha_3 - \alpha_1).$$

Numerically, it has been shown that $\zeta \geq 0$.

Hence, for these vines on five variables with constant partial correlation at each level ℓ, (a) the C-vine has stronger marginal bivariate dependence for level 3 and 4 correlations; (b) the B2-vine has the same level 3 correlation as the four-dimensional D-vine and the same level 4 correlation as the C-vine (this result generalizes to non-Gaussian vine copulae). In standard form, the D-vine leads to Toeplitz matrices (with correlations $\rho_{ij} = \rho_{|i-j|}$ depending only on distance from main diagonal). For the Gaussian C-vine, B0-vine and B2-vine, there are generalized Toeplitz matrices with four distinct correlation values, but the Gaussian B1-vine and B3-vine do not lead to generalized Toeplitz matrices.

For vines with $n \geq 6$ variables, for $\ell \geq 5$, there is no general simplification of the expressions for ρ_ℓ and ρ'_l in the C-vine and D-vine respectively. However, they can be easily computed with a vector of $\boldsymbol{\alpha} = (\alpha_1, \ldots, \alpha_{n-1}) \in (-1,1)^{n-1}$ of partial correlations. Here, α_l is the partial correlation for all conditional (bivariate Gaussian) distributions at level ℓ. In this case, there are simple algorithms for computing $(\rho_1, \ldots, \rho_{n-1})$, where ρ_ℓ is the (unconditional) correlation for level ℓ; see Table 7.11. For the D-vine, the algorithm is based on Durbin's[6] method. From running the algorithms for many (over 10^6) fixed and random inputs of $\boldsymbol{\alpha}$ for dimensions up to 10, the inequalities $\rho_\ell \geq \rho'_\ell$ held for $\ell \geq 1$.

Table 7.11. Algorithms for generalized Toeplitz matrices for C-vine and D-vine.

C-vine: generalized Toeplitz
Input $\boldsymbol{\alpha} = (\alpha_1, \ldots, \alpha_{n-1}) \in (-1, 1)^{n-1}$ with $n \geq 3$.
$\rho_1 \leftarrow \alpha_1$; $\rho_2 \leftarrow \alpha_1^2 + \alpha_2(1 - \alpha_1^2)$;
for $i = 3, \ldots, n - 1$:
  $T \leftarrow \alpha_{i-1}^2 + \alpha_i(1 - \alpha_{i-1}^2)$
  for $j = i - 2, \ldots, 1$: $T \leftarrow \alpha_j^2 + T(1 - \alpha_j^2)$;
  $\rho_i \leftarrow T$.

D-vine: Toeplitz
Input $\boldsymbol{\alpha} = (\alpha_1, \ldots, \alpha_{n-1}) \in (-1, 1)^{n-1}$ with $n \geq 3$.
$\rho_1' \leftarrow \alpha_1$; $\phi_1 \leftarrow \alpha_1$;
for $i = 2, \ldots, n - 1$:
  $T_i \leftarrow \alpha_i$; for $j = 1, \ldots, i - 1$: $T_j \leftarrow \phi_j - \alpha_i \phi_{i-j}$;
  for $j = 1, \ldots, i$: $\phi_j \leftarrow T_j$;
  $\rho_i' \leftarrow \alpha_i$; for $j = 1, \ldots, i - 1$: $\rho_i' \leftarrow \rho_i' + \phi_j \rho_{i-j}'$;

7.6 More Comparisons of Dependence for Different Vines

We consider the five-dimensional vine copulae where there is a fixed bivariate copula $C^{(\ell|}$ at level ℓ, $\ell = 1, \ldots, n - 1$.

We compare bivariate normal (BVN), Plackett,[18] Frank,[7] MTCJ (Mardia–Takahasi–Cook–Johnson[5,15,20]), bivariate t_2 (bivariate t with $\nu = 2$ degrees of freedom) as linking copulae. The bivariate MTCJ copula is the reflection asymmetric representative; it has lower tail dependence. The other families of bivariate copulae are all reflection symmetric.

The parametrization and Kendall tau values for the bivariate copulae are as follows:

(1) BVN: $\tau = (2/\pi) \arcsin(\rho)$.
(2) Plackett: $C(u, v; \delta) = \frac{1}{2}\eta^{-1}\{1 + \eta(u+v) - [(1 + \eta(u+v))^2 - 4\delta\eta uv]^{1/2}\}$, $0 < \delta < \infty$, where $\eta = \delta - 1$. The parameter δ with τ being an integer multiple of 0.1 is given in Table 5.1 of Joe.[10]
(3) Frank: $C(u, v; \delta) = -\delta^{-1} \log([\eta - (1 - e^{-\delta u})(1 - e^{-\delta v})]/\eta)$, $0 \leq \delta < \infty$. The parameter δ with τ being an integer multiple of 0.1 is given in Table 5.1 of Joe.[10]
(4) Bivariate version of MTCJ copula: $C(u, v; \delta) = (u^{-\delta} + v^{-\delta} - 1)^{-1/\delta}$, where $0 < \delta < \infty$. Kendall's tau satisfies $\tau = \delta/(\delta + 2)$; this is a special case of a result on Archimedean copulae, given in Genest and MacKay.[8]
(5) Bivariate t: $\tau = (2/\pi) \arcsin(\rho)$; see Proposition 5.37 in McNeil et al.[16]

Dependence measures that we compare for different bivariate margins are: Kendall's tau, Spearman's rho or rank correlation, Blomqvist's[4,19] beta and a non-limiting tail dependence parameter. Since the tail dependence parameter can not be (easily) computed for all bivariate margins, we consider a non-limiting version $C_{jk}(u,u)/u$ and $\overline{C}_{jk}(1-u, 1-u)/(1-u)$ (for small u near 0 such as 0.1). We use the notation $\lambda_{jk,L}^{(u)}, \lambda_{jk,U}^{(u)}$ for the lower and upper tails respectively. If $u = \frac{1}{2}$, let $\eta_{jk} = \Pr(U_j \leq \frac{1}{2}, U_k \leq \frac{1}{2}) = C_{jk}(\frac{1}{2},\frac{1}{2})$ and let $\beta_{jk} = 4\eta_{jk} - 1$. This is same as Blomqvist's beta because $\overline{C}(\frac{1}{2},\frac{1}{2}) = C(\frac{1}{2},\frac{1}{2})$ for bivariate copulae C.

In a simulation study, using the algorithms in Section 7.3, we used four different parameters $\tau_1, \tau_2, \tau_3, \tau_4$ for the five-dimensional vines. This means level 1 linking copulae have a tau value of τ_1 whatever copula family is used, level ℓ conditional linking copulae have a tau value of τ_ℓ for $\ell = 2, 3, 4$.

We tried several combinations of $(\tau_1, \tau_2, \tau_3, \tau_4)$. Tables 7.12 and 7.13 summarize some representative results, with $\tau_1 = \tau_2 = \tau_3 = \tau_4 = 0.5$ and $\tau_1 = 0.1, \tau_2 = 0.3, \tau_3 = 0.5, \tau_4 = 0.7$ respectively. In Tables 7.12 and 7.13, the column with a heading of 1 has the dependence measure for level 1, so it is the transformation of τ_1 to another dependence measure for the five bivariate copula families. The column with a heading of 2 has the marginal dependence measure for level 2; this is the same for all five-dimensional vines. The column with a heading of 3 has the marginal dependence measure for level 3; there are two possible values (B0 has the same value as the C-vine, B2 has the same value as the D-vine, and the values in columns 3C, 3D can occur for different pairs at level 3 for B1 and B3). The column with a heading of 4 has the marginal dependence measure for level 4; these are different for the different vines, except the dependence value for the B2-vine which is the same as for the C-vine (hence there is no column 4B2). These results are like those in the preceding section on Gaussian vines.

Smaller differences in dependence measures occur at the third and fourth levels if $\tau_1 > \tau_2 \geq \tau_3 \geq \tau_4$, e.g., 0.7, 0.5, 0.3, 0.1 — this case is not included in the tables. For the bivariate normal copulae, the results are based on inverting τs to partial correlation and then applying the inversion to correlations ρ (Section 7.5), from which the dependence measures are computed as a function of ρ. For the Plackett, Frank, MTCJ and bivariate t linking copulae, the results are based on 10^8 replications (using the C programming language).

Tables 7.12 and 7.13 suggest an ordering of non-limiting tail dependence for the three reflection symmetric copula families with no limiting tail dependence: the Plackett copula has less non-limiting tail dependence

Table 7.12. Bivariate marginal dependence measures for $\tau_1 = \tau_2 = \tau_3 = \tau_4 = 0.5$; 10^8 replications.

Copula	1	2	3C	3D	4C	4B0	4B1	4B3	4D
					τ				
BVN	0.500	0.651	0.755	0.651	0.827	0.755	0.699	0.609	0.609
Plackett	0.500	0.628	0.720	0.630	0.788	0.722	0.677	0.630	0.604
Frank	0.500	0.630	0.725	0.625	0.793	0.720	0.671	0.596	0.598
MTCJ	0.500	0.653	0.749	0.634	0.813	0.741	0.673	0.587	0.594
BVT	0.500	0.656	0.764	0.656	0.838	0.764	0.704	0.614	0.613
					ρ_S				
BVN	0.690	0.842	0.920	0.842	0.960	0.920	0.881	0.804	0.804
Plackett	0.680	0.809	0.884	0.817	0.929	0.889	0.861	0.796	0.794
Frank	0.696	0.823	0.897	0.822	0.939	0.896	0.863	0.798	0.780
MTCJ	0.683	0.832	0.903	0.815	0.940	0.899	0.848	0.771	0.778
BVT	0.659	0.815	0.900	0.815	0.946	0.899	0.857	0.777	0.773
					β				
BVN	0.500	0.651	0.755	0.651	0.827	0.755	0.699	0.609	0.609
Plackett	0.543	0.665	0.753	0.659	0.816	0.748	0.699	0.630	0.635
Frank	0.555	0.663	0.754	0.662	0.818	0.745	0.701	0.636	0.636
MTCJ	0.512	0.683	0.785	0.653	0.848	0.770	0.686	0.595	0.612
BVT	0.500	0.658	0.767	0.660	0.841	0.767	0.709	0.619	0.618
					$\lambda_L^{(0.1)}$				
BVN	0.474	0.624	0.733	0.624	0.811	0.733	0.674	0.581	0.581
Plackett	0.433	0.578	0.681	0.569	0.759	0.677	0.616	0.534	0.539
Frank	0.370	0.548	0.662	0.521	0.746	0.653	0.569	0.474	0.489
MTCJ	0.708	0.824	0.883	0.786	0.919	0.872	0.801	0.752	0.775
BVT	0.564	0.705	0.800	0.700	0.864	0.798	0.741	0.663	0.666

dependence: $4C = 4B2 \succ 4B0 \succ 4B1 \succ \{4B3, 4D\}$
λ: MTCJ $\succ$ BVT $\succ$ BVN $\succ$ Plackett $\succ$ Frank

than bivariate normal copula and more non-limiting tail dependence than the Frank copula (see bottom of the tables). The tail behavior follows from the more general bivariate tail analysis in Ledford and Tawn[14] and Heffernan;[9] for the Plackett and Frank copulae, $C(u, u; \delta) = O(u^2)$ as $u \to 0$ (for any dependence parameter) and for the bivariate normal copula, $C(u, u; \rho) = O(u^{2/(1+\rho)}(-\log u)^{-\rho/(1+\rho)})$ as $u \to 0$ for $-1 < \rho < 1$. Also the relative strength of marginal bivariate dependence of the different five-dimensional vines depends on the dependence in the conditional linking copulae.

Table 7.13. Bivariate marginal dependence measures for $\tau_1 = 0.1$, $\tau_2 = 0.3$, $\tau_3 = 0.5$, $\tau_4 = 0.7$; 10^8 replications.

Copula	1	2	3C	3D	4C	4B0	4B1	4B3	4D
				τ					
BVN	0.100	0.310	0.563	0.458	0.814	0.572	0.639	0.374	0.322
Plackett	0.100	0.309	0.552	0.448	0.791	0.565	0.620	0.377	0.326
Frank	0.100	0.309	0.549	0.448	0.782	0.555	0.617	0.368	0.320
MTCJ	0.100	0.311	0.561	0.435	0.795	0.540	0.596	0.334	0.293
BVT	0.100	0.310	0.566	0.457	0.819	0.579	0.639	0.381	0.324
				ρ_S					
BVN	0.150	0.450	0.758	0.641	0.954	0.767	0.831	0.537	0.467
Plackett	0.150	0.446	0.736	0.625	0.929	0.763	0.812	0.543	0.476
Frank	0.150	0.450	0.744	0.633	0.932	0.757	0.814	0.535	0.470
MTCJ	0.150	0.448	0.745	0.607	0.929	0.721	0.783	0.476	0.421
BVT	0.140	0.418	0.714	0.598	0.932	0.731	0.797	0.511	0.438
				β					
BVN	0.100	0.310	0.563	0.458	0.814	0.572	0.639	0.374	0.322
Plackett	0.112	0.342	0.592	0.467	0.821	0.573	0.619	0.368	0.317
Frank	0.113	0.344	0.591	0.470	0.812	0.558	0.615	0.355	0.306
MTCJ	0.097	0.311	0.585	0.469	0.832	0.552	0.636	0.341	0.294
BVT	0.100	0.310	0.567	0.460	0.821	0.583	0.643	0.386	0.328
				$\lambda_L^{(0.1)}$					
BVN	0.154	0.304	0.535	0.434	0.797	0.544	0.611	0.359	0.315
Plackett	0.142	0.270	0.492	0.405	0.758	0.508	0.595	0.355	0.311
Frank	0.140	0.249	0.449	0.373	0.723	0.487	0.582	0.338	0.293
MTCJ	0.219	0.494	0.749	0.504	0.908	0.618	0.553	0.331	0.362
BVT	0.276	0.445	0.653	0.551	0.856	0.653	0.688	0.480	0.446

dependence: $4C = 4B2 \succ \{4B1, 4B0\} \succ \{4B3, 4D\}$

λ: MTCJ,BVT $\succ$ BVN $\succ$ Plackett $\succ$ Frank

7.7 Discussion and Further Research

We started an investigation of the intermediate vines between the C-vine and D-vine in dimensions $n \geq 5$. Because there are only four such vines (called B0, B1, B2, B3) in dimension $n = 5$, we study them in detail, including simulation algorithms, Gaussian vines, generalized Toeplitz matrices and non-Gaussian vines with a common copula for each level of the vine. We obtained some general results from comparing the marginal dependence of C-vines and D-vines when the same set of conditional copulae is used for the two vines. Some of these results extend to the intermediate vines. Under some conditions, the bivariate marginal dependence from the C-vine is the highest of all vines. Some results in Sections 7.5 and 7.6 show that the B2-vine is halfway between the C-vine and D-vine in dimension 5. Understanding this type of comparison should be helpful in the use of vine copulae for modeling multivariate data.

To decide on copula models for multivariate data, a first step in initial data analysis includes measuring bivariate association and assessing the strength of bivariate tail dependence. However, more practical experience is needed for the next step if vine copulae are to be used, that is, the choice of the appropriate vines and the permutation of the variables to indices. It would be useful if there were some heuristics so that modeling with vine copulae can proceed without trying all vines and all permutations. Because of the dependence properties of vine copulae in achieving a wide range of dependence, maybe several different vines can provide equally good fits for multivariate data.

Another further research direction is to figure out in an automated way which backward steps (in Table 7.2) are needed or not needed in the implementation of the algorithm for simulating from an arbitrary regular vine.

Acknowledgments

This research was supported by an NSERC Canada Discovery Grant.

References

1. Aas K., Czado C., Frigessi A. and Bakken H. (2009). Pair-copula constructions of multiple dependence. *Insurance: Mathematics and Economics*, 44:182–198.
2. Bedford T. and Cooke R.M. (2001). Probability density decomposition for conditionally dependent random variables modeled by vines. *Annals of Mathematics and Artificial Intelligence*, 32:245–268.
3. Bedford T. and Cooke R.M. (2002). Vines: A new graphical model for dependent random variables. *Annals of Statistics*, 30:1031–1068.
4. Blomqvist N. (1950). On a measure of dependence between two random variables. *Annals of Mathematical Statistics*, 21:593–600.
5. Cook R.D. and Johnson M.E. (1981). A family of distributions for modelling non-elliptically symmetric multivariate data. *Journal of the Royal Statistical Society: Series B*, 43:210–218.
6. Durbin J. (1960). The fitting of time series models. *Review of the International Statistical Institute*, 28:233–244.

7. Frank M.J. (1979). On the simultaneous associativity of $F(x,y)$ and $x+y-F(x,y)$. *Aequationes Mathematicae*, 19:194–226.
8. Genest C. and MacKay R.J. (1986). Copules archimédiennes et familles de lois bidimensionelles dont les marges sont données. *Canadian Journal of Statistics*, 14: 145–159.
9. Heffernan J.E. (2000). A directory of coefficients of tail dependence. *Extremes*, 3: 279–290.
10. Joe H. (1997). *Multivariate Models and Dependence Concepts*. Chapman & Hall, London.
11. Joe H., Li H. and Nikoloulopoulos A.K. (2010). Tail dependence functions and vine copulas. *Journal of Multivariate Analysis*, 101:252–270.
12. Kurowicka D. and Cooke R. (2006). Completion problem with partial correlation vines. *Linear Algebra and Its Applications*, 418:188–200.
13. Kurowicka D. and Cooke R. (2006). *Uncertainty Analysis with High Dimensional Dependence Modelling*, Wiley, Chichester.
14. Ledford A.W. and Tawn J.A. (1996). Statistics for near independence in multivariate extreme values. *Biometrika*, 83:169–197.
15. Mardia K.V. (1962). Multivariate Pareto distributions. *Annals of Mathematical Statistics*, 33:1008–1015.
16. McNeil A.J., Frey R. and Embrechts P. (2005). *Quantitative Risk Management: Concepts, Techniques, and Tools*. Princeton University Press, Princeton, New Jersey.
17. Morales-Nápoles O., Cooke R.M. and Kurowicka D. (2009). The number of vines and regular vines on n nodes. Submitted to *Discrete Applied Mathematics*.
18. Plackett R.L. (1965). A class of bivariate distributions. *Journal of the American Statistical Association*, 60:516–522.
19. Schmid F. and Schmidt R. (2007). Nonparametric inference on multivariate versions of Blomqvist's beta and related measures of tail dependence. *Metrika*, 66:323–354.
20. Takahasi K. (1965). Note on the multivariate Burr's distribution. *Annals of the Institute of Statistical Mathematics*, 17:257–260.

CHAPTER 8

Tail Dependence in Vine Copulae

Harry Joe

Department of Statistics
University of British Columbia
Vancover BC, Canada, V6T 1Z2

Definitions and properties of conditional and multivariate tail dependence functions are given and applied to vine copulae. We show that vine copulae can have flexible dependence with asymmetry in the joint upper and lower tails, by using appropriate choices of bivariate linking copulae that are reflection asymmetric and have upper/lower tail dependence parameters λ_L, λ_U that independently take values in $(0, 1)$.

8.1	Introduction	165
8.2	Tail Dependence in Different Multivariate Copula Families .	166
8.3	Tail Dependence Parameters and Functions	167
	8.3.1 Bivariate tail dependence	168
	8.3.2 Multivariate tail dependence functions	173
	8.3.3 Conditional tail dependence functions	175
8.4	Main Theorem on Tail Dependence for Vine Copulae	176
8.5	Reflection Asymmetry of Vine Copulae	179
8.6	Choice of Tail Asymmetric Bivariate Linking Copulae . . .	180
8.7	Discussion	184
Appendix		185
References		186

8.1 Introduction

The aim of this chapter is to summarize the definitions and properties of tail dependence functions, and to show how they are applied to vine copulae and other models that involve specified conditional distributions. We show that vine copulae can result in models with flexible dependence including joint upper tail and lower tail asymmetry.

In Aas et al.[2] and Aas and Berg,[1] it is shown that vine copulae constructed from bivariate t-copulae can be better fits to multivariate financial asset return and other data. In Section 2.1.3 of Jondeau et al.,[14] it is mentioned that extreme negative returns in financial assets are more frequent than extreme positive returns, and that correlation or dependence between asset returns tends to increase in high-volatility periods such as crashes. This means that we might expect to have stronger tail dependence in the joint lower tail than upper tail. Vine copulae are a means to empirically assess whether this property holds. More generally, vine copulae are a way to obtain copula models with reflection asymmetry. Although multivariate normal and t-copulae have flexible dependence, they have the property of reflection symmetry.

In Section 8.2, we compare the tail dependence properties of several parametric families of multivariate copulae and point out that vine copula families have much more flexibility for types of tail dependence. In order to investigate tail dependence of vine copulae, we introduce multivariate and conditional tail dependence functions in Section 8.3. The relevant definitions and properties are summarized from Joe et al.[13] The main theorem is given in Section 8.4, with examples of negative tail dependence to show its application. In Section 8.5, we prove some results on reflection symmetry of vine copulae, and give the motivation for the reflection asymmetric bivariate copulae in Section 8.6. Section 8.7 discusses the findings.

To avoid technicalities, limits and derivatives where written are assumed to exist. Generally some conditions involving regular variation are needed to guarantee the existence of the limits used in this chapter.

8.2 Tail Dependence in Different Multivariate Copula Families

In this section, we compare parametric families of multivariate copulae for range of flexible tail dependence and other desirable properties.

(1) Multivariate normal: general dependence but no tail dependence, no closed-form cumulative distribution function (cdf).

(2) Multivariate t: general dependence, no closed-form cdf; flexible tail dependence,[17] but has lower and upper tail dependence parameters $\lambda_{jk,L} = \lambda_{jk,U}$ for all $j \neq k$ because of reflection symmetry.
(3) Mixture of max-infinite divisible (max-id): as proposed in Joe and Hu,[11] this copula family is built from a Laplace transform together with $n(n-1)/2$ bivariate copulae that are each max-id; flexible dependence but not as wide a range as multivariate normal; closed-form cdf; using the Laplace transform family LTI (Joe,[12] p. 376), one can get a multivariate copula with flexible upper and constant lower tail dependence; to get flexible lower tail dependence, one could average with the copula of the reflection, that is, $(C + \widehat{C})/2$, where $\widehat{C}$ is given in (8.1) in the next section.
(4) Vine copula: built from $n(n-1)/2$ bivariate copulae; the multivariate copula has closed-form density (Bedford and Cooke,[3] Aas et al.[2]) but not cdf; general dependence; flexible upper and lower tail dependence if each copula at level 1 of the vine has arbitrarily different upper and lower tail dependence (Joe et al.[13] and Section 8.4 of this chapter).

Having copula families with flexible tail dependence is important for making inferences on joint tail probabilities (which might be representing joint risks). The multivariate normal distribution/copula does not have tail dependence, so using it for modeling will lead to underestimates of joint risks. The multivariate t_ν-copula has tail dependence and reflection symmetry — hence it is not a good choice if one wants a model where the upper and lower tails behave differently for each bivariate margin. Vine copulae have more flexibility, and we think of them mainly as approximations to arbitrary copulae. This is possible because they can achieve a wide range of dependence of different types. Additional details of vines as models based on a set of conditional distributions are given in Bedford and Cooke[3,4] and Kurowicka and Cooke.[15]

8.3 Tail Dependence Parameters and Functions

In this section, we start with definitions of bivariate tail dependence parameters and then go to tail dependence functions. Tail dependence functions are useful as a way to get expressions for bivariate tail dependence parameters for vine copulae or other copula families built from conditional distributions.

8.3.1 *Bivariate tail dependence*

If a bivariate copula C is such that

$$\lim_{u \to 0} C(u,u)/u = \lambda_L$$

exists, C has *lower tail dependence* if $\lambda_L \in (0,1]$ and no lower tail dependence if $\lambda_L = 0$. Similarly, if

$$\lim_{u \to 1} \overline{C}(u,u)/(1-u) = \lambda_U$$

exists, then C has *upper tail dependence* if $\lambda_U \in (0,1]$ and no upper tail dependence if $\lambda_U = 0$.

For two continuous random variables X_1, X_2 with respective cdfs F_1, F_2,

$$\lambda_L = \lim_{u \to 0} \Pr(X_2 \le F_2^{-1}(u) \mid X_1 \le F_1^{-1}(u))$$

$$= \lim_{u \to 0} \Pr(X_1 \le F_1^{-1}(u) \mid X_2 \le F_2^{-1}(u))$$

and

$$\lambda_U = \lim_{u \to 1} \Pr(X_2 > F_2^{-1}(u) \mid X_1 > F_1^{-1}(u))$$

$$= \lim_{u \to 1} \Pr(X_1 > F_1^{-1}(u) \mid X_2 > F_2^{-1}(u)).$$

With $n > 2$ variables and $\binom{n}{2}$ lower and upper bivariate tail dependence values, extra subscripts on λ_L, λ_U will be used, for example, $\lambda_{jk,L}, \lambda_{jk,U}$.

In Aas et al.,[2] some plots of bond returns versus stock market index returns show negative tail dependence (extremes in opposite directions). This is the motivation for the following additional definitions.

We will use the notation $\lambda_{NW}, \lambda_{NE} = \lambda_U, \lambda_{SE}, \lambda_{SW} = \lambda_L$ so that λ_{NW} and λ_{SE} can be used for bivariate negative tail dependence. For a bivariate copula C with $(U_1, U_2) \sim C$, if

$$\lim_{u \to 0} \Pr(1 - U_2 \le u \mid U_1 \le u) = \lim_{u \to 0} [u - C(u, 1-u)]/u = \lambda_{NW}$$

exists and is positive, then there is negative *lower-upper tail dependence*; and if

$$\lim_{u \to 0} \Pr(U_2 \le u \mid 1 - U_1 \le u) = \lim_{u \to 0} [u - C(1-u, u)]/u = \lambda_{SE}$$

exists and is positive, then there is negative *upper-lower tail dependence*.

The bivariate t_ν-copula, with parameters $\nu > 0$ and $-1 < \rho < 1$, is interesting in that all four tail dependence coefficients are non-zero:

$$\lambda_{NE} = \lambda_{SW} = 2T_{\nu+1}\left(-\sqrt{\frac{(\nu+1)(1-\rho)}{(1+\rho)}}\right),$$

$$\lambda_{NW} = \lambda_{SE} = 2T_{\nu+1}\left(-\sqrt{\frac{(\nu+1)(1+\rho)}{(1-\rho)}}\right),$$

where T_η denotes the t-cdf with η degrees of freedom. The usual tail dependence parameter for the t-copula was derived in Embrechts et al.,[8] and the negative tail dependence follows from the following property of the bivariate t_ν-distribution: (X_1, X_2) bivariate t_ν with correlation parameter ρ implies that $(X_1, -X_2)$ bivariate t_ν with correlation parameter $-\rho$.

In order to convert results on positive upper tail dependence to positive lower tail dependence or negative tail dependence, we give some results on the copulae of a vine after reflections of one or more variables on the $U(0, 1)$ random variable space. Some notation is needed to represent these results.

We indicate what happens to a vine copula when some uniform random variables are reflected: variable j is *reflected* if $U_j \to 1 - U_j$ or $X_j \to X_j^* = g_j(X_j)$) where $X_j \sim F_j$ with F_j continuous and g_j is a continuous strictly decreasing real-valued function. In this case, let F_j^* be the distribution of X_j^*.

For $(U_1, \ldots, U_n)$ with copula C, we use the notation $\widehat{C}$ for the copula of $(1 - U_1, \ldots, 1 - U_n)$. For $n = 2$,

$$\widehat{C}(u_1, u_2) = u_1 + u_2 - 1 + C(1 - u_1, 1 - u_2),$$

and for $n > 2$,

$$\widehat{C}(u_1, \ldots, u_n) = 1 - \sum_{j=1}^{n}(1 - u_j)$$
$$+ \sum_{S:S \subset \{1,\ldots,n\}, |S| \geq 2} (-1)^{|S|} C_S(1 - u_j : j \in S). \quad (8.1)$$

For $n = 2$, let $\acute{C}$ be the copula of $(1 - U_1, U_2)$ and $\grave{C}$ be the copula of $(U_1, 1 - U_2)$. The copulae are

$$\acute{C}(u_1, u_2) = u_2 - C(1 - u_1, u_2), \quad \grave{C}(u_1, u_2) = u_1 - C(u_1, 1 - u_2) \quad (8.2)$$

If bivariate negative tail dependence exist, a parametric family

$$u_1 - C(u_1, 1 - u_2; \boldsymbol{\theta}) \quad \text{or} \quad u_2 - C(1 - u_1, u_2; \boldsymbol{\theta})$$

with negative dependence can be considered by converting a family $C(\cdot; \boldsymbol{\theta})$ with positive dependence.

Example 8.1. We next show what happens to the vine representation when some of the variables are reflected. Consider a C-vine on three variables summarized through the distributions $\{F_1, F_2, F_3, C_{12}, C_{13}, C_{23|1}\}$.

(a) For the mapping $(U_1, U_2, U_3) \to (1 - U_1, U_2, U_3)$ with $X_1 \to X_1^* \sim F_1^*$, the C-vine becomes $\{F_1^*, F_2, F_3, \acute{C}_{12}, \acute{C}_{13}, C_{23|1}\}$. If (U_1, U_2) and (U_1, U_3) have lower-upper tail dependence, and (U_2, U_3) has upper tail dependence, then the new vine has bivariate upper tail dependence for all pairs.

(b) For the mapping $(U_1, U_2, U_3) \to (U_1, 1 - U_2, U_3)$ with $X_2 \to X_2^* \sim F_2^*$, the C-vine becomes $\{F_1, F_2^*, F_3, \grave{C}_{12}, C_{13}, \acute{C}_{23|1}\}$. If (U_1, U_2) and (U_3, U_2) have upper-lower tail dependence, and (U_1, U_3) has upper tail dependence, then the new vine has bivariate upper tail dependence for all pairs.

(c) For the mapping $(U_1, U_2, U_3) \to (1 - U_1, 1 - U_2, 1 - U_3)$ with $X_j \to X_j^* \sim F_j^*$, the C-vine becomes $\{F_1^*, F_2^*, F_3^*, \widehat{C}_{12}, \widehat{C}_{13}, \widehat{C}_{23|1}\}$. If the original vine has bivariate upper tail dependence for all pairs, then the new vine has bivariate lower tail dependence for all pairs.

We next introduce some conditional tail dependence functions and show how tail dependence parameters can be obtained from them. Extensions of these functions are given in the subsequent subsections.

Let $\{C_{1|2}(\cdot|u_2)\}$ and $\{C_{2|1}(\cdot|u_1)\}$ be the set of conditional distributions of the differentiable bivariate copula C. Let (j_1, j_2) be equal to $(1, 2)$ or $(2, 1)$. If C has lower tail dependence, then (assuming the limits exist)

$$t_{j_1|j_2}(w_{j_1}|w_{j_2}) = \lim_{u \to 0} C_{j_1|j_2}(uw_{j_1}|uw_{j_2}), \quad \text{for } w_1 > 0, \ w_2 > 0$$

is not the zero function. If C has upper tail dependence, then

$$t_{j_1|j_2}^*(w_{j_1}|w_{j_2}) = \lim_{u \to 0} \overline{C}_{j_1|j_2}(1 - uw_{j_1}|1 - uw_{j_2}), \quad \text{for } w_1 > 0, \ w_2 > 0,$$

is not the zero function. Note that $t_{j_1|j_2}(1|w) = t_{j_1|j_2}(w^{-1}|1)$ is decreasing in w and $t_{j_1|j_2}^*(1|w) = t_{j_1|j_2}^*(w^{-1}|1)$ is decreasing in w.

For the lower tail, with a substitution of $v = uw$ in the integral,

$$\frac{C(u,u)}{u} = u^{-1}\int_0^u C_{1|2}(u|v)\, dv = \int_0^1 C_{1|2}(u|uw)\, dw = \int_0^1 C_{2|1}(u|uw)\, dw$$

$$\to \int_0^1 t_{1|2}(1|w)\, dw = \int_0^1 t_{2|1}(1|w)\, dw, \quad \text{as } u \to 0.$$

Similarly for the upper tail,

$$\frac{\overline{C}(1-u,1-u)}{u} = u^{-1}\int_0^u \overline{C}_{1|2}(1-u|1-v)\,dv = \int_0^1 \overline{C}_{1|2}(1-u|1-uw)\,dw$$

$$\to \int_0^1 t^*_{1|2}(1|w)\,dw = \int_0^1 t^*_{2|1}(1|w)\,dw, \quad \text{as } u \to 0.$$

To conclude this subsection, we show how the above conditional tail dependence functions can be used to obtain the bivariate tail dependence parameters of an n-variate copula constructed from n different bivariate copulae through conditioning. The result of this construction is called a *one-factor copula*.

Consider

$$C(u_1,\ldots,u_n) = \int_0^1 \prod_{j=1}^n C_{j|0}(u_j|v)\,dv, \tag{8.3}$$

where $C_{j0}(u_j,v)$ are bivariate copulae and $C_{j|0}(u_j|v) = \partial C_{j0}(u_j,v)/\partial v$ are conditional distributions. This is the same as a C-vine on variables $X_0, X_1, \ldots, X_n$, with edges $0j$ ($j = 1,\ldots,n$) for the first tree, conditional independence at levels 2 and higher, and X_0 unobserved. A bivariate margin has the form

$$C_{jk}(u_j,u_k) = \int_0^1 C_{j|0}(u_j|v)\,C_{k|0}(u_k|v)\,dv, \quad j \neq k.$$

Suppose C_{j0} has lower tail dependence with conditional tail dependence function $t_{j|0}$, for $j = 1,\ldots,n$. Then, as $u \to 0$,

$$C_{jk}(u,u) = \int_0^1 C_{j|0}(u|v)\,C_{k|0}(u|v)\,dv = u\int_0^{1/u} C_{j|0}(u|uw_0)\,C_{k|0}(u|uw_0)\,dw_0$$

$$\sim u\int_0^{1/w_0} t_{j|0}(1|w_0)\,t_{k|0}(1|w_0)\,dw_0.$$

Hence, by Fatou's lemma,

$$\lim_{u \to 0} \frac{C_{jk}(u,u)}{u} \geq \int_0^\infty t_{j|0}(1|w_0)\,t_{k|0}(1|w_0)\,dw_0,$$

and an inequality for the (j,k) bivariate lower tail dependence parameter is

$$\lambda_{jk,L} \geq \int_0^\infty t_{j|0}(1|w_0)\,t_{k|0}(1|w_0)\,dw_0. \tag{8.4}$$

Similarly if C_{j0} has upper conditional tail dependence function $t^*_{j|0}$ for $j = 1, \ldots, n$, then $\overline{C}_{jk}(u_j, u_k) = \int_0^1 \overline{C}_{j|0}(u_j|v) \overline{C}_{k|0}(u_k|v)\, dv$, and as $u \to 0$,

$$\overline{C}_{jk}(1-u, 1-u) = \int_0^1 \overline{C}_{j|0}(1-u|1-v) \overline{C}_{k|0}(1-u|1-v)\, dv$$

$$= u \int_0^{1/u} \overline{C}_{j|0}(1-u|1-uw_0) \overline{C}_{k|0}(1-u|1-uw_0)\, dw_0$$

$$\sim u \int_0^{1/u} t^*_{j|0}(1|w_0)\, t^*_{k|0}(1|w_0)\, dw_0.$$

Therefore by Fatou's lemma, an inequality for the (j, k) bivariate upper tail dependence parameter is

$$\lambda_{jk,U} \geq \int_0^\infty t^*_{j|0}(1|w_0)\, t^*_{k|0}(1|w_0)\, dw_0. \tag{8.5}$$

These results show that there is upper (lower) tail dependence for all bivariate pairs only if there is upper (lower) tail dependence for all C_{j0} ($j = 1, \ldots, n$).

The above example can be used to show some technicalities that come up in studying tail dependence of vine copulae. The inequalities in (8.4) and (8.5) are equalities in the case where the tail dependence functions $t_{j|0}$ and $t^*_{j|0}$ are distribution functions, and strict inequalities when $t_{j|0}(\cdot|w_0)$ and $t^*_{j|0}(\cdot|w_0)$ are subdistribution functions. Note that $t_{j|0}(w|1) = \lim_{u \to 0} C_{j|0}(uw|u)$ is a limit of distribution functions but can be a subdistribution function (or defective distribution function) if $\lim_{w \to \infty} t_{j|0}(w|1) < 1$.

An intuitive explanation of the inequality is as follows. Instead of conditional independence in (8.3), let us use the conditional Fréchet upper bound. If $C_{j0} = C_{k0}$ and hence $C_{j|0} = C_{k|0}$, then

$$C_{jk}(u_j, u_k) = \int_0^1 \min\{C_{j|0}(u_j|v), C_{k|0}(u_k|v)\}\, dv$$

$$= \int_0^1 C_{j|0}(u_j \wedge u_k|v)\, dv = u_j \wedge u_k$$

is an unconditional Fréchet upper bound copula, and hence has a (lower) tail dependence parameter equal to 1. Suppose C_{j0} has tail dependence and conditional lower tail dependence function $t_{j|0}$. Then

$$\frac{C_{jk}(u,u)}{u} = u^{-1} \int_0^1 C_{j|0}(u|v)\, dv = \int_0^{1/u} C_{j|0}(u|uw)\, dw \to 1,$$

and $C_{j|0}(u|uw) \approx t_{j|0}(1|w)$. But

$$\int_0^\infty t_{j|0}(1|w)\,dw = t_{0|j}(\infty|1); \tag{8.6}$$

this is 1 only if $t_{0|j}(\cdot|1)$ is a distribution function. If C_{j0} is exchangeable, then $t_{j|0} = t_{0|j}$. The proof of the identity in (8.6) is given in the Appendix.

8.3.2 Multivariate tail dependence functions

For tail dependence analysis of vine copulae, we introduce multivariate tail dependence functions. This makes the derivation and analysis of tail dependence easier than in Joe,[10] which had some extreme-value results for D-vines.
Let

$$b(\boldsymbol{w}) = \lim_{u \to 0} C(u\boldsymbol{w})/u, \quad \boldsymbol{w} \in \mathbb{R}_+^n. \tag{8.7}$$

We say that a copula C has *multivariate lower tail dependence* if b is non-zero, and that it has no lower tail dependence if $b \equiv 0$.

Assuming some regular variation conditions on the lower tail of C, then b has the same order of derivatives as C. Assuming that the limit operation and differentiation are commutative, and that C is differentiable to the nth order, then as $u \to 0$, $\to u \to 0$, with $u_i = uw_i$,

$$\frac{\partial C}{\partial u_1}(u\boldsymbol{w}) = \frac{\partial}{\partial w_1}C(u\boldsymbol{w}) \cdot \frac{\partial w_1}{\partial u_1} \sim u\frac{\partial b(\boldsymbol{w})}{\partial w_1} \cdot u^{-1} = \frac{\partial b(\boldsymbol{w})}{\partial w_1},$$

$$\frac{\partial^2 C}{\partial u_1 \partial u_2}(u\boldsymbol{w}) = \frac{\partial^2}{\partial w_1 \partial w_2}C(u\boldsymbol{w}) \cdot \frac{\partial w_1}{\partial u_1}\frac{\partial w_2}{\partial u_2} \sim u\frac{\partial^2 b(\boldsymbol{w})}{\partial w_1 \partial w_2} \cdot u^{-2}$$

$$= u^{-1}\frac{\partial^2 b(\boldsymbol{w})}{\partial w_1 \partial w_2},$$

$$\vdots$$

$$\frac{\partial^n C}{\partial u_1 \cdots \partial u_n}(u\boldsymbol{w}) \sim u^{-(n-1)}\frac{\partial^n b(\boldsymbol{w})}{\partial w_1 \cdots \partial w_n}. \tag{8.8}$$

Using the reflection (8.1), we can get analogous functions for the upper tail. Let

$$b^*(\boldsymbol{w}) = \lim_{u \to 0} \widehat{C}(u\boldsymbol{w})/u, \quad \boldsymbol{w} \in \mathbb{R}_+^n. \tag{8.9}$$

We say that C has *multivariate upper tail dependence* if b^* is non-zero, and that it has no upper tail dependence if $b^* \equiv 0$. Relations like the above for the derivatives of b^* hold if the upper tail of C satisfies regular variation properties.

Example 8.2. We will illustrate the various tail dependence functions using the trivariate MTCJ[7,16,22] copula. The copula is $C(u_1, u_2, u_3; \delta) = (u_1^{-\delta} + u_2^{-\delta} + u_3^{-\delta} - 2)^{-1/\delta}$ for $\delta > 0$. It is straightforward to show that as $u \to 0$, $\to u \to 0$, with $u_i = uw_i$,

$$C(uw_1, uw_2, uw_3) \sim ub(w_1, w_2, w_3), \quad b(w_1, w_2, w_3) = (w_1^{-\delta} + w_2^{-\delta} + w_3^{-\delta})^{-1/\delta}.$$

Some partial derivatives are:

$$\frac{\partial C}{\partial u_1} = (u_1^{-\delta} + u_2^{-\delta} + u_3^{-\delta} - 2)^{-1/\delta - 1} u_1^{-\delta - 1},$$

$$\frac{\partial^2 C}{\partial u_1 \partial u_2} = (1 + \delta)(u_1^{-\delta} + u_2^{-\delta} + u_3^{-\delta} - 2)^{-1/\delta - 2} u_1^{-\delta - 1} u_2^{-\delta - 1},$$

$$\frac{\partial^3 C}{\partial u_1 \partial u_2 \partial u_3} = (1 + \delta)(1 + 2\delta)(u_1^{-\delta} + u_2^{-\delta} + u_3^{-\delta} - 2)^{-1/\delta - 3}$$
$$\times u_1^{-\delta - 1} u_2^{-\delta - 1} u_3^{-\delta - 1}.$$

With $\boldsymbol{w} = (w_1, w_2, w_3)$, it is easy to check that as $u \to 0$,

$$\frac{\partial C(u\boldsymbol{w})}{\partial u_1} \sim (w_1^{-\delta} + w_2^{-\delta} + w_3^{-\delta})^{-1/\delta - 1} w_1^{-\delta - 1} = \frac{\partial b}{\partial w_1},$$

$$\frac{\partial^2 C(u\boldsymbol{w})}{\partial u_1 \partial u_2} \sim u^{-1}(1 + \delta)(w_1^{-\delta} + w_2^{-\delta} + w_3^{-\delta})^{-1/\delta - 2} (w_1 w_2)^{-\delta - 1}$$

$$\sim u^{-1} \frac{\partial^2 b}{\partial w_1 \partial w_2},$$

$$\frac{\partial^3 C(u\boldsymbol{w})}{\partial u_1 \partial u_2 \partial u_3} \sim u^{-2}(1 + \delta)(1 + 2\delta)(w_1^{-\delta} + w_2^{-\delta} + w_3^{-\delta})^{-1/\delta - 3} (w_1 w_2 w_3)^{-\delta - 1}$$

$$\sim u^{-2} \frac{\partial^3 b}{\partial w_1 \partial w_2 \partial w_3}.$$

Some properties on multivariate tail dependence function b in (8.7) are summarized below; see Joe et al.[13] for details. The same properties hold for the function b^* in (8.9) because b^* is the lower tail dependence function of the copula based on the reflection of all variables.

(i) b is increasing.
(ii) $b(\boldsymbol{w})$ is homogeneous of order 1: $b(t\boldsymbol{w}) = tb(\boldsymbol{w})$ for $t > 0$.
(iii) $\partial b(\boldsymbol{w})/\partial w_j$ is homogeneous of order 0.
(iv) $\partial^2 b(\boldsymbol{w})/\partial w_j \partial w_k$ is homogeneous of order -1, etc.
(v) If $b(1, \ldots, 1) > 0$, then $b(\boldsymbol{w}) > 0$ for all $\boldsymbol{w}$ (since b is increasing and homogeneous of order 1).

Next, we state some results on lower-dimensional margins. Let $S \subset \{1,\ldots,n\}$, with marginal C_S. If $S = \emptyset$, then $C_\emptyset = 1$ by definition. If C has lower tail dependence, there are non-negative increasing functions b_S such that

$$C_S(uw_i, i \in S) \sim u\, b_S(w_i, i \in S), \quad u \to 0.$$

Under some conditions (see Proposition 2.4 in Joe et al.[13]), b_S is obtained from b by letting $w_k \to \infty$ for $k \notin S$. Similarly, b_S^* can be defined from $\widehat{C}_S$. From these limits on margins, if a copula C has multivariate lower (upper) tail dependence, then all bivariate and lower-dimensional margins have lower (upper) tail dependence.

We next state the result which links the tail dependence functions $\{b_S\}$ to extreme value limits. The lower extreme value (EV) limit of C comes from the upper EV value limit of the copula of $\boldsymbol{V} = (V_1,\ldots,V_n)$, where $V_j = 1 - U_j$ and $\boldsymbol{U} = (U_1,\ldots,U_n) \sim C$. The copula of $\boldsymbol{V}$ is given in (8.1). Hence, as $u \to 0$, with $u_i = uw_i$,

$$\widehat{C}(1 - u\boldsymbol{w}) = \sum_{S \subset \{1,\ldots,n\}} (-1)^{|S|} C_S(uw_i, i \in S)$$

$$\approx 1 + \sum_{S \subset \{1,\ldots,n\}, S \neq \emptyset} (-1)^{|S|} u\, b_S(w_i, i \in S).$$

Let $a(\boldsymbol{w})$ be defined as

$$a(\boldsymbol{w}) = \sum_{S \subset \{1,\ldots,n\}, S \neq \emptyset} (-1)^{|S|-1} b_S(w_i, i \in S)$$

so that

$$\widehat{C}(1 - u\boldsymbol{w}) = \overline{C}(u\boldsymbol{w}) \approx 1 - u\, a(\boldsymbol{w}), \quad u \to 0.$$

The EV limit is $\lim_{n\to\infty} \widehat{C}^n(u_1^{1/n},\ldots,u_n^{1/n})$. For large n, $u_j^{1/n} = \exp\{n^{-1} \log u_j\} \approx 1 + n^{-1} \log u_j$, so that with $\tilde{u}_j = -\log u_j$,

$$\widehat{C}^n(u_1^{1/n},\ldots,u_n^{1/n}) \approx [1 - n^{-1} a(\tilde{u}_1,\ldots,\tilde{u}_n)]^n \to \exp\{-a(\tilde{u}_1,\ldots,\tilde{u}_n)\}.$$

8.3.3 *Conditional tail dependence functions*

In this subsection, we define more general versions of conditional tail dependence functions. They appear in the theorem in the next section on tail dependence of regular vines.

Let $S = \{k_1, k_2, \ldots, k_m\}$ be a subset of $\{1,\ldots,n\}$ with cardinality m of at least 2, and let C_S be the corresponding margin of the n-dimensional copula

C. Let $(w_{k_1},\ldots,w_{k_m}) \in \mathbb{R}_+^m$. If C_S has multivariate lower tail dependence, then
$$t_{k_1|k_2\cdots k_m}(w_{k_1}|w_{k_2},\ldots,k_m) = \lim_{u\to 0} C_{k_1|k_2\cdots k_m}(uw_{k_1}|uw_{k_2},\ldots,uw_{k_m})$$
is not the zero function. If C_S has multivariate upper tail dependence, then
$$t^*_{k_1|k_2\cdots k_m}(w_{k_1}|w_{k_2},\ldots,k_m)$$
$$= \lim_{u\to 0} C_{k_1|k_2\cdots k_m}(1-uw_{k_1}|1-uw_{k_2},\ldots,1-uw_{k_m})$$
is not the zero function.

It is shown in Joe et al.[13] that the conditional tail dependence functions t and t^* can be obtained from the derivatives of margins of the multivariate tail dependence functions b and b^*. If $b_{\{k_1,k_2,\ldots,k_m\}}$ and $b_{\{k_2,\ldots,k_m\}}$ are lower tail dependence functions, then
$$t_{k_1|k_2\cdots k_m}(w_{k_1}|w_{k_2},\ldots,w_{k_m}) = \frac{\partial^{m-1} b_{\{k_1,k_2,\ldots,k_m\}}}{\partial w_{k_2}\cdots \partial w_{k_m}} \bigg/ \frac{\partial^{m-1} b_{\{k_2,\ldots,k_m\}}}{\partial w_{k_2}\cdots \partial w_{k_m}}.$$

For $m=2$, this simplifies to
$$t_{k_1|k_2}(w_{k_1}|w_{k_2}) = \frac{\partial b_{\{k_1,k_2\}}}{\partial w_{k_2}},$$
since $b_{\{k_2\}}(w) = w$.

Example 8.3. We continue with Example 8.2 to illustrate conditional tail dependence functions. A direct calculation yields
$$C_{3|12}(u_3|u_1,u_2;\delta) = \frac{(u_1^{-\delta}+u_2^{-\delta}+u_3^{-\delta}-2)^{-1/\delta-2}}{(u_1^{-\delta}+u_2^{-\delta}-1)^{-1/\delta-2}}.$$
Therefore,
$$t_{3|12}(w_3|w_1,w_2) = \lim_{u\to 0} C_{3|12}(uw_3|uw_1,uw_2;\delta)$$
$$= \frac{(w_1^{-\delta}+w_2^{-\delta}+w_3^{-\delta})^{-1/\delta-2}}{(w_1^{-\delta}+w_2^{-\delta})^{-1/\delta-2}} = \left(1+\frac{w_3^{-\delta}}{w_1^{-\delta}+w_2^{-\delta}}\right)^{-1/\delta-2}.$$
Note that $t_{3|12}$ is homogeneous of order 0.

8.4 Main Theorem on Tail Dependence for Vine Copulae

The following is a rewriting of Theorem 4.2 in Joe et al.[13] into notation that is valid for any regular vine. The statement of the theorem in this cited paper was given for the D-vine because the proof is less notationally cumbersome for this special case. The theorem shows how the tail dependence of a regular vine depends on its bivariate baseline linking copulae.

The notation used is: a conditional distribution of variables indexed by i_1, i_2, is in level ℓ of the vine if cardinality of conditioning set $D(i_1, i_2, \ell) = \{j_1, \ldots, j_{\ell-1}\}$ is $\ell - 1$ for $\ell = 1, \ldots, n - 1$.

Theorem 8.1. *Consider a regular vine copula C constructed from the linking copulae $\{C_{i_1, i_2 | D(i_1, i_2, \ell)} : 1 \leq i_1 < i_2 \leq n$, with $n - \ell$ pairs at level ℓ, $\ell = 1, \ldots, n - 1\}$. If all the bivariate linking copulae have continuous second-order partial derivatives, then the lower and upper tail dependence functions are given respectively by the recursions. For $\ell > 1$, with $\bm{v} = (v_{j_1}, \ldots, v_{j_{\ell-1}})$ and $D(i_1, i_2, \ell) = \{j_1, \ldots, j_{\ell-1}\}$ as the conditioning set for $\{i_1, i_2\}$,*

$$b_{\{i_1, i_2, j_1, \ldots, j_{\ell-1}\}}(w_{i_1}, w_{i_2}, w_{j_1}, \ldots, w_{j_{\ell-1}})$$

$$= \int_0^{w_{j_1}} \cdots \int_0^{w_{j_{\ell-1}}} C_{i_1, i_2 | j_1, \ldots, j_{\ell-1}}(t_{i_1 | j_1, \ldots, j_{\ell-1}}(w_{i_1} | \bm{v}), t_{i_2 | j_1, \ldots, j_{\ell-1}}(w_{i_2} | \bm{v}))$$

$$\times \frac{\partial^{\ell-1} b_{\{j_1, \ldots, j_{\ell-1}\}}(\bm{v})}{\partial v_{j_1} \cdots \partial v_{j_{\ell-1}}} d\bm{v}.$$

Similar expressions can be obtained for the upper tail with the upper tail functions b^*, t^* replacing the lower tail b, t functions.

If the supports of the bivariate linking copulae are the entire $(0, 1)^2$ and the baseline copulae $\{C_{i_1, i_2} : (i_1, i_2)$ in level 1 tree$\}$ are all lower (upper) tail dependent, then C is lower (upper) tail dependent.

For $n = 3$ and 4, special cases of the theorem for the C-vine are given as follows together with an outline of the derivations.

For $n = 3$, with copulae $C_{12}, C_{13}, C_{23|1}$ and conditional tail dependence functions $t_{2|1}, t_{3|1}$,

$$C_{123}(uw_1, uw_2, uw_3) = \int_0^{uw_1} C_{23|1}(C_{2|1}(uw_2|v_1), C_{3|1}(uw_3|v_1)) \, dv_1$$

$$= u \int_0^{w_1} C_{23|1}(C_{2|1}(uw_1|uv), C_{3|1}(uw_3|uv)) \, dv$$

$$\sim u \int_0^{w_1} C_{23|1}(t_{2|1}(w_2|v), t_{3|1}(w_3|v)) \, dv$$

so that

$$b_{123}(w_1, w_2, w_3) = \int_0^{w_1} C_{23|1}(t_{2|1}(w_2|v), t_{3|1}(w_3|v)) \, dv.$$

C_{123} has trivariate lower tail dependence if $b_{12}, b_{13}, t_{2|1}, t_{3|1}$ are positive and $C_{23|1}$ is positive in $(0,1)^2$. With these conditions, $t_{3|12}$ is also positive.

For $n=4$, with additional copulae $C_{14}, C_{24|1}, C_{34|12}$, the above construction means that b_{124} is positive if $b_{14}, t_{4|1}$ are also positive and $C_{24|1}$ is positive in $(0,1)^2$. With these conditions, $t_{4|12}$ is also positive. If $C_{3|12}$ is a conditional distribution of C_{123} and $C_{4|12}$ is a conditional distribution of C_{124}, then

$$C_{1234}(uw_1,\ldots,uw_4)$$
$$= \int_0^{uw_1}\int_0^{uw_2} C_{34|12}(C_{3|12}(uw_3|z_1,z_2), C_{4|12}(uw_4|z_1,z_2))$$
$$\times c_{12}(z_1,z_2)dz_1 dz_2$$
$$= u^2 \int_0^{w_1}\int_0^{w_2} C_{34|12}(C_{3|12}(uw_3|uv_1,uv_2), C_{4|12}(uw_4|uv_1,uv_2))$$
$$\times c_{12}(uv_1,uv_2)dv_1 dv_2$$
$$\sim u^1 \int_0^{w_1}\int_0^{w_2} C_{34|12}(t_{3|12}(w_3|v_1,v_2), t_{4|12}(w_4|v_1,v_2))$$
$$\times \frac{\partial^2 b_{12}(v_1,v_2)}{\partial v_1 \partial v_2} dv_1 dv_2.$$

The last approximation follows from (8.8). C_{1234} has multivariate lower tail dependence if $b_{123}, b_{124}, t_{1|23}, t_{4|23}$ are positive (and hence b_{12} is positive), and $C_{34|12}$ is positive in $(0,1)^2$.

Example 8.4. This is a continuation of Example 8.1 to show an application of Theorem 8.1. Suppose we start with variables X_1, X_2, X_3 having a joint distribution summarized by the C-vine $\{F_1, F_2, F_3, C_{12}, C_{13}, C_{23|1}\}$.

(a) Let $X_1^* = g_1(X_1)$ be a continuous strictly decreasing transform. Suppose C_{12} and C_{13} have lower-upper tail dependence, so that the copulae $\acute{C}_{12}$ and $\acute{C}_{13}$ from (8.2) have upper tail dependence. Also suppose $C_{23|1}$ has support on $(0,1)^2$. By Theorem 8.1, (X_1^*, X_2, X_3) has upper tail dependence, and hence marginally (X_2, X_3) has upper tail dependence.

(b) Let $X_2^* = g_2(X_1)$ be a continuous strictly decreasing transform. Suppose C_{12} has upper-lower tail dependence and C_{13} has upper tail dependence, so that the copulae $\grave{C}_{12}$ from (8.2) has upper tail dependence. Also suppose $C_{23|1}$ and $\acute{C}_{23|1}$ have support on $(0,1)^2$. By Theorem 8.1, (X_1, X_2^*, X_3) has upper tail dependence, and hence marginally (X_2, X_3) has lower-upper tail dependence.

8.5 Reflection Asymmetry of Vine Copulae

The multivariate uniform vector $\boldsymbol{U} = (U_1, \ldots, U_m)$ is *reflection or complement symmetric* if

$$(U_1, \ldots, U_m) \stackrel{d}{=} (1 - U_1, \ldots, 1 - U_m).$$

If $\boldsymbol{U}$ is reflection symmetric, then any subvector of $\boldsymbol{U}$ is reflection symmetric. If the copula density is $c = c_{1 \cdots m}$, then reflection symmetry implies

$$c(u_1, \ldots, u_m) = c(1 - u_1, \ldots, 1 - u_m).$$

We next prove some results showing that if the copulae $\{C_{i_1, i_2 | j_1, \ldots, j_{\ell-1}} : i_1, i_2 | j_1, \ldots, j_{\ell-1}$ in level ℓ tree; $1 \leq i_1 < i_2 \leq d; d - \ell$ pairs at level $\ell\}$ in a vine are all reflection symmetric, then the vine copula is reflection symmetric. The next section mentions choices of bivariate copulae that can be used in vines to get asymmetry of the upper and lower tails.

Proposition 8.1. *If C is an m-dimensional reflection symmetric copula with a density, then*

$$C_{m|1\ldots m-1}(1 - u_m \mid 1 - u_1, \ldots, 1 - u_{m-1})$$
$$= 1 - C_{m|1\ldots m-1}(u_m \mid u_1, \ldots, u_{m-1}).$$

Similar identities hold for other conditional distributions of one variable on the other $m - 1$ variables.

Proof. The result is immediate from geometry. A calculus-based proof is:

$$C_{m|1\ldots m-1}(1 - u_m \mid 1 - u_1, \ldots, 1 - u_{m-1})$$
$$= \frac{\int_0^{1-u_m} c_{1\ldots m}(1 - u_1, \ldots, 1 - u_{m-1}, v)\, dv}{c_{1\ldots m-1}(1 - u_1, \ldots, 1 - u_{m-1})}$$
$$= \frac{\int_0^{1-u_m} c_{1\ldots m}(u_1, \ldots, u_{m-1}, 1 - v)\, dv}{c_{1\ldots m-1}(u_1, \ldots, u_{m-1})}$$
$$= \frac{\int_{u_m}^{1} c_{1\ldots m-1}(u_1, \ldots, u_{m-1}, v')\, dv'}{c_{1\ldots m-1}(u_1, \ldots, u_{m-1})}$$
$$= 1 - C_{m|1\ldots m-1}(u_m \mid u_1, \ldots, u_{m-1}). \qquad \square$$

Proposition 8.2. *Let $2 \leq \ell < n$ be an integer for level ℓ of the vine. Suppose $C_{i_1, j_1, \ldots j_{\ell-1}}$ and $C_{i_2, j_1 \cdots j_{\ell-1}}$ are reflection symmetric copulae with* densities and $C_{i_1 i_2 | j_1,\ldots,j_{\ell-1}}$ is a reflection symmetric bivariate copula with density $c_{i_1 i_2 | j_1,\ldots,j_{\ell-1}}$, then the function

$$c_{i_1 i_2 | j_1,\ldots,j_{\ell-1}}(C_{i_1 | j_1 \ldots j_{\ell-1}}(u_{i_1} \mid u_{j_1},\ldots,u_{j_{\ell-1}}),$$
$$C_{i_2 | j_1 \ldots j_{\ell-1}}(u_{i_2} \mid u_{j_1},\ldots,u_{j_{\ell-1}})) \qquad (8.10)$$

is reflection symmetric.

Proof. This follows from Proposition 8.1.

$$c_{i_1 i_2 | j_1,\ldots,j_{\ell-1}}(C_{i_1 | j_1 \ldots j_{\ell-1}}(1 - u_{i_1} \mid 1 - u_{j_1},\ldots,1 - u_{j_{\ell-1}}),$$
$$C_{i_2 | j_1 \ldots j_{\ell-1}}(1 - u_{i_2} \mid 1 - u_{j_1},\ldots,1 - u_{j_{-1}}))$$
$$= c_{i_1 i_2 | j_1,\ldots,j_{\ell-1}}(1 - C_{i_1 | j_1 \ldots j_{\ell-1}}(u_{i_1} \mid u_{j_2},\ldots,u_{j_{\ell-1}}),$$
$$1 - C_{i_2 | j_1 \ldots j_{\ell-1}}(u_{i_2} \mid u_{j_1},\ldots,u_{j_{\ell-1}}))$$
$$= c_{i_1 i_2 | j_1,\ldots,j_{\ell-1}}(C_{i_1 | j_1 \ldots j_{\ell-1}}(u_{i_1} \mid u_{j_1},\ldots,u_{j_{\ell-1}}),$$
$$C_{i_2 | j_1 \ldots j_{\ell-1}}(u_{i_2} \mid u_{j_1},\ldots,u_{j_{-1}})). \qquad \square$$

Proposition 8.3. Let $\{C_{i_1,i_2 | D(i_1,i_2,\ell)} : 1 \leq i_1 < i_2 \leq n\}$ be the bivariate linking copulae of a regular vine. If all of the bivariate copulae are reflection symmetric with densities, then the resulting n-dimensional copula $c_{1 \ldots d}$ is reflection symmetric.

Proof. Let $\{c_{i_1,i_2 | D(i_1,i_2,\ell)}\}$ be the set of bivariate copula densities. With $D(i_1, i_2, \ell)$ written generically as $\{j_1 \ldots j_{\ell-1}\}$, then from Bedford and Cooke[3] and Aas et al.,[2] the copula density $c_{1 \ldots n}$ is the product of densities in the vine:

$$c_{i_1 i_2 | j_1 \ldots j_{\ell-1}}(C_{i_1 | j_1 \ldots j_{\ell-1}}(u_{i_1} \mid u_{j_1},\ldots,u_{j_{\ell-1}}),$$
$$C_{i_2 | j_1 \ldots j_{\ell-1}}(u_{i_2} \mid u_{j_1},\ldots,u_{j_{\ell-1}}))$$

of the form (8.10), together with the densities at level 1. From Proposition 8.2, $c_{1 \ldots d}$ is reflection symmetric. $\square$

8.6 Choice of Tail Asymmetric Bivariate Linking Copulae

If reflection symmetric bivariate copulae (e.g., t_ν) are used throughout the construction of the vine, then from results in the preceding two sections, upper and lower bivariate tail dependence occurs for the (j, k) bivariate margin, but $\lambda_{jk,L} = \lambda_{jk,U}$ for any $j < k$. To get a multivariate copula with different upper and lower tail dependence for each bivariate margin, the linking copulae can be chosen to be *reflection asymmetric*.

To get both upper and lower tail dependence for each bivariate margin of $C(u_1, \ldots, u_n)$, we want the $n-1$ bivariate copulae at level 1 to have both upper and lower dependence, and the remaining bivariate copulae should have support on $(0,1)^2$. In the remainder of this section, we discuss possible choices.

Consider bivariate families of the form

$$C(u,v) = \psi_\theta(-\log K(e^{-\psi_\theta^{-1}(u)}, e^{-\psi_\theta^{-1}(v)}); \delta), \tag{8.11}$$

where K is a *max-infinitely divisible* (max-id) copula (meaning all positive powers of K are cdfs) and ψ_θ is a Laplace transform (LT) family. Based on the results in Joe and Hu[11] (or Examples 4.1–4.3, combined with Theorems 4.13 and 4.16 in Joe[12]), the most interesting cases for upper and lower tail dependence are (a) the Gamma LT $\psi_\theta(s) = (1+s)^{-1/\theta}$ ($\theta > 0$) and K having upper tail dependence; (b) the Sibuya[6,21] LT, (labeled as LTC in Joe[12]) $\psi_\theta(s) = 1 - (1-e^{-s})^{1/\theta}$ ($\theta \geq 1$) and K having lower tail dependence.

In (8.11), if K is increasing in concordance as δ increases, then clearly C increases in concordance as δ increases with θ fixed. The concordance ordering for δ fixed and θ varying is harder to check. If K has the form of an Archimedean copula, then C also has the form of an Archimedean copula. That is, if $K(x, y; \delta) = \phi_\delta(\phi_\delta^{-1}(x) + \phi_\delta^{-1}(y))$ for a LT family ϕ_δ, then

$$C(u,v;\theta,\delta) = \psi_\theta(-\log \phi_\delta[\phi_\delta^{-1}(e^{-\psi_\theta^{-1}(u)}) + \phi_\delta^{-1}(e^{-\psi_\theta^{-1}(v)})])$$
$$= \eta_{\theta,\delta}(\eta_{\theta,\delta}^{-1}(u) + \eta_{\theta,\delta}^{-1}(v)), \quad \text{where } \eta_{\theta,\delta}(s) = \psi_\theta(-\log \phi_\delta(s)).$$

The three examples listed below use the classification in Chapter 5 of Joe.[12] Some additional properties of these copula families are mentioned as a guide to their comparisons. The cited bivariate copula families are in Section 5.1 and the cited LT families can be found in Appendix A.1 of Joe.[12]

Family BB1. (Example 5.1 in Joe and Hu[11]). In (8.11), let K be the bivariate Gumbel copula and let ψ be the Gamma LT. The resulting two-parameter copula is

$$C(u,v;\theta,\delta) = \{1 + [(u^{-\theta}-1)^\delta + (v^{-\theta}-1)^\delta]^{1/\delta}\}^{-1/\theta}$$
$$= \eta(\eta^{-1}(u) + \eta^{-1}(v)), \quad \theta > 0, \ \delta \geq 1, \tag{8.12}$$

where $\eta(s) = \eta_{\theta,\delta}(s) = (1+s^{1/\delta})^{-1/\theta}$ is the Mittag-Leffler LT family with a convolution parameter (see Pillai[19]).

Some properties of (8.12) are:

(a) The MTCJ copula is a subfamily when $\delta = 1$, and the Gumbel copula is obtained as $\theta \to 0$. The independence copula C_I obtains as $\theta \to 0$ and $\delta \to 1$ and the Fréchet upper bound copula C_U obtains as $\theta \to \infty$ or $\delta \to \infty$.

(b) The lower tail dependence parameter is $\lambda_L = 2^{-1/(\delta\theta)}$, while the upper tail dependence parameter is $\lambda_U = 2 - 2^{1/\delta}$, independent of θ. λ_U lies in the interval $(0, 1)$. For fixed δ, the lower tail dependence parameter λ_L lies in the entire interval $(0, 1)$ as θ increases from 0 to ∞.

(c) Concordance increases as θ increases because $w(s)/s$ is increasing, where $w(s) = \eta_{\theta_2,\delta}^{-1}(\eta_{\theta_1,\delta}(s)) = [(1 + s^{1/\delta})^\zeta - 1]^\delta$, $\theta_1 < \theta_2$, and $\zeta = \theta_2/\theta_1 > 1$.

Family BB4. (Example 5.3 in Joe and Hu[11]). In (8.11), let K be the bivariate Galambos copula and let ψ be the Gamma LT. The resulting two-parameter copula is

$$C(u, v; \theta, \delta) = (u^{-\theta} + v^{-\theta} - 1 - [(u^{-\theta} - 1)^{-\delta}$$
$$+ (v^{-\theta} - 1)^{-\delta}]^{-1/\delta})^{-1/\theta}, \quad \theta \geq 0, \ \delta > 0. \quad (8.13)$$

Some properties of the family of copulae (8.13) are:

(a) The MTCJ copula is obtained when $\delta \to 0$, and the Galambos family obtains as $\theta \to 0$. The Fréchet upper bound C_U obtains as $\theta \to \infty$ or $\delta \to \infty$.

(b) The lower tail dependence parameter is $\lambda_L = (2 - 2^{-1/\delta})^{-1/\theta}$, while the upper tail dependence parameter is $\lambda_U = 2^{-1/\delta}$, independent of θ. $\lambda_U = 2^{-1/\delta}$ lies in the interval $(0, 1)$. For fixed δ, the lower tail dependence parameter λ_L lies in the entire interval $(0, 1)$ as θ increases from 0 to ∞.

(c) Concordance increases as θ increase if and only if $[x+y-1-((x-1)^{-\delta}+(y-1)^{-\delta})^{-1/\delta}]\log[x+y-1-((x-1)^{-\delta}+(y-1)^{-\delta})^{-1/\delta}]-x\log x-y\log y+[(x-1)^{-\delta}+(y-1)^{-\delta}]^{-1/\delta-1}[(x-1)^{-\delta}x\log x+(y-1)^{-\delta}y\log y] \geq 0$ for all $x, y > 1$ and $\delta > 0$. This condition holds for numerical checks but has not been confirmed analytically.

(d) It can be shown that Blomqvist's[5,20] beta or medial correlation coefficient, which for copulae becomes $4C(\frac{1}{2}, \frac{1}{2}; \theta, \delta) - 1$, is increasing in θ with δ fixed. This requires showing that

$$(2 \cdot 2^\theta - 1 - 2^{-1/\delta}(2^\theta - 1))^{-1/\theta} = ((2 - 2^{-1/\delta})(2^\theta - 1) + 1)^{-1/\theta}$$

is increasing in θ with δ fixed, or $\alpha = 2 - 2^{-1/\delta} \in (1, 2)$ fixed. Let $\zeta = 2^\theta > 1$. We need to show that the derivative of $-\theta^{-1} \log[1 + \alpha(2^\theta - 1)]$ is

$$\theta^{-2} \log[1 + \alpha(2^\theta - 1)] - \theta^{-1} \frac{\alpha 2^\theta \log 2}{1 + \alpha(2^\theta - 1)} \geq 0$$
$$\iff h(\zeta) = [1 + \alpha(\zeta - 1)] \log[1 + \alpha(\zeta - 1)] - \alpha\zeta \log \zeta \geq 0, \ \forall \zeta.$$

Note that $h(1) = 0$ and $h(\zeta) \sim \alpha\zeta \log \alpha$ as $\zeta \to \infty$. Also

$$h'(\zeta) = \alpha \log[1 + \alpha(\zeta - 1)] + \alpha - \alpha \log \zeta - \alpha$$
$$= \alpha[\log[1 + \alpha(\zeta - 1)] - \log \zeta] \geq 0$$

since $1 + \alpha(\zeta - 1) \geq \zeta$ is the same as $\alpha \geq 1$. Hence $h(\zeta) \geq 0$ for all $\zeta \geq 1$ for any fixed $\alpha \geq 1$.

Family BB7. (This family was in the first draft of Joe and Hu[11] but did not appear in the published version.) In (8.11), let K be the bivariate MTCJ family and let ψ be the Sibuya LT family. The resulting two-parameter family is

$$C(u, v; \theta, \delta) = 1 - (1 - [(1 - \bar{u}^\theta)^{-\delta} + (1 - \bar{v}^\theta)^{-\delta} - 1]^{-1/\delta})^{1/\theta}$$
$$= \eta(\eta^{-1}(u) + \eta^{-1}(v)), \quad \theta \geq 1, \quad \delta > 0, \quad \bar{u} = 1 - u, \quad \bar{v} = 1 - v,$$
(8.14)

where $\eta(s) = \eta_{\theta,\delta}(s) = 1 - [1 - (1 + s)^{-1/\delta}]^{1/\theta}$ (family LTI in the Appendix of Joe[12]).

Some properties of the family of copulae (8.14) are:

(a) The MTCJ family is obtained when $\theta = 1$, and the B5 family (Joe[9,12]) is obtained as $\delta \to 0$. The Fréchet upper bound C_U obtains as $\theta \to \infty$ or $\delta \to \infty$.
(b) The lower tail dependence parameter is $2^{-1/\delta}$, independent of θ, and the upper tail dependence parameter is $2 - 2^{1/\theta}$, independent of δ. These can vary independently in the entire interval $(0, 1)$.
(c) Concordance increases as θ increases when $\delta \leq 1$.
(d) The concordance ordering has been shown NOT to hold as θ increases for fixed $\delta \geq 1.5$. The numerical counterexamples are most easily found from

$$\left. \frac{\partial C(u, v; \theta, \delta)}{\partial \theta} \right|_{\theta=1} < 0 \qquad (8.15)$$

for some choices of (u, v).

(e) In fact, (8.15) is negative for $u = v = \frac{1}{2}$ for $\delta \geq 2.23$, so that even Blomqvist's beta, $4C(\frac{1}{2}, \frac{1}{2}; \theta, \delta) - 1$, is not increasing in θ for some fixed δ.

All three families BB1, BB4 and BB7 have flexible upper and lower tail dependence. Because of the known concordance ordering only for BB1, the BB1 family might be preferable over BB4 and BB7 for getting asymmetric tail dependence with vine copulae.

8.7 Discussion

The theorem in Section 8.4 implies that vine copulae with flexible upper and lower tail dependence can be obtained with appropriate choices of bivariate linking copulae that are reflection asymmetric and have upper and lower tail dependence parameters λ_L, λ_U that independently take values in $(0, 1)$. Section 8.6 lists some choices of bivariate copula families with this property.

To decide on copula models for multivariate data, a first step in initial data analysis includes measuring bivariate association and assessing the strength of bivariate tail dependence. Models with flexible upper and lower tail dependence should be useful for multivariate financial asset return and other data where there can be strong dependence in extremes. For modeling multivariate data, by adding extra parameters, one could consider vine copulae for which the conditional linking copulae at levels $2, \ldots, n - 1$ are not constant over the values of the conditioned variables.

The comparison of reflection asymmetric copulae versus t-copulae in vines for financial asset return data is made in Nikoloulopoulos et al.[18] Some initial assessment for such data suggest that the assumption of constant conditional copulae is acceptable. But more experience is needed to go from the initial bivariate analysis to decide on the vine and the indexing of the variables to the vine.

Other directions of research include the study of (a) additional families based on the construction (8.11) and (b) other construction methods for bivariate copulae with reflection asymmetry.

Acknowledgments

This research was supported by an NSERC Canada Discovery Grant. The author is grateful to Haijun Li and Aristidis Nikoloulopoulos for many discussions on tail dependence functions and vine copulae, and to Lei Hua for the concordance ordering counterexample for the BB7 copula. The investigations of tail dependence functions were initiated after questions from Kjersti Aas on conditions for tail dependence in vines.

Appendix

Proposition 8.4. Let $C(u_1, u_2)$ be a bivariate copula with lower tail dependence. Let $b(w_1, w_2)$ be its lower tail dependence function, and let $t_{1|2}(w_1|w_2) = \frac{\partial b(w_1,w_2)}{\partial w_2}$, $t_{2|1}(w_2|w_1) = \frac{\partial b(w_1,w_2)}{\partial w_1}$ be the lower conditional tail dependence functions. Then
$$\int_0^\infty t_{1|2}(1|w)\,dw = t_{2|1}(\infty|1) \le 1.$$

Proof. Since $t_{1|2}(1|w) = \lim_{u \to 0} C_{1|2}(u|uw)$ for $w > 0$ and $\int_0^{1/u} C_{1|2}(u|uw)\,dw = 1$, then by Fatou's lemma, $\int_0^\infty t_{1|2}(1|w)\,dw \le 1$. Because of the integrability of $t_{1|2}(1 \cdot)$, $t_{1|2}(1|w) = o(w^{-1})$ as $w \to \infty$ or $w\, t_{1|2}(1|w) \to 0$ as $w \to \infty$.

To complete the proof, we use homogeneity properties of b (see Ref. 13). Since $t_{1|2}$ is a partial derivative of b,
$$\int_0^\infty t_{1|2}(1|w)\,dw = b(1,\infty) - b(1,0) = b(1,\infty).$$
From homogeneity and Euler's formula on homogeneous functions,
$$b(w_1, w_2) = w_1 t_{2|1}(w_2|w_1) + w_2 t_{1|2}(w_1|w_2)$$
so that
$$b(1, \infty) = t_{2|1}(\infty|1) + \lim_{w \to \infty} w\, t_{1|2}(1|w) = t_{2|1}(\infty|1).$$
The upper bound of 1 follows from $t_{2|1}(w|1) = \lim_{u \to 0} C_{2|1}(uw|u)$. □

To illustrate the proposition, we obtain the conditional tail dependence functions for (a) the bivariate t_ν-copula and (b) the mixture of the independence copula and an Archimedean copula with lower tail dependence. Then we show that they are subdistribution functions.

Example 8.5.

(a) By the bivariate t_ν-copula $C(u_1, u_2; \nu, \rho)$ with $\nu > 0$ and $-1 < \rho < 1$, it follows from results in Nikoloulopoulos et al.[17] that
$$t_{2|1}(w|1) = t_{1|2}(w|1) = T_{\nu+1}\left(\frac{\sqrt{\nu+1}}{\sqrt{1-\rho^2}}(\rho - w^{-1/\nu})\right),$$
and $\lim_{w \to \infty} t_{2|1}(w|1) = T_{\nu+1}(\sqrt{\nu+1}\,\rho/\sqrt{1-\rho^2}) < 1$. Here, T_η is the t cdf with η degrees of freedom.

(b) Let $C_{12}(u_1, u_2; \delta, \beta) = (1-\beta)u_1 u_2 + \beta(u_1^{-\delta} + u_2^{-\delta} - 1)^{-1/\delta}$, where $\delta > 0$ and $0 < \beta \le 1$. Then it is straightforward to obtain

$$b(w_1, w_2) = \beta(w_1^{-\delta} + w_2^{-\delta})^{-1/\delta}$$

and

$$t_{2|1}(w_2|w_1) = \frac{\partial b(w_1, w_2)}{\partial w_1} = \beta\left[1 + \left(\frac{w_1}{w_2}\right)^\delta\right]^{-1-1/\delta},$$

and $t_{2|1}(w|1) = \beta[1 + w^{-\delta}]^{-1-1/\delta}$ is a sub-distribution function $t_{2|1}(\infty|1) = \beta$ so the mass is $(1-\beta)$ at ∞.

References

1. Aas K. and Berg D. (2009). Models for construction of multivariate dependence: A comparison study. *European Journal of Finance*, 15:639–659.
2. Aas K., Czado C., Frigessi A. and Bakken H. (2009). Pair-copula constructions of multiple dependence. *Insurance: Mathematics and Economics*, 44:182–198.
3. Bedford T. and Cooke R.M. (2001). Probability density decomposition for conditionally dependent random variables modeled by vines. *Annals of Mathematics and Artificial Intelligence*, 32:245–268.
4. Bedford T. and Cooke R.M. (2002). Vines: A new graphical model for dependent random variables. *Annals of Statistics*, 30:1031–1068.
5. Blomqvist N. (1950). On a measure of dependence between two random variables. *Annals of Mathematical Statistics*, 21:593–600.
6. Christoph G. and Schreiber K. (2000). Scaled Sibuya distribution and discrete self-decomposability. *Statistics & Probability Letters*, 48:181–187.
7. Cook R.D. and Johnson, M.E. (1981). A family of distributions for modelling non-elliptically symmetric multivariate data. *Journal of the Royal Statistical Society: Series B*, 43:210–218.
8. Embrechts P., McNeil A.J. and Straumann D. (2002). Correlation and dependency in risk management: Properties and pitfalls. In *Risk Management: Value at Risk and Beyond*, M. Dempster (ed.), pp. 176–223. Cambridge University Press, Cambridge.
9. Joe H. (1993). Parametric family of multivariate distributions with given margins. *Journal of Multivariate Analysis*, 46:262–282.
10. Joe H. (1996). Families of m-variate distributions with given margins and $m(m-1)/2$ bivariate dependence parameters. In *Distributions with Fixed Marginals and Related Topics*, L. Rüschendorf, B. Schweizer and M.D. Taylor (eds.), pp. 120–141. IMS Lecture Notes Monograph Series, 28, Institute of Mathematical Statistics, Hayward, CA.
11. Joe H. and Hu T. (1996). Multivariate distributions from mixtures of max-infinitely divisible distributions. *Journal of Multivariate Analysis*, 57:240–265.
12. Joe H. (1997). *Multivariate Models and Dependence Concepts*. Chapman & Hall, London.
13. Joe H., Li H. and Nikoloulopoulos A.K. (2010). Tail dependence functions and vine copulas. *Journal of Multivariate Analysis*, 101:252–270.

14. Jondeau E., Poon S.-H. and Rockinger M. (2007). *Financial Modeling under Non-Gaussian Distributions*. Springer, London.
15. Kurowicka D. and Cooke R. (2006). *Uncertainty Analysis with High Dimensional Dependence Modelling*. Wiley, Chichester.
16. Mardia K.V. (1962). Multivariate Pareto distributions. *Annals of Mathematical Statistics*, 33:1008–1015.
17. Nikoloulopoulos A.K., Joe H. and Li H. (2009). Extreme value properties of multivariate t-copulas. *Extremes*, 12:129–148.
18. Nikoloulopoulos A.K., Joe H. and Li H. (2010). Vine copulas with asymmetric tail dependence and applications to financial return data. *Computational Statistics and Data Analysis*. In Press.
19. Pillai R.N. (1990). On the Mittag-Leffler functions and related distributions. *Annals of the Institute of Statistical Mathematics*, 42:157–161.
20. Schmid F. and Schmidt R. (2007). Nonparametric inference on multivariate versions of Blomqvist's beta and related measures of tail dependence. *Metrika*, 66:323–354.
21. Sibuya M. (1979). Generalized hypergeometric digamma and trigamma distributions. *Annals of the Institute of Statistical Mathematics*, 31:373–390.
22. Takahasi K. (1965). Note on the multivariate Burr's distribution. *Annals of the Institute of Statistical Mathematics*, 17:257–260.

CHAPTER 9

Counting Vines

O. Morales-Nápoles*

Delft Institute of Applied Mathematics
Mekelweg 4, 2628CD Delft, The Netherlands
O.MoralesNapoles@ewi.tudelft.nl

In this chapter, three algorithms for producing and enumerating regular vines are presented. The first one produces all possible vines on n nodes and regular vines are found by inspection. The second one uses the concept of line graphs to produce only regular vines. The third algorithm produces regular vines by extending a regular vine on three nodes to a regular vine on n nodes. The first and second algorithms presented have been used for the construction of a catalogue of labeled regular vines on at most nine nodes and tree-equivalent regular vines on at most seven. This catalogue is presented as an appendix to the chapter.

9.1	Introduction	190
9.2	Basic Definitions	190
9.3	Regular Vines and Prüfer Codes	194
9.4	Regular Vines and Line Graphs	197
9.5	Regular Vines and Regular Vine Arrays	198
9.6	Classifying Regular Vines	203
9.7	Conclusions and Final Comments	205
Appendix		
	A Catalogue of Labeled Regular Vines on at Most Nine Nodes and Tree-Equivalent Regular Vines on at Most Seven Nodes	205
References	. . . .	217

---

*Delft Institute of Applied Mathematics, Faculty of Electrical Engineering, Mathematics and Computer Science, Delft University of Technology.

9.1 Introduction

Regular vines have found application in probability theory and uncertainty analysis. More recently they are becoming popular in statistical analysis of data.[1,2,6,13,16] These last references are concerned with choosing an optimal vine to represent multivariate data sets. Algorithms for enumerating all possible $\frac{n!}{2} \times 2^{\binom{n-2}{2}}$ regular vines on n nodes[19] will be needed for this purpose.

The problem of counting graphs has been undertaken in the past.[11] Trees are the immediate ancestors of vines and were first successfully counted by Cayley.[5] Trees were used in Ref. 9 and Ref. 26 as special cases of graphical models; however, undirected graphs with cycles were also used. Trees were also used in Ref. 7 to infer discrete distributions[a] from data.

After introducing some definitions required in the rest of this chapter (Section 9.2), previous results about the enumeration of trees will be discussed. Algorithms for producing and enumerating regular vines are presented. The first one produces all possible vines on n nodes and regular vines are found by inspection (Section 9.3). The second one uses the concept of line graphs to generate only regular vines (Section 9.4). The third algorithm generates regular vines by extending a regular vine on three nodes up to a regular vine on n nodes (Section 9.5). The first and second algorithms have been used for the construction of a catalogue of labeled regular vines on at most nine nodes and tree-equivalent regular vines on at most seven. This catalogue is presented as an appendix to this chapter. Proofs for the results presented in this chapter have been discussed previously in Ref. 19 and hence are omitted here.

9.2 Basic Definitions

A *tree* is an undirected acyclic graph. The graph isomorphism problem consists of deciding whether there exists a mapping from the nodes of one graph to the nodes of a second graph such that the edge adjacencies are preserved.

Two *labeled graphs* $G_i = (E_i, N_i)$ and $G_j = (E_j, N_j)$ are *isomorphic* if there is a bijection $\varphi : N_i \to N_j$ such that for all pairs $(a,b) \in E_i \iff (\varphi(a), \varphi(b)) \in E_j$. If two graphs are isomorphic they are the same *unlabeled* graph.

Graph isomorphism is important in selecting the regular vine that best fits a given data set. For example, the algorithm proposed on p. 189 of

---

[a]Actually the method presented in Ref. 7 characterized trees as directed graphs and has a close relationship with BBNs (Chapter 14).

Aas et al.[2] suggests that operations to assign the "best" vine to a data set should begin by selecting the first tree and iteratively for selecting subsequent trees of the regular vine. Once the first tree of a regular vine has been selected, only a fixed number of regular vines may be tested in the next steps. Knowing exactly how many regular vines are still possible in the next steps and how to construct them might be of advantage for analysis.

The *line graph* $LG(G)$ of a graph G has as its nodes the edges of G, with two nodes being adjacent in LG if the corresponding edges are adjacent in G.

A connected graph $T = (N, E)$ is called a *labeled tree* with nodes $N = \{1, 2, \ldots, n\}$ and edges E, where E is a subset of pairs of N with no cycle. Every sequence of numbers $R(T) = (A_1, A_2, \ldots, A_{n-2})$ where each A_i is an integer not greater than n is a *Prüfer Code* for some labeled tree T on n nodes.

A *spanning graph* SS of a graph G is a graph with the same set of nodes as G. If SS is a tree, it is called a *spanning tree* of G.

Trees have been used to represent high-dimensional probability distributions[8] and they are often called dependence or Markov-dependence trees. For an account of dependence trees, see Ref. 14. In dependence trees, nodes are associated with random variables with invertible distribution function and arcs are associated with rank correlations realized by bivariate copulae. Figures 9.1 and 9.2 show two different labeled trees with rank correlations attached to their edges. By setting nodes $5 = 2$, $2 = 3$, $4 = 5$ and $3 = 4$ in T_2, it would be transformed into T_1 and hence considered the same unlabeled tree.

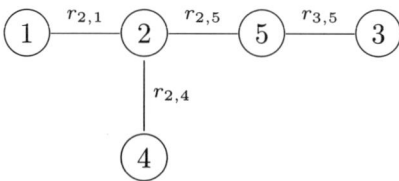

Figure 9.1. T_1, a tree on five nodes with $R(T_1) = (2, 5, 2)$.

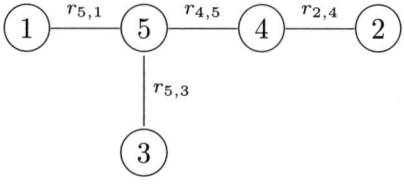

Figure 9.2. T_2, a tree on five nodes with $R(T_2) = (5, 4, 5)$.

A vine[8] is a set of nested trees. Just like labeled trees, vines have been used to represent high-dimensional probability distributions[3,14] with applications in uncertainty analysis. Vines use sequences of conditional distributions to build a multivariate distribution where conditional bivariate constraints are satisfied. The definitions of *vine* and *regular vine* have been provided in Chapter 3 and hence are not repeated here.

As in dependence trees, the nodes of T_1 in a regular vine represent random variables with an invertible distribution function. Edges are associated with rank and conditional rank correlations. Figure 9.3 presents the sequence of trees for a regular vine $V_1(5)$ on five nodes. The conditioned set is separated from the conditioning set by a vertical line "|" in the conditional rank correlations from Fig. 9.3.

Nodes reachable from a given edge in a regular vine are called the *constraint set* of that edge. When two edges are joined by an edge in tree T_i, the intersection of the respective constraint sets form the *conditioning set*. The symmetric difference of the constraint sets is the *conditioned set*.

If node e is an element of node f in a regular vine, we say that e is an *m-child* of f; similarly, if e is reachable from f via the membership relation: $e \in e_1 \in \cdots \in f$, we say that e is an *m-descendant* of f.

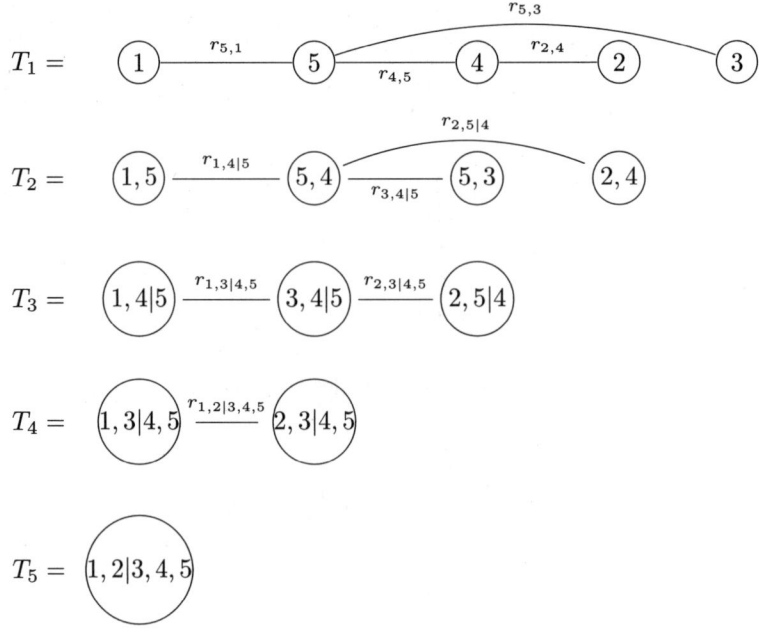

Figure 9.3. $V_1(5)$ (Regular vine on five nodes).

If a bijection may be found for each $T_i \in V_k(n)$ and $T_i \in V_j(n)$, then we speak of the same *tree-equivalent vine* and accordingly the same *tree-equivalent regular vine* when the proximity condition holds. For example, setting nodes $5 = 2$, $2 = 3$, $4 = 5$ and $3 = 4$ in T_2 would generate different labeled regular vines but the same tree-equivalent regular vine. Observe that it is easy to find non-regular vines that are tree-equivalent with regular vines.

If element a occurs with element b as conditioned variables in tree k, then a and b are termed *k-partners*. Nodes A and B are *siblings* if they are m-children of a common parent.

A *natural order* of the elements of a regular vine on n elements is a sequence of numbers $NO(V(n)) = (A_n, A_{n-1}, \ldots, A_1)$ where each A_i is an integer not greater than n obtained as follows: Take one conditioned element of the last tree of a regular vine (a tree with a single node and no edges) and assign it position n; assign the other conditioned element of the top node position $(n-1)$. Element A_{n-1} occurs in one m-child of the top node with an $(n-1)$-partner in the conditioned set. Give this $(n-1)$-partner position $(n-2)$ and iterate this process until all elements have been assigned a position.

Observe that there are two natural orders for every regular vine. Without loss of generality, we will always assign position n to the smallest element of the conditioned set of the last tree of a regular vine. Hence, the natural order of the regular vine in Fig. 9.3 is $NO(V_1(5)) = (1, 2, 3, 4, 5)$.

A *regular vine array* $TA(V(n)) = \{A_{i,j}\}$ for $i, j = 1, \ldots, n$ and $j \geq i$ is a lower triangular matrix with elements in $\{1, \ldots, n\}$ indexed in "reverse order" (see Eq. (9.1)), where $A_{j,j}$ equals the element in position j in $NO(V(n))$ and $A_{j-1,j}$ equals the element in position $j - 1$ in the same natural order. The *echelon* of element $A_{i,j}$ is i and element $A_{i,j}$ codes the node $(A_{j,j}, A_{i,j} | A_{i-1,j}, \ldots, A_{1,j})$. The regular vine array $TA(V_1(5))$ of the regular vine in Fig. 9.3 is presented in Eq. (9.1).

$$TA(V_1(5)) = \begin{pmatrix} A_{5,5} & & & & \\ A_{4,5} & A_{4,4} & & & \\ A_{3,5} & A_{3,4} & A_{3,3} & & \\ A_{2,5} & A_{2,4} & A_{2,3} & A_{2,2} & \\ A_{1,5} & A_{1,4} & A_{1,3} & A_{1,2} & A_{1,1} \end{pmatrix} = \begin{pmatrix} 1 & & & & \\ 2 & 2 & & & \\ 3 & 3 & 3 & & \\ 4 & 5 & 4 & 4 & \\ 5 & 4 & 5 & 5 & 5 \end{pmatrix}$$
(9.1)

The reader may check for example that $A_{2,4} = (5, 2|4)$ and $A_{2,3} = (4, 3|5)$ in (9.1) are children of $A_{3,4} = (3, 2|5, 4)$. $A_{3,4} = (3, 2|5, 4)$ and $A_{3,5} = (3, 1|4, 5)$ are siblings because they are children of the common parent

$A_{4,5} = (2,1|3,4,5)$. Similarly, $A_{2,3} = (4,3|5)$ and $A_{2,5} = (4,1|5)$ are children of $A_{3,5} = (3,1|4,5)$ and hence siblings. Other elements may also be checked by the reader. In Ref. 19 it is shown that *regular vine arrays* represent regular vines. Next, algorithms for producing vines and regular vines are proposed.

9.3 Regular Vines and Prüfer Codes

The first proof about the number of labeled trees on n nodes is due to Cayley.[5] Since then, several proofs have been presented.[18]

Theorem 9.1. *The number of labeled trees on n nodes is n^{n-2}.*

One of various proofs of this theorem due to Ref. 21 provides a very useful result for representing labeled trees. The argument is to notice that there is a one-to-one correspondence between the set of trees with n labeled nodes and the set of Prüfer codes.

In his paper, Prüfer obtains the correspondence by the following procedure: For a given tree, remove the endpoint[b] with the smallest label (other than the root[c]) and let A_1 be the label of the unique node which is adjacent to it. Remove the endpoint and the edge adjacent to it and a tree on $n-1$ nodes is obtained. Repeat the operation with the new tree on $n-1$ nodes to obtain A_2 and so on. The process is terminated when a tree on two nodes has been found. The reader may check that the trees from Figs. 9.1 and 9.2 have Prüfer codes $R(T_1) = (2,5,2)$ and $R(T_2) = (5,5,2)$ respectively. The procedure described above may be easily reversed, that is, suppose you start with a sequence of $(n-2)$-tuples $R(T_k) = (A_1, A_2, \ldots, A_{n-2})$, then to obtain the only tree corresponding to the sequence, one applies Algorithm 9.1:

Algorithm 9.1. Decoding a Prüfer code.

(1) Take a sequence $R(T_k) = (A_1, A_2, \ldots, A_{n-2})$ for $k = 1, 2, \ldots, n^{n-2}$ where each A_i, $i = 1, 2, \ldots, n-2$ is an integer not greater than n.
(2) Write the root in the rightmost position of $R(T_k)$. Notice that $R(T_k)$ has now length $n-1$ which is $|E|$.

---

[b]The endpoints are nodes with degree 1 in the tree; they are sometimes referred to as *leafs*.
[c]Without loss of generality we will choose node n as the root of all labeled trees on n nodes. Choosing any other node as the root makes no difference except that the algorithm and the procedure to find the Prüfer code for a given tree must be modified.

(3) Write another row of integers at the bottom of $R(T_k)$ from left to right. Each entry B_i in this new row is the smallest integer that has not been already written in this new row (the row of $B'_i s$) nor in the first row (the row of $A'_i s$) in the position exactly above it or every other position to the right.
(4) The resulting code $S(T_k)$ is the *extended Prüfer code*. Each column in the extended Prüfer code represents an arc in the unique labeled tree corresponding to it.

$$S(T_k) = \begin{pmatrix} A_1 & A_2 & A_3 & \cdots & n \\ B_1 & B_2 & B_3 & \cdots & B_{n-1} \end{pmatrix}$$

Take the two Prüfer codes $R(T_1) = (2,5,2)$ and $R(T_2) = (5,4,5)$. Apply Algorithm 9.1 to decode each sequence into the extended Prüfer code. The reader may check in Eq. (9.2) that $S(T_1)$ corresponds to Fig. 9.1 and $S(T_2)$ to Fig. 9.2.

$$S(T_1) = \begin{pmatrix} 2 & 5 & 2 & 5 \\ 1 & 3 & 4 & 2 \end{pmatrix}, \quad S(T_2) = \begin{pmatrix} 5 & 4 & 5 & 5 \\ 1 & 2 & 3 & 2 \end{pmatrix} \quad (9.2)$$

Prüfer then gives an induction argument to show that for each $(n-2)$-tuple there is some tree which determines the given sequence by the above procedure. From the code, one can see that a node with degree m would occur exactly $m-1$ times in the code.

Since every labeled tree can be represented by a Prüfer code, then every tree in a vine may also be represented by a Prüfer code and in this way the vine may be generated. A way to write all possible vines on n nodes is presented in Algorithm 9.2.

Algorithm 9.2. Constructing all possible vines on n nodes.

(1) Set $i = 1$.
(2) Construct all Prüfer codes possible for T_i.
(3) The edges of each one of the $n^{n-(i+1)}$ trees in step 2 become nodes in T_{i+1}. Hence, for each tree in step (2):
    (a) Label the $n-i$ edges of each tree, giving label 1 to the edge appearing in the first column in its extended Prüfer code, 2 to the edge in the second column and so on until all edges have been labeled.[d]

---

[d]This labeling is not unique and any other labeling would work equally well as long as all n^{n-2} trees are labeled in the same way.

(b) Construct all Prüfer codes possible for T_{i+1} and connect the new labeled edges (from T_i) as nodes according to these new Prüfer codes.

(4) Set $i := i + 1$ and go to step (3) until two edges must be connected in the last tree. At this point there is only one way to connect them and no Prüfer code is required.

From Algorithm 9.2, it may be observed that to write any vine on n nodes, all that is required are $n - 2$ Prüfer codes. The first one of length $n - 2$, the second one of length $n - 3$ and so on until the last one of length 1. A vine on n nodes may be represented by an upper triangular array of size $(n-2) \times (n-2)$ whose first row represents the Prüfer code of the first tree in the vine, the second row the second tree of the vine and so on. For example, $V_1(5)$ represents the vine from Fig. 9.3:

$$V_1(5) = \begin{pmatrix} 5 & 4 & 5 \\ & 4 & 4 \\ & & 3 \end{pmatrix} \quad (9.3)$$

Representing a vine as an upper triangular array of size $(n-2) \times (n-2)$ provides a convenient way of storing vines. The representation from Eq. (9.3) provides some idea of the unlabeled tree used at each level in the vine. For example, in the first tree node 5 will have degree 3, in the second tree node 4 will also have degree 3 and in the last tree there will be a single node with degree 2 (obviously) which is node 3. A disadvantage is, however, that the array in Eq. (9.3) does not tell us right away which node is node 4 in T_2 while Fig. 9.3 shows that it is node (5,4). Similarly, it is not immediately evident that node 3 in T_3 of Eq. (9.3) corresponds to $(3,4|5)$ in Fig. 9.3. Observe that although a *regular vine array* requires more space for storing, it is clearer regarding the conditioned and conditioning variables used at each level in the vine. Corollary 9.1 follows immediately from the definition of vines and Theorem 9.3.

Corollary 9.1. *The number of vines on n nodes is $\prod_{i=1}^{n} i^{i-2}$.*

Regular vines are most interesting in uncertainty analysis. Implementing Algorithm 9.2 on a computer is very easy and it provides a simple way of constructing all possible regular vines on n nodes by simply discarding those that are not regular. However, this method incurs the excessive burden of searching all vines. According to Corollary 9.1 the number of vines grows extremely fast with n and it could be very restrictive in time to find all regular vines using Algorithm 9.2 even for a modest number of nodes (eight or nine). Another possibility of constructing only regular vines will be discussed in the next section.

9.4 Regular Vines and Line Graphs

The idea is to use the line graph[e] of each tree in the vine. Harary[10] notes that the concept of the line graph of a given graph is so natural that it has been rediscovered independently by many authors.

If the edges of the first tree of Fig. 9.3 are labeled according to the second step in Algorithm 9.2, then the line graph of this tree can be found. This line graph corresponds to Fig. 9.4. If in the same way we label the edges of the second tree in the vine in Fig. 9.3 accordingly, then the line graph in Fig. 9.5 may be obtained.

It is clear that in order to find all regular vines on n nodes, all the spanning trees of the line graphs of all subtrees in the vine must be found. This result is summarized in Algorithm 9.3.

Algorithm 9.3. Constructing all possible regular vines on n nodes based on line graphs.
(1) Set $i = 1$.
(2) Construct all Prüfer codes possible for T_i.

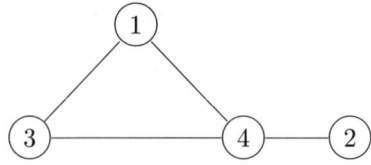

Figure 9.4.   Line graph of the first tree in Fig. 9.3.

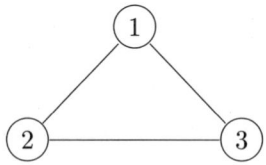

Figure 9.5.   Line graph of the second tree of the vine from Fig. 9.3.

---

[e]Line graphs are also known as derived graphs, interchange graphs, adjoint and edge-to-vertex dual.[4]

(3) The edges of each one of the $n^{n-(i+1)}$ trees in step (2) become nodes in T_{i+1}. Hence, for each tree in step (2):

    (a) Label the edges of each tree, giving the label 1 to the edge appearing in the first column in its extended Prüfer code, 2 to the edge in the second column and so on until all edges have been labeled.[f]

(4) Construct the line graph of each one of the trees from step (2).

(5) For each line graph from step (3) find all possible spanning trees. Connect the edges of each tree in step (1) according to all spanning trees from its line graph. This will give all possible T_{i+1} for each T_i.

(6) Set $i := i + 1$ and go to step (2) until two edges must be connected in the last tree. At this point there is only one way to connect them and no Prüfer code is required.

Notice that the vines generated by this procedure may still be stored as an $(n-2) \times (n-2)$ upper triangular array as in Eq. (9.3) once a way of labeling the edges from each tree in the vine is specified. Algorithm 9.3 does not produce any irregular vine as opposed to Algorithm 9.2. However, it involves a greater programming effort and more operations as all possible spanning trees of the line graphs in all trees in the vine must be found. Several algorithms for finding all spanning trees of a given graph have been proposed and examined.[15,17,22,24,25] In general, finding all possible spanning trees of a given graph other than a complete graph[g] is demanding in terms of time and space.[25] Another algorithm for constructing regular vines without duplication will be presented in the next section.

9.5 Regular Vines and Regular Vine Arrays

One disadvantage of using a triangular array such as the one in Section 14.1 is that the information regarding the label of variables in the first tree of a regular vine is lost when assigning new labels to its edges as they become nodes of the next tree. The same happens as more trees are added to a regular vine. This means that conditioned and conditioning sets are not immediately visible anymore. However, a regular vine array preserves the

---

[f]As before, this labeling is not unique and any other labeling would work equally well as long as all n^{n-i+1} are uniquely labeled.

[g]For a complete graph, all possible spanning trees are the n^{n-2} Prüfer codes.

information concerning the labels of the first tree as lower trees in the vine are added.

The *regular vine array* defined in Section 9.2 was used in Ref. 19 to show that the number of regular vines possible with n nodes is $\frac{n!}{2} \times 2^{\binom{n-2}{2}}$. Example 9.5.1 shows how to construct all possible regular vines on five nodes with the natural order $NO(V(5)) = (1, 2, 3, 4, 5)$. This example is useful in showing how to arrive at a general result about the number of labeled regular vines on n nodes.

Example 9.5.1. Constructing regular vines with natural order $NO(V(5)) = (1, 2, 3, 4, 5)$.

Observe that the diagonal and off-diagonal elements of the *regular vine array* are fixed from the natural order and the definition of *regular vine array*. Element $A_{1,3}$ will also always be fixed by the choices of $A_{3,3}$ and $A_{2,3}$. This means that we start with a regular vine on three nodes. The objective is to extend this regular vine on three nodes to a regular vine on five nodes in all possible ways that preserve regularity within our natural order. Hence, the *regular vine array* with the natural order $NO(V(5)) = (1, 2, 3, 4, 5)$ will look as in Eq. (9.4) in the beginning.

$$TA(V(5)) = \begin{pmatrix} A_{5,5} & & & & \\ A_{4,5} & A_{4,4} & & & \\ A_{3,5} & A_{3,4} & A_{3,3} & & \\ A_{2,5} & A_{2,4} & A_{2,3} & A_{2,2} & \\ A_{1,5} & A_{1,4} & A_{1,3} & A_{1,2} & A_{1,1} \end{pmatrix} = \begin{pmatrix} 1 & & & & \\ 2 & 2 & & & \\ A_{3,5} & 3 & 3 & & \\ A_{2,5} & A_{2,4} & 4 & 4 & \\ A_{1,5} & A_{1,4} & 5 & 5 & 5 \end{pmatrix}$$
(9.4)

We will start filling in $TA(V(5))$ from top to bottom and from right to left. Hence $A_{2,4}$ will be the next element to be filled in $TA(V(5))$. Observe that element $A_{1,4}$ will be fixed by the choice of $A_{2,4}$. Since we are filling in column 4 of $TA(V(5))$, from the definition of *regular vine array*, $A_{2,4} \in \{A_{3,3}, A_{2,2}, A_{1,1}\}$. However, also from the definition of *regular vine array*, $A_{3,3} = A_{3,4}$ and hence $A_{2,4} \in \{A_{3,3}, A_{2,2}, A_{1,1}\} \setminus A_{3,3} = \{A_{2,2}, A_{1,1}\}$. In order to preserve regularity, node $A_{3,4} = (3, 2|4, 5)$ must have two children in the lower tree of the vine. These siblings must also have a common child in one tree lower in the vine. By the definition of a regular vine array, the first child of node $A_{3,4} = (3, 2|4, 5)$ must be node $A_{2,4}$ and its sibling must be in some column $h < 4$ and some row $k \leq 2$, that is, in the known part of the *regular vine array*. In this case nodes $A_{2,4}$ and $A_{2,3} = (3, 4|5)$

must be siblings and must have common child $A_{1,2} = (5,4)$. Observe that if element $A_{2,4} = 4$, then nodes $A_{2,4} = (4,2|5)$ and $A_{2,3} = (3,4|5)$ will be siblings and have common child $A_{1,2} = (5,4)$. If element $A_{2,4} = 5$, then nodes $A_{2,4} = (5,2|4)$ and $A_{2,3} = (3,4|5)$ will be siblings and have common child $A_{1,2} = (5,4)$. Hence, either $A_{2,4} = A_{2,2}$ or $A_{2,4} = A_{1,1}$ preserves regularity and immediately fixes element $A_{1,4}$. Then both *regular vine arrays* in Eq. (9.5) are possible.

$$TA_a(V(5)) = \begin{pmatrix} 1 & & & & \\ 2 & 2 & & & \\ A_{3,5} & 3 & 3 & & \\ A_{2,5} & 5 & 4 & 4 & \\ A_{1,5} & 4 & 5 & 5 & 5 \end{pmatrix} \quad TA_b(V(5)) = \begin{pmatrix} 1 & & & & \\ 2 & 2 & & & \\ A_{3,5} & 3 & 3 & & \\ A_{2,5} & 4 & 4 & 4 & \\ A_{1,5} & 5 & 5 & 5 & 5 \end{pmatrix}$$
(9.5)

Next, element $A_{3,5}$ must be found for both $TA_a(V(5))$ and $TA_b(V(5))$ in Eq. (9.5). From a similar argument as before, $A_{3,5} \in \{A_{3,3}, A_{2,2}, A_{1,1}\}$ for $TA_a(V(5))$ and $TA_b(V(5))$. Consider first $TA_a(V(5))$. Nodes $A_{3,5}$ and $A_{3,4} = (3,2|4,5)$ have common parent $A_{2,5} = (2,1|3,4,5)$ and hence are siblings. Nodes $A_{3,5}$ and $A_{3,4}$ must have a common child in the lower tree in order to keep regularity. Possible candidates are nodes $A_{2,4} = (5,2|4)$ or $A_{2,3} = (4,3|5)$. However, element $A_{4,4}$ is not an element of node $A_{3,5}$, hence $A_{2,3} = (4,3|5)$ must be the sibling of node $A_{2,5}$. And they must have a common child in the lower tree. The three possible choices for $A_{3,5}$ are listed next.

(1) $A_{3,5} = A_{3,3} = 3$, then either $A_{2,5} = (1,4|5)$ or $A_{2,5} = (1,5|4)$. $A_{2,5} = (4,5|1)$ is not a valid choice because element $A_{5,5} = 1$ and it must be in the conditioned set of every node in column 5. Nodes $A_{2,3} = (4,3|5)$ and $A_{2,5}$ must have a common child in the lower tree. This must be either $A_{1,3} = (5,3)$ or $A_{1,2} = (5,4)$. Element $A_{3,3} = 3$ is not an element of node $A_{2,5}$ from the definition of a *regular vine array* and the assumption that $A_{3,5} = A_{3,3} = 3$. Hence, node $A_{1,2} = (5,4)$ must be the common child. If node $A_{2,5} = (1,4|5)$, then nodes $A_{1,5} = (5,1)$ and $A_{1,2} = (5,4)$ are its children and regularity is preserved. On the other hand, if node $A_{2,5} = (1,5|4)$, then nodes $A_{1,5} = (4,1)$ and $A_{1,2} = (5,4)$ are its children and regularity is again preserved. Hence, element $A_{3,3} = 3$ is a valid choice for $A_{3,5}$ and both matrices in Eq. (9.6) are possible. Observe also that given element $A_{3,5} = 3$, there are two possible choices for element

$A_{2,5}$. That is, either $A_{2,5} = 4$ or $A_{2,5} = 5$ and either choice maintains regularity. $A_{1,5}$ is fixed by our previous choices.

$$TA_1(V(5)) = \begin{pmatrix} 1 & & & & \\ 2 & 2 & & & \\ 3 & 3 & 3 & & \\ 4 & 5 & 4 & 4 & \\ 5 & 4 & 5 & 5 & 5 \end{pmatrix} \quad TA_2(V(5)) = \begin{pmatrix} 1 & & & & \\ 2 & 2 & & & \\ 3 & 3 & 3 & & \\ 5 & 5 & 4 & 4 & \\ 4 & 4 & 5 & 5 & 5 \end{pmatrix}$$
(9.6)

(2) $A_{3,5} = A_{2,2} = 4$, then either $A_{2,5} = (1,5|3)$ or $A_{2,5} = (1,3|5)$. $A_{2,5} = (3,5|1)$ is not a valid choice because element $A_{5,5} = 1$ and it must be in the conditioning set of every node in column 5. Nodes $A_{2,3} = (4,3|5)$ and $A_{2,5}$ must have a common child in the lower tree. This must be either $A_{1,3} = (5,3)$ or $A_{1,2} = (5,4)$. Element $A_{2,2} = 4$ is not an element of node $A_{2,5}$ from the definition of a *regular vine array* and the assumption that $A_{3,5} = A_{3,3} = 4$. Hence, node $A_{1,3} = (5,3)$ must be the common child. If node $A_{2,5} = (1,5|3)$, then nodes $A_{1,5} = (3,1)$ and $A_{1,3} = (5,3)$ are its children and regularity is preserved. On the other hand, if node $A_{2,5} = (1,3|5)$, then nodes $A_{1,5} = (5,1)$ and $A_{1,3} = (5,3)$ are its children and regularity is again preserved. Hence, element $A_{2,2} = 4$ is a valid choice for $A_{3,5}$ and both matrices in Eq. (9.7) are possible. Observe also that given element $A_{3,5} = 4$, there are two possible choices for element $A_{2,5}$. That is, either $A_{2,5} = 3$ or $A_{2,5} = 5$ and either choice maintains regularity. $A_{1,5}$ is fixed by our previous choices.

$$TA_3(V(5)) = \begin{pmatrix} 1 & & & & \\ 2 & 2 & & & \\ 4 & 3 & 3 & & \\ 5 & 5 & 4 & 4 & \\ 3 & 4 & 5 & 5 & 5 \end{pmatrix} \quad TA_4(V(5)) = \begin{pmatrix} 1 & & & & \\ 2 & 2 & & & \\ 4 & 3 & 3 & & \\ 3 & 5 & 4 & 4 & \\ 5 & 4 & 5 & 5 & 5 \end{pmatrix}$$
(9.7)

(3) $A_{3,5} = A_{1,1} = 5$, then there are two possibilities, either $A_{2,5} = (3,1|4)$ or $A_{2,5} = (4,1|3)$. Nodes $A_{2,3} = (4,3|5)$ and $A_{2,5}$ must have a common child in the lower tree. This must be either $A_{1,3} = (5,3)$ or $A_{1,2} = (5,4)$; however, element $A_{1,3} = A_{1,2} = 5$ is not an element of node $A_{2,5}$ from the definition of a *regular vine array* and the assumption that $A_{3,5} = A_{1,1} = 5$. Hence, element $A_{1,1} = 5$ is not a valid choice for $A_{3,5}$.

By a similar procedure as described earlier, the reader may check that $TA_b(V(5))$ in Eq. (9.5) may be extended to the four regular vines in Eqs. (9.8) and (9.9).

$$TA_5(V(5)) = \begin{pmatrix} 1 & & & & \\ 2 & 2 & & & \\ 3 & 3 & 3 & & \\ 4 & 4 & 4 & 4 & \\ 5 & 5 & 5 & 5 & 5 \end{pmatrix} \quad TA_6(V(5)) = \begin{pmatrix} 1 & & & & \\ 2 & 2 & & & \\ 3 & 3 & 3 & & \\ 5 & 4 & 4 & 4 & \\ 4 & 5 & 5 & 5 & 5 \end{pmatrix} \quad (9.8)$$

$$TA_7(V(5)) = \begin{pmatrix} 1 & & & & \\ 2 & 2 & & & \\ 4 & 3 & 3 & & \\ 5 & 4 & 4 & 4 & \\ 3 & 5 & 5 & 5 & 5 \end{pmatrix} \quad TA_8(V(5)) = \begin{pmatrix} 1 & & & & \\ 2 & 2 & & & \\ 4 & 3 & 3 & & \\ 3 & 4 & 4 & 4 & \\ 5 & 5 & 5 & 5 & 5 \end{pmatrix} \quad (9.9)$$

To summarize, we may see that for every one of the two choices for $A_{2,4}$ that keep regularity, there are two choices for $A_{3,5}$ that will keep regularity. Again, for each of the two choices of $A_{3,5}$, there will also be two possible choices for $A_{2,5}$ that will keep regularity. In other words, there are $2 \times 2 \times 2 = 2^3 = 8$ regular vines possible with the natural order $NO(V(5)) = (1, 2, 3, 4, 5)$.

In Ref. 19, an argument similar to the one presented in Example 9.5.1 (using induction on n) is used to show that the number of regular vines possible with a fixed natural order $NO(V(n)) = A_{n,n}, A_{n-1,n-1}, \ldots, A_{1,1}$ is: $\prod_{j=1}^{n-3} 2^j = 2^{\binom{n-2}{2}}$. It is easy to see that there are $\binom{n}{2}$ ways of choosing the pair $A_{n,n}, A_{n-1,n-1}$ in a natural order and $(n-2)!$ ways of permuting elements $A_{n-2,n-2}, \ldots, A_{1,1}$. Hence, there are $\frac{n!}{2} \times 2^{\binom{n-2}{2}}$ labeled regular vines on n nodes.

The procedure to find regular vines from *regular vine arrays* presented in Example 9.5.1 also provides another algorithm for constructing regular vines without duplication. The algorithm is presented next. The basic idea is first to construct all possible natural orders on n nodes and then use them to build all *regular vines arrays* possible.

Algorithm 9.4. Constructing all possible regular vine arrays on n nodes.

(1) For $n \leq 3$, constructing regular vines is trivial, hence consider $n \geq 4$.
(2) Create the $\binom{n}{2} \times (n-2)! = \frac{n!}{2}$ natural orders possible on n nodes.

(3) For each of the natural orders in step (1):

   (a) Generate the regular vine on three nodes that corresponds to the natural order as in Eq. (9.1).
   (b) Set $c := 4$ $r := 2$
   (c) Find the two possibilities of selecting each one of $A_{c-2,r}, \ldots, A_{2,c}$ that would preserve regularity, given the previous choices. Each choice should be in $\{A_{c-2,c-2}, \ldots, A_{1,1}\}$ in the natural order.
   (d) If $c = n$, stop; else, set $c := c + 1$ and $r := r + 1$ and go to (b)
   (e) The $\prod_{j=1}^{n-3} 2^j = 2^{\binom{n-2}{2}}$ possibilities of building a regular vine array given the natural order have been found.

Algorithm 9.4 does not require the additional operations that Algorithm 9.3 requires for building line graphs and finding spanning trees for each tree in each level of the regular vine. However, it still requires a search in $A_{c-2,c-2}, \ldots, A_{1,1}$ in the natural order for selecting choices for each $A_{c-2,c}, \ldots, A_{2,c}$ in the regular vine array. Additionally, the construction of the regular vine array provides a natural way of enumerating regular vines. Next, the classification of regular vines is discussed.

9.6 Classifying Regular Vines

Organizing regular vines in a systematic way may be of advantage. One natural way to start classifying regular vines is according to their equivalence class as shown in Chapter 7 by Joe. Another natural way to classify them is according to the unlabeled trees used in their construction. The appendix is presented as a first step towards a better organization of regular vines. This appendix presents a catalogue of labeled regular vines on at most nine nodes, organized according to the unlabeled tree used in the first tree of the regular vine. It also presents tree-equivalent regular vines on at most seven nodes. In principle, any one of the three algorithms for generating regular vines presented in this chapter may be used to classify regular vines. Algorithms 9.2 and 9.3 were used for the construction of the catalogue presented here.

The names of trees from Table A.1 used in each level of each regular vine in Tables A.4 and A.5 will be displayed in order after the + sign. There is one tree-equivalent regular vine on three nodes: V3 = T3 + T2 + T1. Every regular vine on n nodes for $n > 3$ must necessarily use V3 in its construction. For this reason, T3 + T2 + T1 will be omitted when indicating the sequence of trees used in the construction of different

Table 9.1. Apportioning regular vines from Example 9.5.1 to tree-equivalence classes.

Tree sequence	Vines from Example 9.5.1 within given tree sequence
T6+T4	TA_3
T7+T4	TA_4, TA_7
T7+T5	TA_1, TA_2, TA_6
T8+T4	TA_8
T8+T5	TA_5

tree-equivalent regular vines. For example, the D-vine on four nodes will be V4 = T4 + V3 = T4.

The eight regular vines generated in Example 9.5.1 may be classified according to their equivalence and tree-equivalence classes. According to Table A.4 in the appendix, there are five possible tree sequences for regular vines on five nodes. These five sequences are shown in the first column of Table 9.1. The second column apportions the regular vines from Example 9.5.1 to these five tree sequences. The names of trees from Table A.1 used in each level of each regular vine in Example 9.5.1 are displayed as explained before.

The reader may check that TA_3 corresponds to a D-vine and TA_5 to a C-vine. TA_8 has one node with maximal degree in its first tree and a D-vine on four nodes is attached to this first tree. It is evident that these three vines, besides corresponding to different tree-equivalence classes, correspond to different equivalence classes.

Besides being tree-equivalent TA_4 and TA_7, are in the same equivalence class. The reader may check this by permuting nodes $4 \leftrightarrows 3$ and $2 \leftrightarrows 1$ in either TA_4 or TA_7 to obtain the same vine.

TA_1, TA_2 and TA_6 are tree-equivalent. By making node $2 = 1$, $5 = 4$, $3 = 2$ and $1 = 3$ in TA_6, it becomes TA_2 and hence these two are in the same equivalence class. However, TA_1 cannot be transformed into either TA_2 or TA_6 by a permutation of nodes in the first tree. Hence, TA_1 falls in a different equivalence class than TA_2 and TA_6 despite the fact that the three are tree-equivalent. Observe that the six equivalence classes for regular vines on five nodes mentioned in Chapter 7 are represented in Example 9.5.1.

According to the results from the previous section, there are $\frac{5!}{2} = 60$ other possible natural orders than the one used in Example 9.5.1. Hence, there must be 60×1 D-vines and C-vines. Also, 60×2 regular vines in the class of TA_4 and TA_7 must be observed. Finally, there must be 60×3 regular vines with tree-sequence T7 + T5. This result may also be observed in V6-V10 in Table A.4 in the appendix. Observe that after the classification of the *regular vine arrays* from Example 9.5.1, from the 180 regular vines with tree-sequence T7+T5, 60 must be in the same equivalence class as TA_1. The appendix was elaborated without the implementation of *regular vine arrays*. This verifies that it is possible to arrive at the same conclusions with the three different methods proposed in this chapter. Finally, observe that it is sufficient to investigate one natural order to classify regular vines within equivalence and tree-equivalence classes.

9.7 Conclusions and Final Comments

A way to efficiently code and store vines on n nodes based on the Prüfer code is proposed. This consists of an upper triangular matrix of size $(n-2)\times(n-2)$. An algorithm for building vines and two others for building regular vines on n nodes have been presented. Algorithm 9.2 is easy to implement and efficient if regular vines on less than six nodes are required. Algorithms 9.3 and 9.4 would produce only regular vines at the cost of greater programming effort and a larger number of arithmetic operations. Tables A.1 to A.3 present the number of labeled trees, regular vines per labeled tree and tree-equivalent regular vines according to unlabeled trees on at most nine nodes. Table A.1 presents the 25 trees on seven nodes or less. These trees will be used to present pictures of tree-equivalent regular vines on at most six nodes in Table A.4. Finally, Table A.5 present tree-equivalent regular vines on seven nodes.

We have made a first step towards organizing vines and regular vines in a more systematic way. We believe that this task is necessary in order to progress more rapidly the space of applications to vines and make them more accessible to people interested in the subject. Hence, our recommendation is to enhance efforts for a more systematic organization of vines, including algorithms for generating and storing them.

Appendix
A Catalogue of Labeled Regular Vines on at Most Nine Nodes and Tree-Equivalent Regular Vines on at Most Seven Nodes

The purpose of this catalogue is to classify regular vines according to their graphical structure. We hope that this catalogue will help researchers interested in regular vines with their investigations. Like the authors of Ref. 23, this author has "tried that the data are free of errors, but accept[s] no responsibility for any loss of time, money, patience or temper occurring as a result of any mistakes that may have crept into the pages of this [catalogue]. Furthermore, [the author] wishes it to be understood that any mistakes are entirely the fault of the other author".

Tables A.1–A.3 present the 95 trees on seven nodes or less. Catalogues of trees on at most 12 nodes have been presented before. In Ref. 18, pictures of trees on at most five nodes are presented. In Ref. 12, a catalogue of trees on at most eight nodes may be found.[h] Harary[10] presents trees on at most ten nodes.[i] The 987 trees on at most 12 vertices (together with about 10,000 other graphs and many tables of interest for graph theorists) may be found in Ref. 23. None of the above catalogues present results concerning vines.

Vines will be presented in pictures in the next section and the names of the trees from Table A.1 used in each level of each regular vine in Tables A.4 and A.5 will be displayed in order after the + sign. There is one tree-equivalent regular vine on three nodes V3 = T3 + T2 + T1. Every regular vine on n nodes for $n > 3$ must necessarily use V3 in its construction. For this reason, T3 + T2 + T1 will be omitted when indicating the sequence of trees used in the construction of different tree-equivalent regular vines. For example, the D-vine on four nodes will be V4 = T4 + V3 = T4. Next, the catalogue is presented.

Table A.1. Trees with at most seven nodes.

Prüfer code example			1	12	11	123
	T1	T2	T3	T4	T5	T6
# Labeled trees	1	1	3	12	4	60
# Regular vines per labeled tree	1	1	1	1	3	1
# Tree-equivalent reg. vines / tree	1	1	1	1	1	1

---

[h]This catalogue repeats a tree on eight nodes, neglecting, another one. In the same reference, tables with the number of non-isomorphic trees on less than 26 nodes may be found.

[i]Harary refers to Ref. 20 for diagrams of trees on at most 12 nodes. However this reference is not available to the author at the time of the publication of this catalogue.

Table A.1. (*Continued*)

Prüfer code example	112	111	1234	1123	1213	2244
	T7	T8	T9	T10	T11	T12
# Labeled trees	60	5	360	360	360	90
# Regular vines per labeled tree	5	24	1	7	11	48
# Tree-equivalent reg. vines / tree	2	2	1	3	3	5
Prüfer code example	1112	1111	12345	12344	12234	12324
	T13	T14	T15	T16	T17	T18
# Labeled trees	120	6	2,520	2,520	5,040	840
# Regular vines per labeled tree	75	480	1	9	19	33
# Tree-equivalent reg. vines / tree	5	5	1	4	7	3

Table A.1. (*Continued*)

Prüfer code example	11233	11223	11123	12223
	T19	T20	T21	T22
# Labeled trees	630	2,520	840	1,260
# Regular vines per labeled tree	80	168	168	342
# Tree-equivalent reg. vines / tree	9	17	12	17
Prüfer code example	11122	11112	11111	
	T23	T24	T25	
# Labeled trees	420	210	7	
# Regular vines per labeled tree	1,452	2,928	23,040	
# Tree-equivalent reg. vines / tree	22	22	22	

Table A.2. Trees with at most eight nodes.

Prüfer code example	123456	123455	122345	123345	123435	112324
	T26	T27	T28	T29	T30	T31
# Labeled trees	20,160	20,160	40,320	20,160	20,160	10,080
# Regular vines per labeled tree	1	11	29	39	71	820
# Tree-equivalent reg. vines / tree	1	5	12	8	10	44

Prüfer code example	112344	122344	122334	123344	112233	122324
	T32	T33	T34	T35	T36	T37
# Labeled trees	5,040	20,160	20,160	20,160	5,040	6,720
# Regular vines per labeled tree	120	315	815	423	4,520	2,181
# Tree-equivalent reg. vines / tree	14	38	55	41	72	44

Table A.2. (*Continued*)

Prüfer code example	244466	123444	123334	112333	122333	111222
	T38	T39	T40	T41	T42	T43
# Labeled trees	10,080	6,720	20,160	3,360	6,720	560
# Regular vines per labeled tree	11,246	315	1,046	3,384	8,667	89,712
# Tree-equivalent reg. vines / tree	114	24	61	72	111	133

Prüfer code example	122223	123333	112222	122222	222222	
	T44	T45	T46	T47	T48	
# Labeled trees	3,360	1,680	840	336	8	
# Regular vines per labeled tree	27,222	11,160	117,072	279,000	2,580,480	
# Tree-equivalent reg. vines / tree	114	83	136	136	136	

Table A.3. Trees with nine nodes.

Prüfer code example	2345678	2345578	2345668	2345677	2345658	2345478
	T49	T50	T51	T52	T53	T54
# Labeled trees	181,440	362,880	362,880	181,440	181,440	181,440
# Regular vines on each tree	1	69	41	13	129	181
# Tree-equivalent reg. vines / tree	1	21	18	6	22	18

Prüfer code example	2345477	2335658	2343677	2335668	2344668	2245677
	T55	T56	T57	T58	T59	T60
# Labeled trees	181,440	181,440	90,720	181,440	362,880	45,360
# Regular vines on each tree	2,651	5,390	1,708	1,646	2,708	168
# Tree-equivalent reg. vines / tree	164	203	104	125	221	20

Prüfer code example	2335677	2344677	2345577	2344478	2345558	2345666
	T61	T62	T63	T64	T65	T66
# Labeled trees	181,440	181,440	181,440	90,720	181,440	60,480
# Regular vines on each tree	528	887	887	4,202	2,567	528
# Tree-equivalent reg. vines / tree	70	105	91	147	162	42

Table A.3. (*Continued*)

Prüfer code example	2345448	2343638	2245577	2335577	2245477	2344438
	T67	T68	T69	T70	T71	T72
# Labeled trees	181,440	15,120	90,720	181,440	45,360	30,240
# Regular vines on each tree	8,738	18,504	11,296	34,417	36,892	72,546
# Tree-equivalent reg. vines / tree	275	99	287	628	350	428
Prüfer code example	2343377	2225668	2333668	2344666	2225677	2333677
	T73	T74	T75	T76	T77	T78
# Labeled trees	90,720	60,480	181,440	60,480	30,240	90,720
# Regular vines on each tree	120,444	20,904	99,028	34,143	6,756	32,812
# Tree-equivalent reg. vines / tree	724	332	840	439	166	516
Prüfer code example	2344477	2345555	2344448	2333637	2244666	2244477
	T79	T80	T81	T82	T83	T84
# Labeled trees	90,720	15,120	60,480	30,240	30,240	22,680
# Regular vines on each tree	54,004	32,688	149,901	360,084	428,388	680,576
# Tree-equivalent reg. vines / tree	607	245	765	724	980	1,034

Table A.3. (*Continued*)

Prüfer code example	2225666	2333666	2245555	2333377
	T85	T86	T87	T88
# Labeled trees	5,040	30,240	7,560	30,240
# Regular vines on each tree	262,080	1,232,820	414,432	1,919,610
# Tree-equivalent reg. vines / tree	465	1,328	735	1,328
Prüfer code example	2335555	2344444	2333338	2225555
	T89	T90	T91	T92
# Labeled trees	15,120	3,024	7,560	2,520
# Regular vines on each tree	1,232,340	1,869,120	5,255,904	14,889,744
# Tree-equivalent reg. vines / tree	1,195	901	1,328	1,464
Prüfer code example	2244444	2333333	1111111	
	T93	T94	T95	
# Labeled trees	1,512	504	9	
# Regular vines on each tree	23,334,480	62,523,360	660,602,880	
# Tree-equivalent reg. vines / tree	1,464	1,464	1,464	

Table A.4. Tree-equivalent regular vines with at most six nodes.

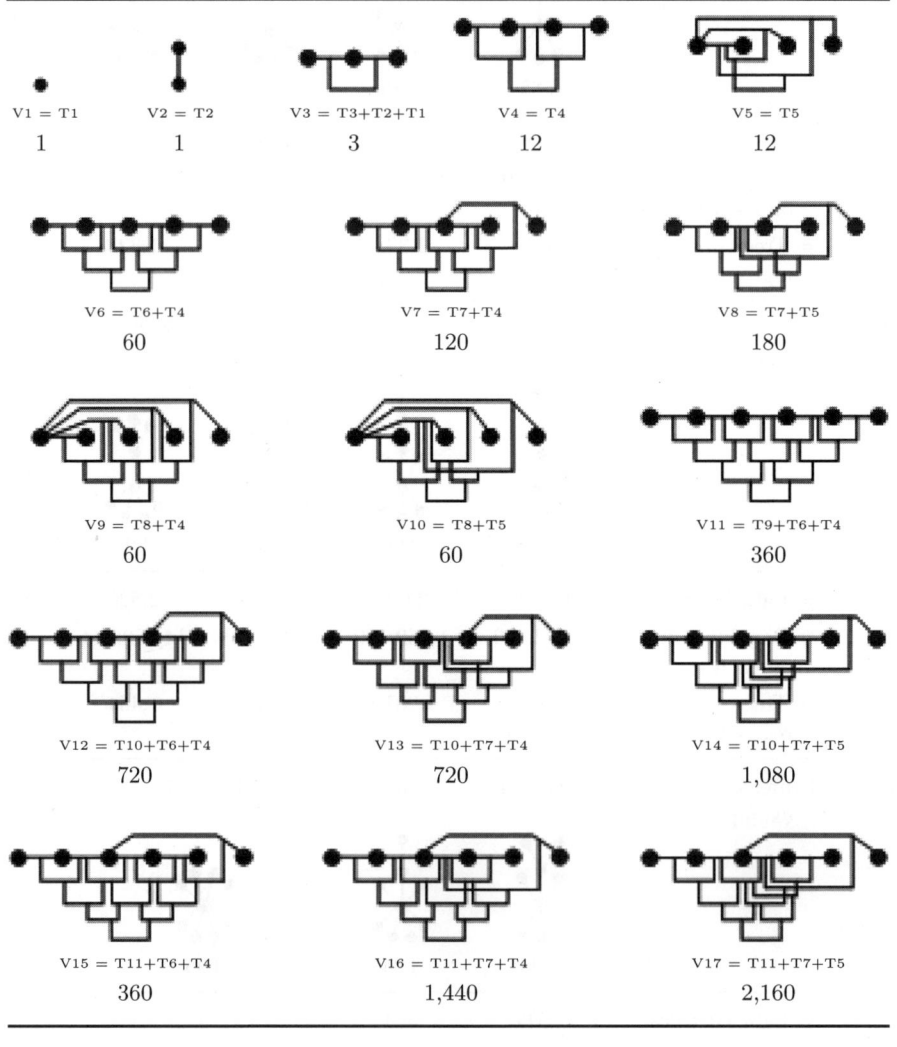

Table A.4. (*Continued*)

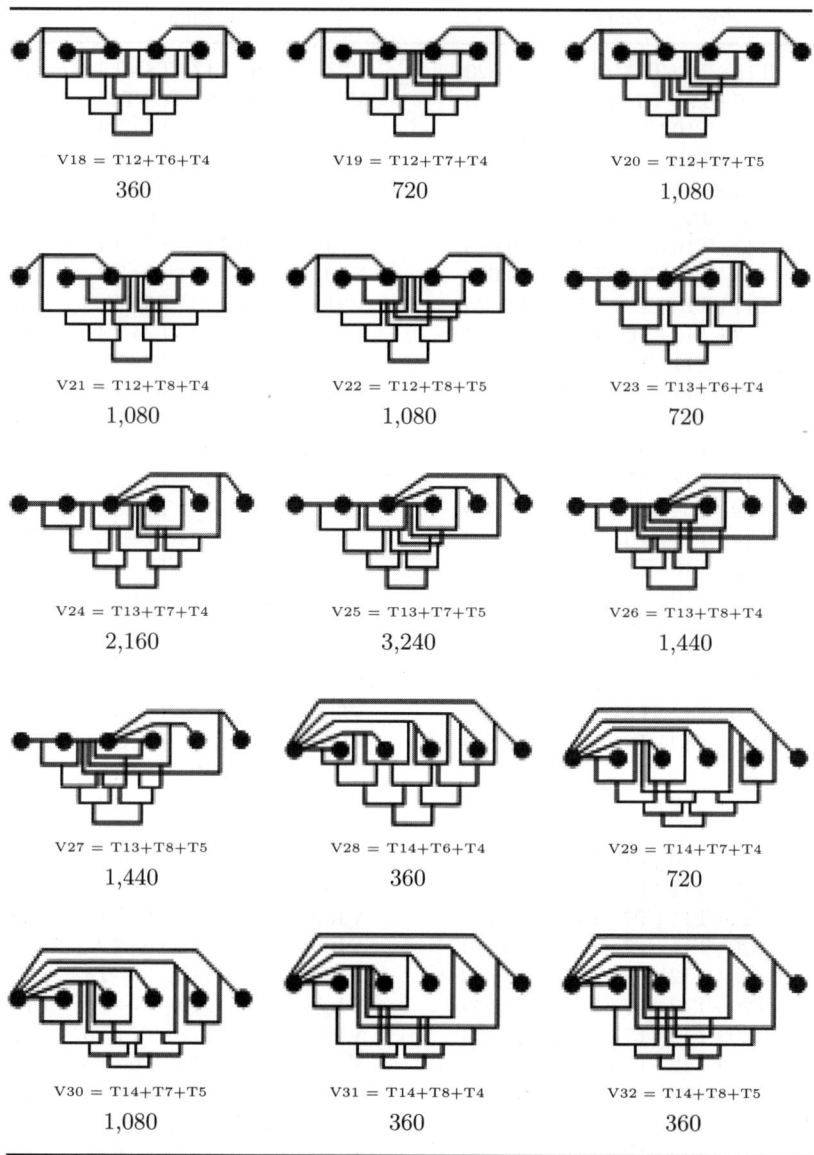

V18 = T12+T6+T4 360	V19 = T12+T7+T4 720	V20 = T12+T7+T5 1,080
V21 = T12+T8+T4 1,080	V22 = T12+T8+T5 1,080	V23 = T13+T6+T4 720
V24 = T13+T7+T4 2,160	V25 = T13+T7+T5 3,240	V26 = T13+T8+T4 1,440
V27 = T13+T8+T5 1,440	V28 = T14+T6+T4 360	V29 = T14+T7+T4 720
V30 = T14+T7+T5 1,080	V31 = T14+T8+T4 360	V32 = T14+T8+T5 360

Table A.5. Tree-equivalent regular vines with seven nodes.

Tree sequence & # Tree-equivalent labeled regular vines		Tree sequence & # Tree-equivalent labeled regular vines	
V33 = T15+T9+T6+T4	2,520	V73 = T20+T13+T8+T5	30,240
V34 = T16+T9+T6+T4	5,040	V74 = T21+T9+T6+T4	5,040
V35 = T16+T10+T6+T4	5,040	V75 = T21+T10+T6+T4	5,040
V36 = T16+T10+T7+T4	5,040	V76 = T21+T10+T7+T4	5,040
V37 = T16+T10+T7+T5	7,560	V77 = T21+T10+T7+T5	7,560
V38 = T17+T9+T6+T4	5,040	V78 = T21+T11+T6+T4	5,040
V39 = T17+T10+T6+T4	10,080	V79 = T21+T11+T7+T4	20,160
V40 = T17+T10+T7+T4	10,080	V80 = T21+T11+T7+T5	30,240
V41 = T17+T10+T7+T5	15,120	V81 = T21+T13+T6+T4	5,040
V42 = T17+T11+T6+T4	5,040	V82 = T21+T13+T7+T4	15,120
V43 = T17+T11+T7+T4	20,160	V83 = T21+T13+T7+T5	22,680
V44 = T17+T11+T7+T5	30,240	V84 = T21+T13+T8+T4	10,080
V45 = T18+T11+T6+T4	2,520	V85 = T21+T13+T8+T5	10,080
V46 = T18+T11+T7+T4	10,080	V86 = T22+T9+T6+T4	2,520
V47 = T18+T11+T7+T5	15,120	V87 = T22+T10+T6+T4	10,080
V48 = T19+T9+T6+T4	2,520	V88 = T22+T10+T7+T4	10,080
V49 = T19+T10+T6+T4	5,040	V89 = T22+T10+T7+T5	15,120
V50 = T19+T10+T7+T4	5,040	V90 = T22+T11+T6+T4	7,560
V51 = T19+T10+T7+T5	7,560	V91 = T22+T11+T7+T4	30,240
V52 = T19+T12+T6+T4	2,520	V92 = T22+T11+T7+T5	45,360
V53 = T19+T12+T7+T4	5,040	V93 = T22+T12+T6+T4	10,080
V54 = T19+T12+T7+T5	7,560	V94 = T22+T12+T7+T4	20,160
V55 = T19+T12+T8+T4	7,560	V95 = T22+T12+T7+T5	30,240
V56 = T19+T12+T8+T5	7,560	V96 = T22+T12+T8+T4	30,240
V57 = T20+T9+T6+T4	5,040	V97 = T22+T12+T8+T5	30,240
V58 = T20+T10+T6+T4	15,120	V98 = T22+T13+T6+T4	15,120
V59 = T20+T10+T7+T4	15,120	V99 = T22+T13+T7+T4	45,360
V60 = T20+T10+T7+T5	22,680	V100 = T22+T13+T7+T5	68,040
V61 = T20+T11+T6+T4	5,040	V101 = T22+T13+T8+T4	30,240
V62 = T20+T11+T7+T4	20,160	V102 = T22+T13+T8+T5	30,240
V63 = T20+T11+T7+T5	30,240	V103 = T23+T9+T6+T4	5,040
V64 = T20+T12+T6+T4	10,080	V104 = T23+T10+T6+T4	10,080
V65 = T20+T12+T7+T4	20,160	V105 = T23+T10+T7+T4	10,080
V66 = T20+T12+T7+T5	30,240	V106 = T23+T10+T7+T5	15,120
V67 = T20+T12+T8+T4	30,240	V107 = T23+T11+T6+T4	5,040
V68 = T20+T12+T8+T5	30,240	V108 = T23+T11+T7+T4	20,160
V69 = T20+T13+T6+T4	15,120	V109 = T23+T11+T7+T5	30,240
V70 = T20+T13+T7+T4	45,360	V110 = T23+T12+T6+T4	5,040
V71 = T20+T13+T7+T5	68,040	V111 = T23+T12+T7+T4	10,080
V72 = T20+T13+T8+T4	30,240	V112 = T23+T12+T7+T5	15,120

Table A.5. (Continued)

Tree sequence & # Tree-equivalent labeled regular vines		Tree sequence & # Tree-equivalent labeled regular vines	
V113 = T23+T12+T8+T4	15,120	V141 = T24+T13+T8+T5	40,320
V114 = T23+T12+T8+T5	15,120	V142 = T24+T14+T6+T4	12,600
V115 = T23+T13+T6+T4	20,160	V143 = T24+T14+T7+T4	25,200
V116 = T23+T13+T7+T4	60,480	V144 = T24+T14+T7+T5	37,800
V117 = T23+T13+T7+T5	90,720	V145 = T24+T14+T8+T4	12,600
V118 = T23+T13+T8+T4	40,320	V146 = T24+T14+T8+T5	12,600
V119 = T23+T13+T8+T5	40,320	V147 = T25+T9+T6+T4	2,520
V120 = T23+T14+T6+T4	25,200	V148 = T25+T10+T6+T4	5,040
V121 = T23+T14+T7+T4	50,400	V149 = T25+T10+T7+T4	5,040
V122 = T23+T14+T7+T5	75,600	V150 = T25+T10+T7+T5	7,560
V123 = T23+T14+T8+T4	25,200	V151 = T25+T11+T6+T4	2,520
V124 = T23+T14+T8+T5	25,200	V152 = T25+T11+T7+T4	10,080
V125 = T24+T9+T6+T4	5,040	V153 = T25+T11+T7+T5	15,120
V126 = T24+T10+T6+T4	15,120	V154 = T25+T12+T6+T4	2,520
V127 = T24+T10+T7+T4	15,120	V155 = T25+T12+T7+T4	5,040
V128 = T24+T10+T7+T5	22,680	V156 = T25+T12+T7+T5	7,560
V129 = T24+T11+T6+T4	7,560	V157 = T25+T12+T8+T4	7,560
V130 = T24+T11+T7+T4	30,240	V158 = T25+T12+T8+T5	7,560
V131 = T24+T11+T7+T5	45,360	V159 = T25+T13+T6+T4	5,040
V132 = T24+T12+T6+T4	10,080	V160 = T25+T13+T7+T4	15,120
V133 = T24+T12+T7+T4	20,160	V161 = T25+T13+T7+T5	22,680
V134 = T24+T12+T7+T5	30,240	V162 = T25+T13+T8+T4	10,080
V135 = T24+T12+T8+T4	30,240	V163 = T25+T13+T8+T5	10,080
V136 = T24+T12+T8+T5	30,240	V164 = T25+T14+T6+T4	2,520
V137 = T24+T13+T6+T4	20,160	V165 = T25+T14+T7+T4	5,040
V138 = T24+T13+T7+T4	60,480	V166 = T25+T14+T7+T5	7,560
V139 = T24+T13+T7+T5	90,720	V167 = T25+T14+T8+T4	2,520
V140 = T24+T13+T8+T4	40,320	V168 = T25+T14+T8+T5	2,520

References

1. Aas K. and Berg D. (2009). Models for construction of multivariate dependence: A comparison study. *European Journal of Finance*, 15:639–659.
2. Aas K., Czado C., Frigessi A. and Bakken H. (2009). Pair-copula constructions of multiple dependence. *Insurance: Mathematics and Economics*, 44(2):182–198.
3. Bedford T. J. and Cooke R. M. (2002). Vines: A new graphical model for dependent random variables. *Annals of Statistics*, 30(4):1031–1068.
4. Beineke L. (2006). Derived graphs with derived complements. *Recent Trends in Graph Theory: Proceedings of the First New York City Graph Theory Conference held on June 11, 12, and 13, 1970*. Springer.

5. Cayley A. (1889). A theorem on trees. *The Quarterly Journal of Pure and Applied Mathematics*, 23:376–378.
6. Chollete L., Heinen A. and Valdesogo A. (2009). Modeling international financial returns with a multivariate regime switching copula. *Journal of Financial Econometrics*, 7(4):437–480.
7. Chow C. and Liu C. (1968). Approximating discrete probability distributions with dependence trees. *IEEE Transactions on Information Theory*, 14(3):462–467.
8. Cooke R.M. (1997). Markov and entropy properties of tree and vine-dependent variables. *Proceedings of the ASA Section on Bayesian Statistical Science*. American Statistical Association, Washington.
9. Darroch J.N., Lauritzen S. and Speed T.P. (1980). Markov fields and log-linear interaction models for contingency tables. *Annals of Statistics*, 8(3):522–539.
10. Harary F. (1969). *Graph Theory*. Addison-Wesley, Reading, MA.
11. Harary F. and Palmer E.M. (1973). *Graphical Enumeration*. Academic Press, New York.
12. Kasyanov V.N. and Evstigneev V.A. (2000). *Graph Theory for Programmers: Algorithms for Processing Trees*. Kluwer Academic Publishers, Norwell, MA.
13. Kolbjornsen O. and Stien M. (2008). D-vine creation of non-Gaussian random field. *Proceedings of the Eight International Geostatistics Congress*, 399–408.
14. Kurowicka D. and Cooke R.M. (2006). *Uncertainty Analysis with High Dimensional Dependence Modelling*. Wiley, New York.
15. Mayeda W. and Seshu S. (1967). Generation of trees without duplication. *IEEE Transactions on Circuit Theory*, 12:181–185.
16. Min A. and Czado C. (2010). Bayesian inference for multivariate copulas using pair copula constructions. *Journal of Financial Econometrics*. In press.
17. Minty G.J. (1965). A simple algorithm for listing all the trees of a graph. *IEEE Transactions on Circuit Theory*, 12:120–120.
18. Moon J. (1967). Various proofs of Cayley's formula for counting trees. In: *A Seminar on Graph Theory*, F. Harary (ed.). Holt, Rihehat and Winston, New York.
19. Morales-Nápoles O., Cooke R.M. and Kurowicka D. (2009). The number of vines and regular vines on n nodes. Submitted to *Discrete Applied Mathematics*.
20. Prins G. (1957). On the automorphism group of a tree. PhD thesis, University of Michigan.
21. Prüfer V.H. (1918). Neuer Beweis Eines Satzes Über Permutationen. *Archiv der Mathematik und Physik*, 27:742–744.
22. Read R.C. and Tarjan R.E. (1975). Bounds on backtrack algorithms for listing cycles, paths, and spanning trees. *Networks*, 5:678–692.
23. Read R.C. and Wilson R.J. (2005). *An Atlas of Graphs*. Oxford University Press, Oxford.
24. Shioura A., Tamura A. and Uno T. (1994). An optimal algorithm for scanning all spanning trees of undirected graphs. *SIAM Journal on Computing*, 26:678–692.
25. Smith M.J. (1997). Generating spanning trees. Master's thesis, University of Victoria.
26. Speed T.P. and Kiiveri H.T. (1986). Gaussian Markov distributions over finite graphs, *Annals of Statistics*, 14(1):138–150.

CHAPTER 10

Regular Vines: Generation Algorithm and Number of Equivalence Classes

Harry Joe*, Roger M. Cooke[†] and Dorota Kurowicka[†]

*Department of Statistics, University of British Columbia
[†]Department of Mathematics, Delft University of Technology

A natural order for a regular vine on n variables is an assignment of indices to the variables such that variables indexed with j and $j+1$ occur as conditioned variables in a node of tree j, $j = 1, \ldots, n-1$. Regular vines V and U on n variables are equivalent if there is a permutation $\pi \in n!$ such that $\pi(V) = U$. U and V are equivalent if and only if the regular vines in natural order corresponding to U and V are equivalent. The number of equivalence classes for regular vines is obtained by counting the number of equivalence classes for regular vines in natural order.

10.1	Introduction	219
10.2	Naming Convention for Vines	221
10.3	Number of Equivalence Classes	222
10.4	Examples	230
10.5	Discussion	231
	Reference	231

10.1 Introduction

Morales-Nápoles et al.[1] introduced the notion of a natural order for regular vines. Roughly, this is a method for assigning indices to the variables of a regular vine on n variables, such that a conditioned variable in the top node (the single node in tree $n-1$) gets index n, and its partner in this conditioned set gets index $n-1$. Further, if indices $n, \ldots, j+1$ have been assigned, index j is assigned to the conditioned-set partner of the variable index $j+1$ in the node of tree j which is an m-child of the node with conditioned set $(j+1, j+2)$. Hence, a natural ordering for a regular vine is a permutation of the indices defined with respect to that vine. For any regular vine, there are two natural orderings, according to the variable in the top node which is assigned index n. There are two regular vines in natural order corresponding to the original vine, and in some cases these two vines actually coincide. The two natural orderings for a C-vine and D-vine always yield the same regular vine.[a] Let $NO(n)$ denote the set of regular vines on n variables in a natural order. Morales-Nápoles et al.[1] show that there are 2^{n-3} ways of extending a regular vine on $n-1$ variables in natural order, to a regular vine in natural order on n elements; and that the set $NO(n)$ of regular vines in natural order has cardinality $N_n = 2^{(n-2)(n-3)/2}$.

A useful tool in deriving this result is the representation of a regular vine as a triangular array, or vine array. An $n \times n$ upper triangular matrix A is a vine array if (i) $A_{ii} = i$ for $i = 1, \ldots, n$, (ii) $A_{i-1,i} = i - 1$ for $i = 2, \ldots, n$, (iii) $A_{1,i}, \ldots, A_{i-2,i}$ is a permutation of $\{1, \ldots, i-2\}$ and (iv) another condition based on binary vectors is satisfied and it restricts the number of possible permutations to 2^{i-3} in columns $i = 4, \ldots, n$.

Regular vines V and U on n variables are equivalent if there is a permutation $\pi \in n!$ such that $\pi(V) = U$. This means that $\{i, j | k \cdots m\}$ is a node of V if and only if $\{\pi(i), \pi(j) | \pi(k) \cdots \pi(m)\}$ is a node of U. U and V are equivalent if and only if the regular vines in natural order corresponding to U and V are equivalent (or indeed identical). The purpose of this chapter is to count the number of equivalence classes for regular vines by counting the number of equivalence classes for regular vines in natural order.

Based on the theory in Morales-Nápoles et al.,[1] an algorithm for generating all vine arrays $A \in NO(n)$, $n \geq 4$, is the following.

1. Input $b_4, \ldots, b_n$ where $b_i = b_i(\cdot)$ is a binary vector of length i or a mapping from $(1, \ldots, i)$ to $\{0, 1\}^i$. Assume $b_i(1) = b_i(i-1) = b_i(i) = 1$ for $i = 4, \ldots, n$. [b_1, b_2, b_3, with these properties, can also be added but they would be fixed over all $A \in NO(n)$.]
2. For an $n \times n$ matrix, set $A_{i,i} = i$ for $i = 1, \ldots, n$, $A_{i-1,i} = i - 1$ for $i = 2, \ldots, n$, $A_{1,3} = 1$.
3. For columns $d = 4, \ldots, n$,

    - ac = active column $\leftarrow d - 2$;
    - For j from $d - 2$ to 1

---

[a]The B1-vine in Chapter 7 is an example in which the two natural orderings yield different regular vines.

- If $b_d(j) = 1$, then $A_{j,d} = A_{ac,ac}$ and the new ac is the largest value among $1, \ldots, d-2$ not yet assigned in column j;
- Else if $b_d(j) = 0$, then $A_{j,d} = A_{j-1,ac}$.

4. Return A.

Next is an example to illustrate the algorithm and the b_i notation for a regular vine A in natural order. In (10.1) below, $b_4(2) = 1$, $b_5(3) = 1$, $b_5(2) = 0$, $b_6(4) = 0$, $b_6(3) = 1$, $b_6(2) = 0$ are set and applied sequentially in the algorithm to get respectively $A_{2,4} = 2$, $A_{3,5} = 3$, $A_{2,5} = 1$, $A_{4,6} = 3$, $A_{3,6} = 4$, $A_{2,6} = 1$:

$$\begin{bmatrix} b_1 & b_2 & b_3 & b_4 & b_5 & b_6 \\ 1 & 1 & 1 & 1 & 1 & 1 \\ & 1 & 1 & 1 & 0 & 0 \\ & & 1 & 1 & 1 & 1 \\ & & & 1 & 1 & 0 \\ & & & & 1 & 1 \\ & & & & & 1 \end{bmatrix}, \quad A = \begin{bmatrix} 1 & 1 & 1 & 1 & 2 & 2 \\ & 2 & 2 & 2 & 1 & 1 \\ & & 3 & 3 & 3 & 4 \\ & & & 4 & 4 & 3 \\ & & & & 5 & 5 \\ & & & & & 6 \end{bmatrix} \quad (10.1)$$

With nodes in the dth column being:

$$A_{1,d}A_{d,d}, A_{2,d}A_{d,d}|A_{1,d}, \ldots, A_{i,d}A_{d,d}|A_{1,d}\cdots A_{i-1,d}, \ldots$$

the nodes of this vine are

$$\begin{bmatrix} 12 & 13 & 14 & 25 & 26 \\ & 23|1 & 24|1 & 15|2 & 16|2 \\ & & 34|12 & 35|12 & 46|12 \\ & & & 45|123 & 36|124 \\ & & & & 56|1234 \end{bmatrix}.$$

Note that this vine is invariant to transpositions of 3,4 and 5,6 (at the same time).

10.2 Naming Convention for Vines

In this section, we use a notation for vines of dimenions $n \geq 5$ based on the binary vectors $b_4, \ldots, b_n$. If $b_4(2) = 1$, we start the vine with the letter "C", and if $b_4(2) = 0$, we start the vine with the letter "D". Then for columns $j \in 5, \ldots, n$, we consider $b_j(2) \cdots b_j(j-2)$ as a binary number and convert it to decimal. The decimal form of the columns are separated by dots. That is, the notation of the vine has the form D.$i_5.\cdots.i_n$ or C.$i_5.\cdots.i_n$, where i_d is an integer between 0 and $2^{d-3} - 1$ inclusive for $d = 5, \ldots, n$.

Specifically, for $n = 5$, the eight arrays in $NO(5)$ are denoted as D.0, D.1, D.2, D.3, C.0, C.1, C.2, C.3. For $n = 6$, the 64 arrays in $NO(6)$ are denoted as D.0.0, ..., D.0.7, ..., D.3.7, C.0.0, ..., C.0.7, ..., C.3.7. For $n = 7$, the 1024 arrays in $NO(7)$ are denoted as D.0.0.0, ..., D.0.0.15, ..., D.3.7.15, C.0.0.0, ..., C.0.0.15, ..., C.3.7.15. In this notation, D.0, D.0.0, D.0.0.0, etc. are the D-vines and C.3, C.3.7, C.3.7.15, etc. are the C-vines, and everything else is in-between. That is, for the D-vines, $b_d(2), \ldots, b_d(d-2)$ are 0s for $d = 4, \ldots, n$, and for the C-vines, $b_d(2), \ldots, b_d(d-2)$ are 1s for $d = 4, \ldots, n$. This notation also shows that the D-vines and C-vines are the boundary cases of regular vines.

This compact notation of the binary vectors in columns 5 to n in terms of decimal equivalents is convenient for the generation of the vine arrays in $NO(n)$.

10.3 Number of Equivalence Classes

We assume $n \geq 5$ below. For dimension n, we consider permutations of $\{1, \ldots, n\}$ in a vine such that the new vine array A^* can be put into the natural order. This will enable us to count the number of equivalence classes of regular vines in dimension n. The number of such permutations will be shown to be $[n/2]$, that is, $n/2$ for n even and $(n-1)/2$ for n odd.

Note that in $NO(n)$, the conditional distributions or nodes $23|1; 34|12; \ldots; n-1, n|1, \ldots, n-2$ must exist. The first observation is that it is only necessary to check the subgroup of permutations generated from the transpositions:

(a) $(1\ 2), (3\ 4), \ldots, (n-1\ n)$ if n is even,
(b) $(2\ 3), (4\ 5), \ldots, (n-1\ n)$ if n is odd.

For $n = 5$, in order that a permutation of $1, 2, 3, 4, 5$ leaves the node $45|123$ invariant, it can transpose $4, 5$ and permute $1, 2, 3$; then to leave $23|1$ invariant, permutations of $1, 2, 3$ can only transpose $2, 3$. For $n = 6$, in order that a permutation of $1, 2, 3, 4, 5, 6$ leaves the node $56|1234$ invariant, it can transpose $5, 6$ and permute $1, 2, 3, 4$; then to leave $34|12$ invariant, permutations of $1, 2, 3, 4$ can only transpose $3, 4$ and $1, 2$. Generalizing these two cases leads to the above.

But there are more restrictions. For $n = 5$, with nodes $34|12$, $j_3 5|j_1 j_2$ and $45|123$, if only $4, 5$ are transposed we get nodes $35|12$, $j_3 4|j_1 j_2$ and $45|123$. Therefore $j_3 = 3$ and $\{j_1, j_2\} = \{1, 2\}$. Hence $35|12$ is a node of A in order that the (4 5) transposition can be considered. This corresponds to $b_5(3) = 1$ in generating the vine array. For $n > 5$, with nodes $n-2, n-1|1\cdots n-3, j_{n-2}n|j_1\cdots j_{n-3}$ and $n-1,n|1\cdots n-2$, if only $n-1,n$ are transposed we get nodes $n-2, n|1\cdots n-3$, $j_{n-2}, n-1|j_1\cdots j_{n-3}$ and $n-1, n|1\cdots n-2$. Therefore $j_{n-2} = n-2$ and $\{j_1,\ldots,j_{n-3}\} = \{1,\ldots,n-3\}$. Hence $n-2, n|1\cdots n-3$ is a node of A in order that the $(n-1\ n)$ transposition can be considered. This corresponds to $b_n(n-2) = 1$ in generating the vine array.

Next, consider $n = 5$ and transposing $2,3$. With the permutation denoted as π, this means the nodes $12, 13, 23|1, j_35|j_1j_2, 34|12$ map to $13, 12, 32|1, \pi(j_3)\pi(5)|\pi(j_1)\pi(j_2), 24|13$. In order that the permutation includes $34|12$, it is necessary that 5 be permuted and the above implies that it is transposed with 4. Hence if $2,3$ are transposed, it is necessary that $4,5$ be transposed. For $n = 6$, if $3,4$ are transposed, then it is necessary that $5,6$ be transposed; if $1,2$ are transposed, then it is necessary that pairs $3,4$ and $5,6$ be transposed.

The generalization is that the only permutations to considered for equivalent classes are:

(a) $(n-1\ n)$, $(n-3\ n-4)(n-1\ n)$, ..., $(1\ 2)\cdots(n-1\ n)$ if n is even,
(b) $(n-1\ n)$, $(n-3\ n-4)(n-1\ n)$, ..., $(2\ 3)\cdots(n-1\ n)$ if n is odd.

For dimension n, denote P_n as the set of these $[(n-1)/2]$ permutations.

Let us next look more carefully at vectors $b_4,\ldots,b_n$, which define $NO(n)$, that are consistent with each of the permutations in P_n, in the sense that the vine with permuted indices is still in $NO(n)$. The above shows that the transposition $(n-1\ n)$ corresponds to $b_n(n-2) = 1$ and this covers $N_n/2$ members of $NO(n)$.

The next permutation $(n-3\ n-2)(n-1\ n)$ corresponds to $b_{n-2}(n-4) = 1$ and $b_n(n-2) = 0$, and this covers $N_n/4$ members of $NO(n)$ if $n \geq 6$ and $N_n/2$ members if $n = 5$. We show the idea of the proof of this with $n = 5$. If $2,3$ are transposed and $4,5$ are transposed, then $34|12$ is converted to $25|13$, and there must be $25|13$ in the vine V to convert to $34|12$. This means that $A_{35} = 2$ and $b_5(3) = b_n(n-2) = 0$. For $n = 5$, $b_{n-2}(n-4) = b_3(1) = 1$ always. For $n > 5$, the argument leading to $b_n(n-2) = 1$ for the transposition $(n-1\ n)$ means that $b_{n-2}(n-4) = 1$ for $(n-3\ n-2)(n-1\ n)$.

The pattern extends to other permutations in P_n. For $(n-2j-1\ n-2j)\cdots(n-1\ n)$ [a permutation that involves $j+1$ transpositions] with $1 \leq j < [n/2]-1$, the constraints are $b_n(n-2) = \cdots = b_{n-2j+2}(n-2j) = 0$ and $b_{n-2j}(n-2j-2) = 1$. This corresponds to $N_n/2^{j+1}$ members of

$NO(n)$. If n is even, then $(1\ 2)\cdots(n-1\ n)$ corresponds to constraints $b_n(n-2) = \cdots = b_4(2) = 0$ and $N_n/2^{n/2-1}$ members of $NO(n)$. If n is odd, then $(2\ 3)\cdots(n-1\ n)$ corresponds to constraints $b_n(n-2) = \cdots = b_5(3) = 0$ (and $b_3(1) = 1$ by definition) and $N_n/2^{(n-1)/2-1}$ members of $NO(n)$.

Consider a permutation $\pi \in P_n$ applied to a regular vine V with array $A \in NO(n)$. It can lead to a vine V^* with array A^* satisfying one of the following: (i) $A^* \notin NO(n)$, (ii) $A^* = A$ and (iii) $A^* \in NO(n)$ but $A^* \neq A$. From the above properties of the binary vectors $b_4, \ldots, b_n$, for any vine array $A \in NO(n)$ and corresponding regular vine V, there is exactly one of the permutations $\pi \in P_n$ leading to either (ii) or (iii). Then for case (ii), the vine/array V, A is in an equivalence class with one member of $NO(n)$, and for case (iii), the vine/array V, A is in an equivalence class with two members of $NO(n)$. It is not possible for an equivalence class to contain more than two members of $NO(n)$.

Now we explain the remaining combinatorial arguments, with ideas illustrated with $n = 5$ or $n = 6$. For $n = 5$, suppose the transposition $(4\ 5)$ does not change the vine with $i_14, i_24|i_1, j_15, j_25|j_1, 35|12$. Then it is necessary and sufficient that $i_i = j_1$ and $i_2 = j_2$ (and $\{i_1, i_2\} = \{1, 2\}$). This means that after b_4 is set, there is no degree of freedom in choosing $b_5(2)$; the condition in terms of the b's is: $b_5(k) = b_4(k)$ for $k = 2$. With the similar argument, for $n > 5$, if the transposition $(n-1\ n)$ does not change the vine, then after b_{n-1} is set, there is no degree of freedom in choosing $b_n(k)$ for $k = 2, \ldots, n-3$; the condition is: $b_n(k) = b_{n-1}(k)$ for $k = 2, \ldots, n-3$.

Next, for $n = 6$, consider the permutation $(3\ 4)(5\ 6)$. If this leaves the vine unchanged, then there is no freedom for $b_4(2)$ and $b_6(k)$, $k = 2, 3$ after b_3, b_5 are set; the additional conditions are that $b_4(k) = b_3(k)$ for $k = 2$ (but note that $b_3(2) = 1$ by definition) and $b_6(k) = b_5(k)$, $k = 2, 3$. For $n > 6$, if the permutation $(n-3\ n-2)(n-1\ n)$ leaves the vine unchanged, then there is no freedom for $b_{n-2}(k)$, $k = 2, \ldots, n-5$ and $b_n(k)$ for $k = 2, \ldots, n-3$ after b_{n-3}, b_{n-1} are set; the additional conditions are that $b_{n-3}(k) = b_{n-2}(k)$ for $k = 2, \ldots, n-5$ and $b_n(k) = b_{n-1}(k)$ for $k = 2, \ldots, n-3$.

This pattern in terms of freedom to set the b_i vectors extends to all other permutations in P_n. But the conditions that relate consecutive b_i vectors are more complicated than above. For n even, if the permutation $(1\ 2)\cdots(n-1\ n)$ leaves the vine unchanged, $b_4(2) = 0$, $b_6(4) = \cdots = b_n(n-2) = 0$ and there is no freedom in other elements of $b_6, b_8, \ldots, b_n$ after $b_5, b_7, \ldots, b_{n-1}$ are set. For $n \geq 7$ odd, if the permutation $(2\ 3)\cdots(n-1\ n)$ leaves the vine unchanged, $b_5(3) = 0$, $b_7(5) = \cdots = b_n(n-2) = 0$ and there is no freedom in other elements of $b_5, b_7, \ldots, b_n$ after $b_4, b_6, \ldots, b_{n-1}$ are set.

But the conditions that relate consecutive b_i vectors are more complicated than above.

This leads to the following tables; Table 10.1 has equivalence classes by $\pi \in P_n$ for $n = 5, 6, 7, 8$ and Table 10.2 is for general n to show how to get the count of equivalence classes.

Table 10.1. Counts of equivalence classes in $NO(n)$ (those with one member, those with two members, all) by permutation π in P_n. The column "number" indicates the number of $A \in NO(n)$ such that the permuted vine has $A^* \in NO(n)$ with the given π; the column "ec1" is the number such that $A = A^*$; the column "ec2" is half the number such that $A \neq A^*$; and the column "ec" is the sum of "ec1" and "ec2".

		$n=5$		
permutation π	number	ec1	ec2	ec
(45)	4	2	1	3
(23)(45)	4	2	1	3
total	8	4	2	6

		$n=6$		
permutation π	number	ec1	ec2	ec
(56)	32	8	12	20
(34)(56)	16	4	6	10
(12)(34)(56)	16	4	6	10
total	64	16	24	40

		$n=7$		
permutation π	number	ec1	ec2	ec
(67)	512	64	224	288
(45)(67)	256	16	120	136
(23)(45)(67)	256	16	120	136
total	1024	96	464	560

		$n=8$		
permutation π	number	ec1	ec2	ec
(78)	16384	1024	7680	8704
(56)(78)	8192	128	4032	4160
(34)(56)(78)	4096	64	2016	2080
(12)(34)(56)(78)	4096	64	2016	2080
total	32768	1280	15744	17024

Table 10.2. Annotated version of Table 10.1.

π	ok	ec1	ec2
		$n=5$	
(45)	4 $[b_5(3)=1]$		2 $[C.3,D.1, b_4(2)=b_5(2)]$ 1 $[C.0 \equiv D.2]$
(23)(45)	4 $[b_5(3)=0]$		2 $[D.0,C.2, b_4(2)=b_5(2)]$ 1 $[C.1 \equiv D.3]$
total	8	4	2
		$n=6$	
π	number	ec1	ec2
(56)	32 $[b_6(4)=1]$	8 $[b_5(k)=b_6(k), k=2,3]$ $C.i_5.(2i_5+1), D.i_5.(2i_5+1), i_5=0,1,2,3$	12
(34)(56)	16 $[b_6(4)=0, b_4(2)=1]$	4 $[b_5(k)=b_6(k), k=2,3]$ $C.i_5.(2i_5), i_5=0,1,2,3$	6
(12)(34)(56)	16 $[b_6(4)=0, b_4(2)=0]$	4 $[b_5(k)=b_6(k), k=2,3]$ $D.0.0, D.1.6, D2.4, D3.2$	6
total	64	16	24
		$n=7$	
π	number	ec1	ec2
(67)	512 $[b_7(5)=1]$	64 $[b_6(k)=b_7(k), k=2,3,4]$ $C.i_5, i_6.(2i_6+1), D.i_5.i_6.(2i_6+1), i_6=0,\ldots,7$	224
(45)(67)	256 $[b_7(5)=0, b_5(3)=1]$	16 $[b_6(k)=b_7(k), k=2,3,4; b_4(2)=b_5(2)]$ $C.3.i_6.(2i_6), D.1.i_6.(2i_6)$	120
(23)(45)(67)	256 $[b_7(5)=0, b_5(3)=0]$	16 $[C.2, D.0].[0.0, 1.6, 2.4, 3.2, 4.8, 5.14, 6.12, 7.10]$	120
total	1024	96	464
		$n=8$	
π	number	ec1	ec2
(78)	16384 $[b_8(6)=1]$	1024 $[b_7(k)=b_8(k), k=2,\ldots,5]$ $C.i_5.i_6.i_7.(2i_7+1), D.i_5.i_6.i_7.(2i_7+1)$	7680
(56)(78)	8192 $[b_8(6)=0, b_6(4)=1]$	128 $[b_7(k)=b_8(k), k=2,\ldots,5; b_5(k)=b_6(k), k=2,3]$ $C.i_5.(2i_5+1).i_6.(2i_6), D.i_5.(2i_5+1).i_6.(2i_6)$	4032
(34)(56)(78)	4096 $[b_8(6)=0, b_6(4)=0, b_4(2)=1]$	64 $C.i_5.(2i_5).(2i).(4i), i=0,\ldots,7,$ $C.i_5.(2i_5).[1.6,3.2,5.14,7.10,9.22,11.18,13.30,15.26]$	2016
(12)(34)(56)(78)	4096 $[b_8(6)=0, b_6(4)=0, b_4(2)=0]$	64 $D.[0.0,1.6,2.4,3.2].[1.6,3.2,5.30,7.26,9.22,11.18,13.14,15.10],$ $D.[0.0,2.4].[0.0,2.4,4.8,6.28,8.16,10.20,12.24,14.12],$ $D.[1.6,3.2].[0.0,2.4,4.24,6.28,8.16,10.20,12.8,14.12]$	2016
total	32768	1280	15744

The cases satisfying invariance (column "ec1") in Table 10.2 are based on the summaries from running the algorithm for generating all A arrays in $NO(n)$ based on binary vectors $b_4, \ldots, b_n$.

Table 10.2 suggests that for a permutation π with three or more transpositions, it may not be easy to characterize $b_4, \ldots, b_n$ for the A's counted in column "ec1". However the A matrix has simple form for pairs of columns $n-2j-1, n-2j, \ldots, n-1, n$ with $j \geq 2$. If $A_{k,n-2i-1} = d$, then $A_{k,n-2i} = \pi(d)$ for $0 \leq i \leq j$ and $1 \leq k \leq n - 2i - 3$. In Table 10.2, we use the notation for vines based on the binary vectors $b_4, \ldots, b_n$ that is introduced in Section 10.2.

In Table 10.3, we show the pattern of Table 10.2 for general n and indicate the two-way classification of number of equivalence classes by permutation π and the number of members of the equivalence class.

Summing the cases in Table 10.3 leads to the following formula for the number of equivalence classes of regular vines in dimension n:

$$E_n = (N_n + E_{1n})/2,$$

where

$$E_{1n} = \sum_{k=1}^{[n/2]-1} N_n \ell_k \cdot 2^{-k} \cdot 2^{-\sum_{i=0}^{k-1}(n-4-2i)}, \quad N_n = 2^{(n-2)(n-3)/2},$$

and $\ell_k = 1$ for all k except $\ell_{[n/2]-1} = 2$.

Table 10.4 has the counts of equivalence classes for $n = 4$ to 11. The numbers have been confirmed numerically up to $n = 10$ based on the enumeration of all members of $NO(n)$.

We next mention some comments on equivalence classes. If the equivalence class is based on $(n - 1\ n)$ $[n \geq 6]$ or $(n - 3\ n - 2)(n - 1\ n)$ $[n \geq 7]$, then there is a simple form for $b_{n-3}, b_{n-2}, b_{n-1}, b_n$.

For $(n - 1\ n)$, if A, A^* are in the same equivalence class and have binary vectors b_i, b_i^*, then $b_{n-1}(k) = b_n^*(k)$ and $b_n(k) = b_{n-1}^*(k)$ for $k = 2, \ldots, n-3$, $b_n(n-2) = b_n^*(n-2) = 1$ and $b_i = b_i^*$ for $i = 4, \ldots, n-2$. The equivalence class has just one member if the transposition of b_{n-1}, b_n does not lead to something different.

For $(n - 3\ n - 2)(n - 1\ n)$, if A, A^* are in same equivalence class and have binary vectors b_i, b_i^*, then $b_{n-3}(k) = b_{n-2}^*(k)$ and $b_{n-2}(k) = b_{n-3}^*(k)$ for $k = 2, \ldots, n-5$, $b_{n-2}(n-4) = b_{n-2}^*(n-4) = 1$, $b_n(n-2) = b_n^*(n-2) = 0$ and $b_i = b_i^*$ for $i = 4, \ldots, n-4$. The equivalence class has just one member if the transpositions of b_{n-1}, b_n and b_{n-3}, b_{n-2} do not lead to something different.

Table 10.3. Summary of counts of equivalence classes in $NO(n)$ (those with one member, those with two members, all) by permutation in P_n; separated by n even/odd. The number of equivalence classes in $NO(n)$ is equal to $E_n = (N_n + E_{1n})/2 = E_{1n} + E_{2n}$.

π	number	ec1	ec2
		n even	
$(n-1\ n)$	$m_n = N_n/2$	$e_{1,n} = m_n/2^{n-4}$	$(m_n - e_{1,n})/2$
$(n-3\ n-2)(n-1\ n)$	$m_{n-2} = N_n/4$	$e_{1,n-2} = m_{n-2}/2^{(n-4)+(n-6)}$	$(m_{n-2} - e_{1,n-2})/2$
$\vdots$			
$(3\ 4)\cdots(n-1\ n)$	$m_4 = N_n/2^{n/2-1}$	$e_{1,4} = m_4/2^{(n-4)+(n-6)+\cdots+2}$	$(m_4 - e_{1,4})/2$
$(1\ 2)\cdots(n-1\ n)$	$m_2 = N_n/2^{n/2-1}$	$e_{1,2} = m_2/2^{(n-4)+(n-6)+\cdots+2}$	$(m_2 - e_{1,2})/2$
total	$N_n = 2^{(n-2)(n-3)/2}$	E_{1n}	$E_{2n} = (N_n - E_{1n})/2$

perm	number	ec1	ec2
		n odd	
$(n-1\ n)$	$m_n = N_n/2$	$e_{1,n} = m_n/2^{n-4}$	$(m_n - e_{1,n})/2$
$(n-3\ n-2)(n-1\ n)$	$m_{n-2} = N_n/4$	$e_{1,n-2} = m_{n-2}/2^{(n-4)+(n-6)}$	$(m_{n-2} - e_{1,n-2})/2$
$\vdots$			
$(4\ 5)\cdots(n-1\ n)$	$m_5 = N_n/2^{(n-1)/2-1}$	$e_{1,5} = m_5/2^{(n-4)+(n-6)+\cdots+1}$	$(m_5 - e_{1,5})/2$
$(2\ 3)\cdots(n-1\ n)$	$m_3 = N_n/2^{(n-1)/2-1}$	$e_{1,3} = m_3/2^{(n-4)+(n-6)+\cdots+1}$	$(m_3 - e_{1,3})/2$
total	$N_n = 2^{(n-2)(n-3)/2}$	E_{1n}	$E_{2n} = (N_n - E_{1n})/2$

Table 10.4. Number of equivalence classes of regular vines in dimension n.

n	E_n
4	2
5	6
6	40
7	560
8	17024
9	1066496
10	135307264
11	34496249856

However, the form of an equivalence class is much more complex when it is based on permutations with three or more transpositions (see Table 10.2 for the form of equivalence classes with one member). This is because transpositions in b_{n-5}, b_{n-4} can have bigger effects on columns b_{n-1}, b_n.

The above theory and Table 10.2 show how to easily find the permutation π for an $A \in NO(n)$ based on its $b_4, \ldots, b_n$. If this π has three or more transpositions, the other member of the equivalence class can be found as follows. Transpose appropriate elements in the A matrix based on π and obtain the A^* matrix as a member of $NO(n)$. Then A^* can be inverted to obtain $b_4^*, \ldots, b_n^*$.

Below is the algorithm for the inversion — it also includes a check of whether an $n \times n$ matrix A^* is a member of $NO(n)$.

1. Input A^*: assume $A_{i,i}^* = i$ for $i = 1, \ldots, n$, $A_{i-1,i}^* = i-1$ for $i = 2, \ldots, n$, $A_{1,3} = 1$ and $(A_{1,j}^*, \ldots, A_{j-2,j}^*)$ is a permutation of $(1, \ldots, j-2)$ for $j = 4, \ldots, n$. Otherwise, return -1 to indicate that $A^* \notin NO(n)$.
2. For column 4: $b_4^*(2) = 1$ if $A_{2,4}^* = 2$ and $b_4^*(2) = 0$ if $A_{2,4}^* = 1$.
3. For columns $d = 5, \ldots, n$,
    - $b_d^*(1) = b_d^*(d-1) = b_d^*(d) = 1$.
    - $ac =$ active column $\leftarrow d - 2$;
    - For j from $d - 2$ to 1
    - If $A_{j,d}^* = A_{ac,ac}^*$, then $b_d^*(j) = 1$ and the new ac is the largest value among $A_{1,d}^*, \ldots, A_{j-1,d}^*$;
    - Else if $A_{j,d}^* = A_{j-1,ac}^*$, then $b_d^*(j) = 0$;
    - Else A^* is not in $NO(n)$ and return -1.
4. Return $b_4^*, \ldots, b_n^*$.

10.4 Examples

In this section, we comment on the example in (10.1) and show another example to illustrate an equivalence class of two members for $NO(6)$.

Note that (10.1) satisfies $b_4(2) = 1$ and $b_6(4) = 0$, and so from Table 10.2, this means that the relevant permutation is (3 4)(5 6). Because the vine is invariant under this permutation, in $NO(6)$, it is in an equivalence class of size 1. In the notation of Section 10.2, the vine in (10.1) is called C.1.2.

For another example, we start with an array A constructed from b_4, b_5, b_6 with $b_4(2) = 0$ and $b_6(4) = 0$. From Table 10.2, this means that the relevant permutation is (1 2)(3 4)(5 6).

$$\begin{bmatrix} b_1 & b_2 & b_3 & b_4 & b_5 & b_6 \\ 1 & 1 & 1 & 1 & 1 & 1 \\ & 1 & 1 & 0 & 0 & 1 \\ & & 1 & 1 & 1 & 0 \\ & & & 1 & 1 & 0 \\ & & & & 1 & 1 \\ & & & & & 1 \end{bmatrix}, \quad A = \begin{bmatrix} 1 & 1 & 1 & 2 & 2 & 2 \\ & 2 & 2 & 1 & 1 & 4 \\ & & 3 & 3 & 3 & 1 \\ & & & 4 & 4 & 3 \\ & & & & 5 & 5 \\ & & & & & 6 \end{bmatrix} \quad (10.2)$$

The nodes of the vine are

$$\begin{bmatrix} 12 & 13 & 24 & 25 & 26 \\ & 23|1 & 14|2 & 15|2 & 46|2 \\ & & 34|12 & 35|12 & 16|24 \\ & & & 45|123 & 36|124 \\ & & & & 56|1234 \end{bmatrix}.$$

In the notation used previously, the vine in (10.2) is called D.1.4. From the first level nodes of the vine, one can see that it is not invariant with respect to (1 2)(3 4)(5 6). After this permutation, the vine and its representation in $NO(6)$ are:

$$\begin{bmatrix} 12 & 24 & 13 & 16 & 15 \\ & 14|2 & 23|1 & 26|1 & 35|1 \\ & & 34|12 & 46|12 & 25|13 \\ & & & 36|124 & 45|123 \\ & & & & 56|1234 \end{bmatrix}, \quad \begin{bmatrix} 12 & 13 & 24 & 15 & 16 \\ & 23|1 & 14|2 & 35|1 & 26|1 \\ & & 34|12 & 25|13 & 46|12 \\ & & & 45|123 & 36|124 \\ & & & & 56|1234 \end{bmatrix}.$$

The array A^* for the other member of the equivalence class is given below and its corresponding b vectors can be obtained with the inverse algorithm.

$$A^* = \begin{bmatrix} 1 & 1 & 1 & 2 & 1 & 1 \\ & 2 & 2 & 1 & 3 & 2 \\ & & 3 & 3 & 2 & 4 \\ & & & 4 & 4 & 3 \\ & & & & 5 & 5 \\ & & & & & 6 \end{bmatrix}, \begin{bmatrix} b_1 & b_2 & b_3 & b_4 & b_5 & b_6 \\ 1 & 1 & 1 & 1 & 1 & 1 \\ & 1 & 1 & 0 & 1 & 1 \\ & & 1 & 1 & 0 & 1 \\ & & & 1 & 1 & 0 \\ & & & & 1 & 1 \\ & & & & & 1 \end{bmatrix}. \quad (10.3)$$

In the notation of Section 10.2, the vine in (10.3) is called D.2.6. That is, D.1.2 and D.2.6 are in an equivalence class with two members.

10.5 Discussion

For all of these results on equivalence classes, a key component is the algorithm for generating all members of $NO(n)$. The patterns seen for $n = 5, 6, 7, 8$ point to the direction of proof for the formula for the number of equivalence classes in $NO(n)$.

Reference

1. Morales-Nápoles O., Cooke R.M. and Kurowicka D. (2009). The number of vines and regular vines on n nodes. Submitted to *Discrete Applied Mathematics*.

CHAPTER 11

Optimal Truncation of Vines

Dorota Kurowicka

Delft University of Technology
Mekelweg 4, 2628CD Delft, the Netherlands
D.Kurowicka@tudelft.nl

The vine copulae representation is a very flexible model of n-dimensional random vectors. However when n becomes too large, some simplifying assumptions have to be made as fitting this model becomes too cumbersome. In this chapter truncations of vines are discussed. We could reduce a vine to a Markov tree structure but Markov trees allow $n-1$ copulae to be specified out of $\binom{n}{2}$ possible for vines. Hence trees may be too restrictive for a given set of data. Another possibility would be to model subsets of variables with vines and connect these smaller vines in a tree structure. We suggest one more strategy of choosing the "most suitable" vine for the correlation matrix. The "best vine" is the one whose nodes of top trees (tree with the most conditioning) correspond to the smallest absolute values of partial correlations. To search for the "best vine" we developed a new algorithm of generating a regular vine. We start building the vine from the top node (node in tree n-1) and progress to the lower trees, ensuring that the regularity condition is satisfied and that the partial correlations corresponding to these nodes are the smallest. If we assume that we can assign the independent copula to nodes of the vine with small absolute values of partial correlations, then this algorithm can be used to find an optimal truncation of a vine structure. We advocate using it as a preprocessing step of fitting a vine to the data.

11.1	Introduction	234
11.2	Vines	236
11.3	Vine Distributions	237
	11.3.1 Markov trees	238
	11.3.2 Vines in trees	239
11.4	Optimal Truncation	240
	11.4.1 Generating regular vines	240
	11.4.2 Best vine	242

11.5 Optimal Truncation: Results 242
    11.5.1 Example 243
    11.5.2 Comparison 244
11.6 Conclusions 247
References 247

11.1 Introduction

The vine copula representation is a very flexible model of a joint n-dimensional distribution. It allows not only information in marginal distributions to be separated from information about dependencies in the joint distribution; it also gives a way of specifying $\binom{n}{2}$ different bivariate copulae as "building blocks" of the joint distribution. These bivariate building blocks can control correlation structure as well as other features of joint distribution, e.g., tail dependence.[7] Choosing copulae with different tail dependence parameters and different correlations gives us a very rich set of models.

Vine distributions can be quantified with data.[2] When their performance was compared with other existing models,[1,2] it was shown that they outperform other models significantly. If data are not available, vines can be quantified with structured expert judgment.[12]

In Aas et al.[2] only two types of regular vines, namely C-vines and D-vines (see Figs. 11.1 and 11.2), were used. Up to the fourth dimension, all possible vines are of these two types. For higher dimensions, other different types of vines are available (see Chapters 7 and 10). In many cases it can be more advantageous to fit different types of regular vines to data.[9] Fitting all possible vines is not feasible as the number of vines grows rapidly with

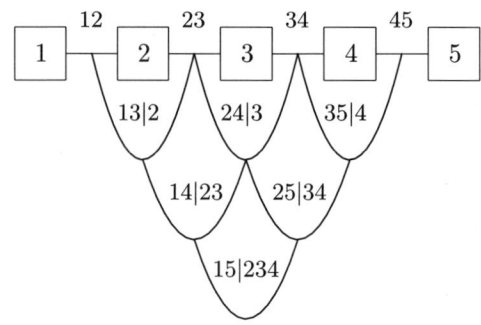

Figure 11.1. A D-vine on five elements showing conditioned and conditioning sets.

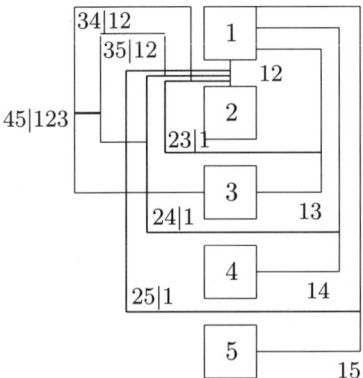

Figure 11.2. A C-vine on five elements showing conditioned and conditioning sets.

dimension.[11] There are three vines on three variables, 24 vines on four variables, 480 on five variables and more than 23,000 on six variables.

In high-dimensional cases where we are unable to fit all possible vines, simplified dependence structures must be considered. We could reduce the model to a Markov tree structure which is a special case of a vine. But as Markov trees allow $n-1$ out of $n(n-1)/2$ copulae to be specified, trees may be too restrictive for a given data. Another possibility would be to model subsets of variables with vines and connect these smaller vines in a tree structure.

In this chapter we suggest a different strategy for choosing the "most suitable" vine for the correlation matrix. The "best vine" is the one whose nodes of top trees (tree with the most conditioning) correspond to the smallest absolute values of partial correlations. To search for the "best vine" we developed a new algorithm of generating a regular vine. We start building the vine from the top node (node in tree $n-1$) and progress to the lower trees, ensuring that the regularity condition is satisfied and that the partial correlations corresponding to these nodes are the smallest. If we assume that we can assign the independent copula to nodes of the vine with small absolute values of partial correlations, then this algorithm will be useful in finding an optimal truncation of a vine structure. In Ref. 10 it is shown that such truncations of vines allow us to build distributions for which the dependence structure corresponds to a chordal graph. We advocate finding an appropriate vine truncation as a preprocessing step of fitting a vine to the data.

This chapter is organized as follows: We first summarize some known facts of regular vines and extend this exposition with a few new properties.

Next, known truncations of vines, namely Markov trees and vines in trees, are discussed. In Section 11.4 a new algorithm for generating a regular vine is presented, which is later used in a heuristic search for "the best" regular vine for the correlation matrix. Finally some results and conclusions are presented.

11.2 Vines

A *vine* on n elements $\mathcal{V} = (T_1, \ldots, T_{n-1})$ is a nested set of trees where the edges of tree j are nodes of tree $j+1$ and each tree has the maximum number of edges. A *regular* vine on n elements is one in which two edges in tree j are joined by an edge in tree $j+1$ only if these edges share a common node. Two edges a, b joined by an edge c in the next tree are called *m-children* of c and c is the *m-parent* of a and b. a and b are called *siblings*. Hence, a regularity property can be formulated: all siblings have a common child.

For each edge of a vine we define *constraint, conditioned* and *conditioning* sets of this edge as follows: The nodes of the first tree reachable from a given edge via the membership relation are called the constraint set of that edge. When two edges are joined by an edge in the next tree, the intersection of the respective constraint sets form the conditioning set, and the symmetric difference of the constraint sets is the conditioned set of this edge. Formal definitions can be found in any of the references above and in Chapter 3. We adopt the following notation: for each edge e of $\mathcal{V}$, let C_e and D_e denote conditioned and conditioning sets of e. Moreover we will exchangeably denote the constraint set of e as $\{C_e|D_e\}$ to indicate conditioned and conditioning sets or simply show variables that this set contains as $C_e \cup D_e$. If $C_e = \{x, y\}$, then we will call x and y *partners* and we will denote that y is a partner of x as $y = pt(x)$.

In Figs. 11.1 and 11.2, two special types of the regular vines, the C-vine and the D-vine, on five elements with conditioned sets and conditioning sets assigned to their edges are shown.

For vines in Figs. 11.1 and 11.2 we can easily check some general properties of regular vines (for proofs and rigorous formulation see Ref. 8 and Chapter 3).

Properties:

(1) There are $n-1$ trees and $\binom{n}{2}$ edges in a regular vine on n elements;
(2) Conditioned sets are doubletons;
(3) Each pair appears once as a conditioned set of an edge;

(4) There are $i-1$ and $i+1$ elements in the conditioning and constraint sets of an edge of the ith tree, respectively;
(5) If two nodes have the same constraint sets, they are the same node;
(6) If element i is a member of the conditioned set of an edge e of a regular vine, then i is a member of the conditioned set of exactly one of the m-children of e and the conditioning set of an m-child is a subset of D_e.

The following two lemmas can be added to vine properties.

Lemma 11.1. *Let $A \subset \{1,\ldots,n\}$ and $x_1, x_2 \notin A$, $x_1 \neq x_2$ and $y_1, y_2 \in A$. Let $N_1 = \{x_1, y_1 | A \setminus \{y_1\}\}$ and $N_2 = \{x_2, y_2 | A \setminus \{y_2\}\}$ be nodes of tree T_i of regular vine on n variables; then N_1 and N_2 have a common m-child. Moreover if $y_1 \neq y_2$, then this common m-child is: $\{y_1, y_2 | A \setminus \{y_1, y_2\}\}$.*

Proof. Node N_1 has two m-children with constraint sets $\{x_1, A \setminus \{y_1\}\}$ and A and N_2 has two m-children whose constraint sets are $\{x_2, A \setminus \{y_2\}\}$ and A. From property [5], N_1 and N_2 have a common m-child.

If $y_1 \neq y_2$, then from property [6], y_1 and y_2 have to be in the conditioned set of the m-child. □

Lemma 11.2. *Let $A \subset \{1,\ldots,n\}$, $x_1, x_2, \ldots, x_k \notin A$, $x_j \neq x_r$, for $j \neq r$ and $y_1, y_2, \ldots, y_k \in A$. Let $N_1 = \{x_1, y_1 | A \setminus \{y_1\}\}$, $N_2 = \{x_2, y_2 | A \setminus \{y_2\}\}$, $\ldots$, $N_k = \{x_k, y_k | A \setminus \{y_k\}\}$ be nodes of tree T_i of regular vine on n variables, then*

$$y_j = s \text{ or } y_j = t, \quad s, t \in \{y_1, \ldots, y_k\}.$$

Proof. Each N_j has two m-children. One of them has the constraint set A. Hence all N_j's have a common m-child. By Lemma 11.1 if $y_j \neq y_r$, then they both have to belong to the conditioned set which concludes the proof. □

11.3 Vine Distributions

In Bedford and Cooke[3] the following representation theorem for joint density in terms of product of (conditional) copulae assigned to the edges of a vine and the marginal densities is proven.

Theorem 11.1. *Let $(F, \mathcal{V}, B)$ be a copula vine specification where: $F = (F_1, \ldots, F_n)$ and each F_i has density f_i, $i = 1, \ldots, n$, $\mathcal{V}$ is a regular vine on n elements and $B = \{C_{jk} \mid e(j,k)$ where $e(j,k)$ is the unique edge with conditioned set $\{j,k\}$, and C_{jk} is a copula for $\{X_j, X_k\}$ conditional*

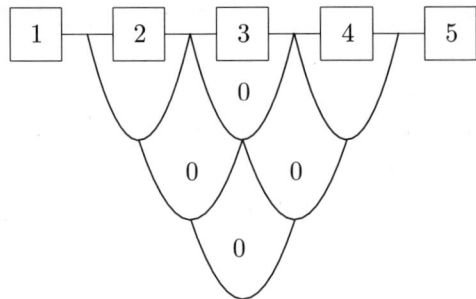

Figure 11.3. A D-vine on five variables with four conditional independent copulae assigned to top nodes.

on $D_{e(j,k)}$ with density $c_{jk|D_e}$}. Then the vine-dependent distribution for $(F, \mathcal{V}, B)$ is uniquely determined and has a density given by

$$f_{1\cdots n} = f_1 \cdots f_n \prod_{i=1}^{n-1} \prod_{e(j,k) \in E_i} c_{jk|D_e}(F_{j|D_e}, F_{k|D_e}). \qquad (11.1)$$

Assigning independent copula to an edge e of the vine ensures that variables in the conditioned set of e are conditionally independent given variables in D_e. When the independent conditional copulae are assigned to top nodes (with the most conditioning) of the vine, then the density (11.1) simplifies significantly. For the D-vine in Fig. 11.3 with four independent conditional copulae (shown as "0"), the density is of the form:

$$f_{1\cdots 5} = f_1 \cdots f_5 c_{12}(F_1, F_2) c_{23}(F_2, F_3) c_{34}(F_3, F_4) c_{45}(F_4, F_5)$$
$$c_{13|2}(F_{1|2}, F_{3|2}) c_{35|4}(F_{3|4}, F_{5|4}).$$

Independent conditional copulae assigned to top edges of the vine make the dependence structure much simpler. This process can be seen as a "truncation" of the vine to a more constrained model. Below, some special cases of vines are briefly discussed.

11.3.1 *Markov trees*

If all conditional copulae are assumed to be the independent copula, then the vine reduces to a Markov tree. The first trees of the D-vine and the C-vine in Figs. 11.1 and 11.2 are shown in Fig. 11.4. For both models, only four out of ten copulae can be specified: $c_{12}, c_{23}, c_{34}, c_{45}$ for the D-vine and $c_{12}, c_{13}, c_{14}, c_{15}$ for the C-vine.

To choose the best tree for the data we can use one of the minimum spanning tree algorithms (see, e.g., Ref. 6). Weights assigned to edges of the

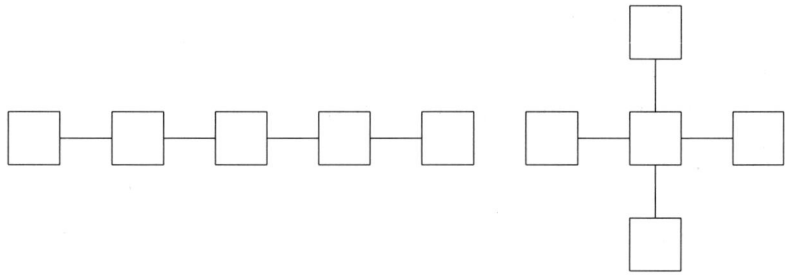

Figure 11.4. First trees of the D-vine (left) and the C-vine (right).

saturated graph might be chosen equal to correlations, tail indices, combinations of those or other features of the joint data.

11.3.2 Vines in trees

Markov trees allow $n-1$ out of $n(n-1)/2$ copulae to be specified. Hence trees may be too restrictive for a given set of data. Another possibility is to model subsets of variables with vines and connect these smaller vines in a tree structure. A simple example is presented in Fig. 11.5.

If the structure in Fig. 11.5 is such that variable 3 from the first subvine is connected with variable 4 from the second one, then this structure represents a truncation of the D-vine on six variables in which only the first tree and two conditional copulae in the second tree can be different from the independent copula.

If we understand the edge between two vines as connecting distributions of variables $\{1, 2, 3\}$ and $\{4, 5, 6\}$, then the structure in Fig. 11.5 will correspond to a chain graph with two chain components $\{1, 2, 3\}$ and $\{4, 5, 6\}$ (see, e.g., Ref. 5). Distributions represented as chain graphs, however, are not so flexible as vines in terms of sampling and quantification.

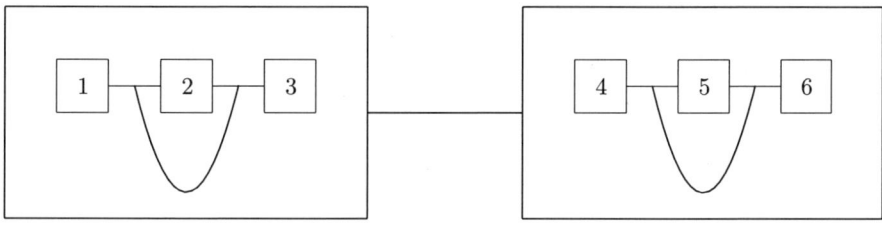

Figure 11.5. Vines in a tree.

11.4 Optimal Truncation

In this section a strategy for choosing the "most suitable" truncation of the vine for the correlation matrix is described. For this purpose we propose a new way of generating a regular vine.

11.4.1 *Generating regular vines*

A regular vine on n variables can be generated in different ways. One way is to follow its definition: choose the first tree; for $j = 2, \ldots, n-1$, build T_j by connecting two edges in T_{j-1} if they share a common node.

A different algorithm is presented in Morales-Nápoles et al.[11] and Chapters 9 and 10, based on extending a vine on $j-1$ variables by adding the variable j. Choices for these extensions have to satisfy a certain condition which assures the regularity. It is shown that there are 2^{j-3} possible choices for the extension.

Our algorithm will start building a vine from the top tree, T_{n-1}, that has only one edge. We can choose any pair of variables to be in the conditioned set of the top edge. If we have chosen n and $n-1$, then the constraint set of this edge is $\{n, n-1|1, \ldots, n-2\}$. Constraint sets of its two m-children are of the form $\{n, 1, \ldots, n-2\}$ and $\{n-1, 1, \ldots, n-2\}$. We must choose partners of n and $n-1$ in T_{n-2} such that regularity will be satisfied. Hence we first develop conditions that by Lemma 11.1 and 11.2 are necessary for regularity and then show that when applied recursively in all trees they will insure regularity.

Consider all sets of the form $B_x = \{x\} \cup A_x$ where $x \in C_e$ and $A_x = D_e$ for some edge e of tree T_j, $j = 2, \ldots, n-2$.

Condition 1.

Suppose $B_x = B_y$ and $x \neq y$, then there exists an edge f of T_{j-1} such that $C_f = \{x, y\}$ and $D_f = A_x \setminus \{y\}$.

Condition 2.

For all $B_{i_1}, \ldots, B_{i_k}$ such that $|B_{i_p} \triangle B_{i_u}| = 2$,

$$B_{i_p} = \{i_p, s | A_{i_p} \setminus \{s\}\} \quad \text{or} \quad B_{i_p} = \{i_p, t | A_{i_p} \setminus \{t\}\}, \ s, t \in A_{i_p}.$$

where $U \triangle V$ denotes the symmetric difference of sets U and V.

The algorithm that builds a regular vine can now be stated as follows:

Algorithm A

Step 1. Choose two variables, say $x, y \in \{1, \ldots, n\} = I$, to be in the conditioned set of the top edge; constraint sets of its m-children are $B_x = \{x\} \cup \{I\backslash\{x,y\}\}$ and $B_y = \{y\} \cup \{I\backslash\{x,y\}\}$; choose partners of x and y in T_{n-2} from the set $I\backslash\{x,y\}$. Then there are two edges in T_{n-2}:

$$E_{n-2} = \{\{x, pt(x)|I\backslash\{x, y, pt(x)\}\}, \ \{y, pt(y)|I\backslash\{x, y, pt(y)\}\}\}.$$

For all $j = n-2, \ldots, 1$

Step 2. Set $B_x = \{x\} \cup A_x$ such that $x \in C_e$ and $A_x = D_e$ for each edge e of tree T_j;

Step 3. Remove all sets for which $x_i = x_k$, for $i \neq k$;

Step 4. Apply Condition 1;

Step 5. Choose partners of variables x such that Condition 2 is satisfied;

Edge set of T_{j-1} is the set containing elements of the form $\{x, pt(x)|A_x \backslash \{pt(x)\}\}$.

We prove that the above algorithm always produces a regular vine.

Theorem 11.2. *Algorithm A produces a regular vine.*

Proof. We prove that the recursive application of Algorithm A to the jth level produces the jth tree of a regular vine. It is enough to show that the procedure insures that all siblings in tree T_j have a common m-child in tree T_{j-1}. Obviously the statement is true for initiating Step 1 as Condition 1 insures that two siblings in T_{n-2} have a common m-child in T_{n-3}. Suppose for all trees $j = n-1, \ldots, k+1$ that all siblings in T_j have a common child in T_{j-1}. Step 2 of the algorithm creates $2(n-k)$ constraint sets of m-children of edges of T_k. These sets are indexed by variables that were in conditioned sets of edges of T_k. Multiple instances of these sets are removed in Step 3. At this point some sets can be equal but they are indexed by different variables. We combine them in Step 4 to satisfy Condition 1 as by Property [5] they are constraint sets of the same edge. Since both indexing variables were in the conditioning set of an edge in T_k, then by Lemma 11.1 they have to be in the conditioned set of their common m-child. We get now that all constraint sets of m-children of edges in T_k are different. The symmetric difference of constraint sets of siblings in

T_k has two elements and the symmetric difference of constraint sets of their m-children has also two elements. Condition 2 applied to all sets B such that $|B_v \triangle B_u| = 2$, insures by Lemma 2 that all siblings in T_k have a common m-child in T_{k-1}. □

Remark 11.1. Notice that the edges of T_1 are already obtained after Step 3 of Algorithm A.

We will use the algorithm described in this section in finding the "best vine" for the correlation matrix.

11.4.2 Best vine

Quantifying a vine with data is performed sequentially by first fitting copulae on the first tree, then transforming data through the fitted copulae to find copula parameters on the second tree, etc.[2] In fitting high-dimensional vine distributions it would be of interest to first find a vine structure with the maximum number of independent copulae in the top nodes that offers the best approximation of the data. We propose to base this choice on a partial correlation vine corresponding to the correlation matrix obtained from data. Partial correlations[13] are calculated from the correlation matrix as follows:

$$\rho_{ij;I\setminus\{i,j\}} = \frac{C_{i,j}}{\det(I)}$$

where $\det(I)$ denotes the determinant of the correlation matrix of variables in I and $C_{i,j}$, its (i,j)th cofactor. Partial correlations are assigned to the edges of a vine such that conditioning variables are equal to the conditioning set and the conditioned variables are equal to the conditioned set.

For elliptically contoured distributions, partial correlations are equal to conditional correlations. For the normal distribution, zero partial correlation corresponds to conditional independence. In general, however, zero partial correlation does not have to indicate conditional independence. Nevertheless, in our algorithm for finding the best vine, we will choose partners of variables x in Algorithm A such that Condition 2 is satisfied and such that the absolute values of partial correlations $\rho_{x,pt(x);A_x \setminus pt(x)}$ are the smallest.

11.5 Optimal Truncation: Results

We start this section with a simple example to show how the algorithm works and then test its performance. We conclude with an application of the algorithm to the data matrix analyzed in Ref. 9. A comparison of the performance of our algorithm and the algorithm based on the majorization principle proposed in Ref. 9 is performed.

11.5.1 *Example*

Consider the following correlation matrix:

$$M = \begin{bmatrix} 1 & 0.2 & 0.4 & 0.5 & 0.7 \\ 0.2 & 1 & 0.3 & 0.6 & 0.7 \\ 0.4 & 0.3 & 1 & 0.8 & 0.5 \\ 0.5 & 0.6 & 0.8 & 1 & 0.8 \\ 0.7 & 0.7 & 0.5 & 0.8 & 1 \end{bmatrix}.$$

The normalized inverse matrix of M is:

$$\begin{bmatrix} 1 & 0.5309 & -0.2034 & 0.2246 & -0.7437 \\ 0.5309 & 1 & 0.0914 & -0.0827 & -0.6021 \\ -0.2034 & 0.0914 & 1 & -0.7930 & 0.3236 \\ 0.2246 & -0.0827 & -0.7930 & 1 & -0.5613 \\ -0.7437 & -0.6021 & 0.3236 & -0.5613 & 1 \end{bmatrix}.$$

We choose the smallest absolute partial correlation which is 0.0827. This is the partial correlation $\rho_{24;13}$. Then we follow Algorithm A and obtain that the best vine, with the smallest partial correlations in the top nodes, is:

$$V = \begin{bmatrix} & C_e & D_e & \rho_{C_e;D_e} \\ 2 & 4\ |\ 1 & 3\ \ 5 & = 0.0827 \\ 2 & 3\ |\ 1 & 5 & = 0.0425 \\ 4 & 1\ |\ 3 & 5 & = 0.3179 \\ 3 & 1\ |\ 5 & & = 0.0808 \\ 2 & 1\ |\ 5 & & = 0.5686 \\ 4 & 3\ |\ 5 & & = 0.7698 \\ 3 & 5 & & = 0.5 \\ 2 & 5 & & = 0.7 \\ 1 & 5 & & = 0.7 \\ 4 & 5 & & = 0.8 \end{bmatrix}.$$

The vine obtained is neither a D-vine nor a C-vine (see Fig. 11.6). If we assume that we can assign the independent copula to nodes of the vine with small absolute partial correlations, then we see that only seven out of ten copulae $c_{15}, c_{25}, c_{35}, c_{45}$ and $c_{43|5}, c_{12|5}, c_{41|35}$ have to be fitted. Moreover, because of the sequential fitting of copula in vines, the estimates for copulae

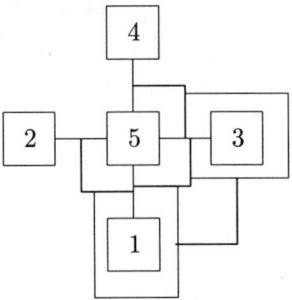

Figure 11.6. Vine corresponding to matrix M.

with more conditioned variables are not so accurate. Removing them may be of benefit.

If we decided to search for a vine with the highest correlations in lower trees, then we would end up with the following vine:

$$V^* = \begin{bmatrix} & C_e & & D_e & \rho_{C_e;D_e} \\ 2 & 3 \mid 1 & 4 & 5 & = -0.0914 \\ 2 & 4 \mid 1 & & 5 & = 0.0169 \\ 1 & 3 \mid 4 & & 5 & = 0.2985 \\ 3 & 5 \mid 4 & & & = -0.3889 \\ 2 & 1 \mid 5 & & & = -0.5686 \\ 4 & 1 \mid 5 & & & = -0.1400 \\ 3 & 4 & & & = 0.8 \\ 2 & 5 & & & = 0.7 \\ 1 & 5 & & & = 0.7 \\ 4 & 5 & & & = 0.8 \end{bmatrix}.$$

We see that in this case the top node is also associated with small partial correlation but in general it does not have to be the case.

Table 11.1 presents sums and average partial correlations in trees of V and V^*. We see that the partial correlation of the top edge of V is slightly smaller than the partial correlation of the top edge of V^*. The choice of slightly smaller correlations in T_4 leads to significantly higher average correlations in T_3.

11.5.2 Comparison

We consider an example treated in Ref. 4. We have eight variables corresponding to weather monitoring stations in Europe. The original data has

Table 11.1. Sum and average absolute values of partial correlations in trees of V and V^*.

	T_4	T_3	T_2	T_1
sum V	0.0827	0.3604	1.4192	2.7
average V	0.0827	0.1802	0.4731	0.675
sum V^*	0.0914	0.3154	1.0975	3
average V^*	0.0914	0.1577	0.3658	0.75

a sample correlation matrix:

$$\begin{bmatrix} 1 & .35 & .50 & .49 & .68 & .38 & .50 & .59 \\ & 1 & .79 & .69 & .12 & .64 & .62 & .49 \\ & & 1 & .72 & .18 & .61 & .58 & .43 \\ & & & 1 & .05 & .46 & .47 & .43 \\ & & & & 1 & .33 & .51 & .71 \\ & & & & & 1 & .97 & .77 \\ & & & & & & 1 & .90 \\ & & & & & & & 1 \end{bmatrix}.$$

In Kurowicka *et al.*[9] a heuristic search of a vine was adopted for which the logarithm of one minus squared partial correlations assigned to its edges majorizes all others based on minimizing the entropy function.

The heuristic works as follows:

(1) Choose an ordering of the variables.
(2) Start with subvine consisting of variables 1 and 2 in the ordering. For $j = 3, \ldots, n$, find the subvine extending the current subvine by adjoining variable $j+1$, so as to minimize entropy function of $\log(1-\rho^2_{C_e;D_e})$. Store the vine obtained for $j = n$.
(3) Go to 1.
(4) Choose the optimal partial correlation vine minimizing entropy among all those stored.

In general it is not feasible to search all permutations; heuristic search methods or Monte Carlo sampling must be used. The optimal vine V^* obtained with this procedure is shown below. We can see that there are many small partial correlations in this vine but they are not necessarily in the top trees. The vine V was obtained with Algorithm A. We can see that there are many more small correlation values in trees with higher indices.

Table 11.2. Sum and average absolute values of partial correlations in trees of V and V^*.

	T_7	T_6	T_5	T_4	T_3	T_2	T_1
sum V	0.0109	0.2382	1.1841	1.3341	2.3696	1.5308	4.3384
average V	0.0109	0.1191	0.3947	0.3335	0.4739	0.2551	0.6198
sum V^*	0.8883	1.4657	0.2035	1.3987	1.9273	1.5745	3.6045
average V^*	0.8883	0.7328	0.0678	0.3497	0.3855	0.2624	0.5149

$$V = \begin{bmatrix} C_e & D_e & & \rho_{C_e;D_e} \\ 2 & 5 & |\ 134678 & 0.0109 \\ 2 & 8 & |\ 13467 & 0.0823 \\ 5 & 7 & |\ 13468 & 0.1559 \\ 8 & 7 & |\ 1346 & 0.9499 \\ 2 & 6 & |\ 1347 & 0.0377 \\ 5 & 3 & |\ 1468 & 0.1965 \\ 7 & 6 & |\ 134 & 0.9666 \\ 8 & 3 & |\ 146 & -0.3310 \\ 2 & 1 & |\ 347 & 0.2756 \\ 5 & 6 & |\ 148 & 0.4229 \\ 6 & 3 & |\ 14 & 0.4250 \\ 7 & 1 & |\ 34 & 0.2854 \\ 8 & 6 & |\ 14 & 0.7166 \\ 2 & 4 & |\ 37 & 0.2802 \\ 5 & 4 & |\ 18 & 0.6624 \\ 3 & 1 & |\ 4 & 0.2495 \\ 7 & 4 & |\ 3 & 0.0936 \\ 8 & 4 & |\ 1 & 0.1944 \\ 6 & 1 & |\ 4 & 0.2037 \\ 2 & 7 & |\ 3 & 0.3236 \\ 5 & 1 & |\ 8 & 0.4660 \\ 7 & 3 & & 0.5822 \\ 6 & 4 & & 0.4602 \\ 2 & 3 & & 0.7888 \\ 5 & 8 & & 0.7140 \\ 3 & 4 & & 0.7155 \\ 1 & 4 & & 0.4890 \\ 8 & 1 & & 0.5886 \end{bmatrix} \quad V^* = \begin{bmatrix} C_e & D_e & & \rho_{C_e;D_e} \\ 7 & 8 & |\ 123456 & 0.8883 \\ 6 & 7 & |\ 12345 & 0.9783 \\ 4 & 8 & |\ 12356 & 0.4874 \\ 4 & 6 & |\ 1235 & 0.0171 \\ 4 & 7 & |\ 1235 & 0.1086 \\ 1 & 8 & |\ 2356 & 0.0778 \\ 1 & 7 & |\ 235 & -0.0591 \\ 1 & 6 & |\ 235 & -0.0961 \\ 1 & 4 & |\ 235 & 0.4632 \\ 5 & 8 & |\ 236 & 0.7803 \\ 2 & 4 & |\ 35 & 0.2928 \\ 1 & 5 & |\ 23 & 0.6998 \\ 5 & 6 & |\ 23 & 0.3068 \\ 5 & 7 & |\ 23 & 0.5450 \\ 3 & 8 & |\ 26 & -0.0829 \\ 1 & 3 & |\ 2 & 0.3935 \\ 3 & 4 & |\ 5 & 0.7190 \\ 2 & 5 & |\ 3 & -0.0309 \\ 3 & 6 & |\ 2 & 0.2208 \\ 3 & 7 & |\ 2 & 0.1918 \\ 2 & 8 & |\ 6 & -0.0185 \\ 1 & 2 & & 0.3491 \\ 2 & 3 & & 0.7888 \\ 3 & 5 & & 0.1762 \\ 4 & 5 & & 0.0487 \\ 2 & 6 & & 0.6424 \\ 2 & 7 & & 0.6210 \\ 6 & 8 & & 0.9783 \end{bmatrix}$$

11.6 Conclusions

The new algorithm for generating a regular vine presented in this chapter allows us to build a vine starting from the edge in tree $n-1$, progressing to lower trees, making sure that the regularity condition is satisfied. We proposed applying this algorithm in a heuristic search for a vine with small absolute values of partial correlations assigned to its top edges that corresponds to a given correlation matrix. It was observed that the choice of a small partial correlation in tree T_j may severely constrain the choices that we have in tree T_{j-1}. We could improve the heuristic by basing our choices on more than one tree at a time.

References

1. Aas K. and Berg D. (2009). Models for construction of multivariate dependence: A comparison study. *European Journal of Finance*, 15:639–659.
2. Aas K., Czado C., Frigessi A. and Bakken H. (2009). Pair-copula constructions of multiple dependence. *Insurance: Mathematics and Economics*, 44(2):182–198.
3. Bedford T.J. and Cooke R.M. (2001). Probability density decomposition for conditionally dependent random variables modeled by vines. *Annals of Mathematics and Artificial Intelligence*, 32:245–268.
4. Callies U., Kurowicka D. and Cooke R. (2003). Graphical models for the evaluation of multisite temperature forecasts: Comparison of vines and independence graphs. *Proceedings of ESREL 2003, Safety and Reliability*, 1:363–371.
5. Cowell R., David A., Lauritzen S. and Spiegelhalter D. (1999). *Probabilistic Networks and Expert Systems, Statistics for Engineering and Information Sciences*. Springer-Verlag, New York.
6. Even S. (1979). *Graph Algorithms*. Computer Science Press, Potomac, MD.
7. Joe H. (1997). *Multivariate Models and Dependence Concepts*. Chapman & Hall, London.
8. Kurowicka D. and Cooke R.M. (2006). *Uncertainty Analysis with High Dimensional Dependence Modelling*. Wiley, Chichester.
9. Kurowicka D., Cooke R.M. and Callies U. (2006). Vines inference. *Brazilian Journal of Probability and Statistics*, 20:103–120.
10. Kurowicka D. and Cooke R.M. (2006). Completion problem with partial correlation vines. *Linear Algebra and Its Applications*, 418(1):188–200.
11. Morales-Nápoles O., Cooke R.M. and Kurowicka D. (2009). The number of vines and regular vines on n nodes. Submitted to *Discrete Applied Mathematics*.
12. Morales-Nápoles O., Kurowicka D. and Roelen A. (2007). Elicitation procedures for conditional and unconditional rank correlations. *Reliability Engineering and Systems Safety*, 95(5):699–710.
13. Yule G.U. and Kendall M.G. (1965). *An Introduction to the Theory of Statistics*, 14th ed. Charles Griffin & Co., Belmont, California.

CHAPTER 12

Bayesian Inference for D-Vines: Estimation and Model Selection

Claudia Czado[*,†] and Aleksey Min[*,‡]

Technische Universität München, Zentrum Mathematik
Boltzmannstr. 3, 85747 Garching, Germany
[†]*cczado@ma.tum.de*
[‡]*aleksmin@ma.tum.de*

In the last two decades the advent of fast computers has made Bayesian inference, based on Markov chain Monte Carlo (MCMC) methods, very popular in many fields of science. These Bayesian methods, are good alternatives to traditional maximum likelihood (ML) methods since they can often estimate complicated statistical models for which an ML approach fails. In this chapter we review available MCMC estimation and model selection algorithms as well as their possible extensions for D-vine pair-copula constructions (PCC) based on bivariate t-copulae. However the discussed methods can easily be extended for an arbitrary regular vine PCC based on any bivariate copulae. A Bayesian inference for Australian electricity loads demonstrates the addressed algorithms at work.

12.1	Introduction	250
12.2	D-Vine	251
12.3	D-Vine PCC Based on t-Copulae	253
12.4	Bayesian Inference for D-Vine PCC Based on t-Copulae . .	255
12.5	Application: Australian Electricity Loads	257
12.6	Bayesian Model Selection for Australian Electricity Loads	259
12.7	Summary and Discussion	260
	References	262

---

*Technische Universität München, Zentrum Mathematik, Boltzmannstr. 3, 85747 Garching, Germany.

12.1 Introduction

Pair-copula constructions (PCC) for multivariate copulae have been successful in extending the class of available multivariate copulae (see Refs. 8 and 1). Estimation of the corresponding copula parameters has been done so far using maximum likelihood (ML). However, the number of parameters of a PCC model to be estimated can be considerable. So far it has been facilitated by numerical optimization of the log-likelihood to obtain ML estimates.

For inference purposes, one needs to have reliable standard error estimates for the estimated parameters. The standard approach for this is to impose regularity conditions such that asymptotic normality of the parameter estimates holds and to approximate the estimated standard errors by evaluating numerically the Hessian matrix. However, for large parameter vectors this evaluation is time-consuming and its reliability is uncertain. In addition, numerical estimates of the Hessian matrix might result in non-positive definite matrices yielding negative variance estimates. For these reasons Min and Czado[19] started to investigate Bayesian inference for PCC models based on Markov chain Monte Carlo (MCMC) methods (see Refs. 18 and 13). Bayesian inference has the advantage of providing natural interval estimates based on the posterior distribution and does not rely on asymptotic normality. Besides, the Bayesian approach is able to incorporate prior information which might be available from the data context or previous data analyses.

Min and Czado[19] developed and implemented Bayesian MCMC algorithms for D-vines based on pair t-copulae. While this solved the problem of obtaining reliable interval estimates for parameters needed for inference purposes, the problem of model selection needed to be approached. The gain of flexibility by using PCC is huge, however, the problem of which PCC model to choose becomes important. In particular Morale-Nápoles et al.[21] have shown that the number of PCC models, even in small dimensions, can be enormous, so it is impossible to fit all models and compare them. Therefore efficient model selection strategies are needed. While Heinen and Valdesogo[14] approached this problem by using truncated PCC constructions, Min and Czado[20] approached the problem of reducing a chosen PCC by using reversible jump (RJ) MCMC methods suggested by Green[12] and successfully applied to search large model spaces as is needed in PCCs. The purpose of this chapter is to give an overview of these Bayesian estimation and model selection procedures and illustrate their usefulness in a data set involving Australian electricity loads from Chapter 13.

The proposed methodology is developed for data transformed to the copula level, i.e., for data living on the multivariate unit cube. We will use a parametric and nonparametric approach to create the copula data used for illustration. We have chosen the above data set to facilitate comparison of two-step estimation procedures to the joint estimation procedure from Chapter 13 and in Ref. 9. For this data set, it turns out that two-step estimation procedures are nearly as efficient as the joint estimation procedure, thus making the extra effort required for the joint estimation less necessary.

The chapter is organized as follows. In Section 12.2 we briefly consider a general D-vine decomposition for a multivariate density. Section 12.3 presents the likelihood of D-vine PCCs based on t-copulae. In Section 12.4 we survey MCMC estimation and model selection algorithms, as well as their possible extensions for a D-vine PCC based on t-copulae. Sections 12.5 and 12.6 illustrate the discussed MCMC methods in the Bayesian analysis of Australian electricity loads from Chapter 13. In Section 12.7 we summarize our findings and discuss further open problems as well as future research directions.

12.2 D-Vine

Using Sklar's theorem in n dimensions, multivariate distributions on $\mathbb{R}^n$ with given margins can be easily constructed. However, this general approach does not provide a solution for the construction of flexible multivariate distributions. In this section we give such a construction proposed first by Joe,[15] organized by Bedford and Cooke[3] and applied to Gaussian copulae only. Later, Aas et al.[2] used bivariate Gaussian, t, Gumbel and Clayton copulae as building blocks to increase model flexibility.

Let $f(x_1, \ldots, x_n)$ be an n-dimensional density function and $c(u_1, \ldots, u_n)$ be the corresponding copula density function. For a pair of integers r and s $(1 \leq r \leq s \leq n)$, a set $r : s$ denotes all integers between r and s, namely $r : s := \{r, \ldots, s\}$. If $r > s$, then $r : s = \emptyset$. Further, let $\mathbf{X}_{r:s}$ denote the set of variables $\{X_r, \ldots, X_s\}$, $u_{i|r:s}$ denote the conditional cumulative distribution function (cdf) $F_{i|r:s}(x_i|\mathbf{x}_{r:s})$ and u_i denote the unconditional cdf $F_i(x_i)$. It is well-known that the density $f(x_1, \ldots, x_n)$ can be factorized as

$$f(x_1, \ldots, x_n) = f_n(x_n) \cdot f_{n-1|n}(x_{n-1}|x_n) \cdot f_{n-2|(n-1)n}(x_{n-2}|x_{n-1}, x_n) \cdots$$
$$\cdot f_{1|2\cdots n}(x_1|x_2, \ldots, x_n). \tag{12.1}$$

The above factorization is a simple consequence from the definition of conditional densities and is invariant under permutation of the variables.

The second factor $f_{n-1|n}(x_{n-1}|x_n)$ on the right-hand side of (12.1) can be represented as a product of a copula density and the marginal density $f_{n-1}(x_{n-1})$ in the following way. Consider the bivariate density function $f_{(n-1)n}(x_{n-1}, x_n)$ with marginal densities $f_{n-1}(x_{n-1})$ and $f_n(x_n)$, respectively. Using Sklar's theorem for $n = 2$, we have that the conditional density $f_{n-1|n}(x_{n-1}|x_n)$ is given by

$$f_{n-1|n}(x_{n-1}|x_n) = \frac{f_{(n-1)n}(x_{n-1}, x_n)}{f_n(x_n)}$$
$$= c_{(n-1)n}(u_{n-1}, u_n) \cdot f_{n-1}(x_{n-1}). \quad (12.2)$$

Similarly, the conditional density $f_{n-2|(n-1)n}(x_{n-2}|x_{n-1}, x_n)$ is given by

$$f_{n-2|(n-1)n}(x_{n-2}|x_{n-1}, x_n)$$
$$= \frac{f_{(n-2)n|n-1}(x_{n-2}, x_n|x_{n-1})}{f_{n|n-1}(x_n|x_{n-1})}$$
$$= c_{(n-2)n|n-1}(u_{n-2|n-1}, u_{n|n-1}) \cdot f_{n-2|n-1}(x_{n-2}|x_{n-1})$$
$$= c_{(n-2)n|n-1}(u_{n-2|n-1}, u_{n|n-1})$$
$$\times c_{(n-2)(n-1)}(u_{n-2}, u_{n-1}) \cdot f_{n-2}(x_{n-2}). \quad (12.3)$$

The copula density $c_{(n-2)n|n-1}(\cdot, \cdot)$ is the conditional copula density corresponding to the conditional distribution $F_{(n-2)n|n-1}(x_{n-2}, x_n|x_{n-1})$ and, in general, it depends on the given conditioning value x_{n-1}. By induction, the $(j+1)$th factor $(j = 3, \ldots, n-1)$ in (12.1) is given by

$$f_{n-j|(n-j+1):n}(x_{n-j}|\mathbf{x}_{(n-j+1):n})$$
$$= c_{n-j,n|(n-j+1):(n-1)}(u_{n-j|(n-j+1):(n-1)}, u_{n|(n-j+1):(n-1)})$$
$$\times f_{n-j|(n-j+1):(n-1)}(x_{n-j}|\mathbf{x}_{(n-j+1):(n-1)})$$
$$= \prod_{t=1}^{j-1} c_{n-j,n-t+1|(n-j+1):(n-t)}(u_{n-j|(n-j+1):(n-t)}, u_{n-t+1|(n-j+1):(n-t)})$$
$$\times c_{n-j,n-j+1}(u_{n-j}, u_{n-j+1}) \cdot f_{n-j}(x_{n-j}). \quad (12.4)$$

Thus we can represent each term on the right-hand side of (12.1) as the product of the corresponding marginal density and copula density terms.

Combining (12.2)–(12.4), expression (12.1) can be rewritten as

$$f(x_1,\ldots,x_n) = \prod_{t=1}^{n} f(x_t) \times \prod_{j=1}^{n-1} c_{n-j,n-j+1}(u_{n-j}, u_{n-j+1})$$

$$\times \prod_{j=2}^{n-1}\prod_{t=1}^{j-1} c_{n-j,n-t+1|\mathbf{i}(j,t)}(u_{n-j|\mathbf{i}(j,t)}, u_{n-t+1|\mathbf{i}(j,t)}), \quad (12.5)$$

where $\mathbf{i}(j,t) := (n-j+1) : (n-t)$. The density $f(x_1,\ldots,x_n)$ is the product of n marginal densities and $n(n-1)/2$ pair-copula density terms. The pair-copula density terms are unconditional copulae evaluated at marginal distribution function values or conditional copulae evaluated at univariate conditional distribution function values. The above construction was defined in Ref. 2 and was called the D-vine pair-copula construction (PCC) for multivariate distributions.

12.3 D-Vine PCC Based on t-Copulae

From now on, we use as the building pair-copulae of the PCC model (12.5) bivariate t-copulae. However, the estimation and model selection methodology is generic and applies much more widely. Further we assume that the margins of **X** are uniform. This is motivated by the standard two-step semi-parametric copula estimation procedure suggested by Ref. 10, where approximate uniform margins are obtained by applying the empirical probability integral transformation to standardized fitted residuals based on specified marginal models.

The bivariate t-copula (see, e.g., Ref. 7) has two parameters: the association parameter $\rho \in (-1, 1)$ and the degrees of freedom (df) parameter $\nu \in (0, \infty)$. Its density is given by

$$c(u_1, u_2|\nu, \rho) = \frac{\Gamma\left(\frac{\nu+2}{2}\right)\Gamma\left(\frac{\nu}{2}\right)}{\sqrt{1-\rho^2}\left[\Gamma\left(\frac{\nu+1}{2}\right)\right]^2} \cdot \frac{\left(\left[1+\frac{x_1^2}{\nu}\right]\left[1+\frac{x_2^2}{\nu}\right]\right)^{-\frac{\nu+1}{2}}}{\left(1+\frac{x_1^2+x_2^2-2\rho x_1 x_2}{\nu(1-\rho^2)}\right)^{-\frac{\nu+2}{2}}},$$

where $x_i := t_\nu^{-1}(u_i)$ for $i = 1, 2$ and $t_\nu^{-1}(\cdot)$ is a quantile function of a t-distribution with ν degrees of freedom. Specifying the pair-copulae and assuming uniform margins, the conditional distribution function for a bivariate t-copula is needed. It is called the h-function for the t-copula with parameters ρ and ν, and Ref. 2 derives it as

$$h(u_1|u_2,\rho,\nu) = t_{\nu+1}\left(\frac{t_\nu^{-1}(u_1) - \rho\, t_\nu^{-1}(u_2)}{\sqrt{\frac{(\nu+(t_\nu^{-1}(u_2))^2)(1-\rho^2)}{\nu+1}}}\right). \tag{12.6}$$

The D-vine PCC (12.5) with building bivariate t-copulae depends now on an $n(n-1)$ dimensional parameter vector $\boldsymbol{\theta}$ given by

$$\boldsymbol{\theta} = (\rho_{12}, \nu_{12}, \rho_{23}, \nu_{23}, \ldots, \rho_{1n|2:(n-1)}, \nu_{1n|2:(n-1)})^t,$$

where $\rho_{tj|(t+1):(j-1)}$ and $\nu_{tj|(t+1):(j-1)}$ are the parameters of the t-copula density $c_{tj|(t+1):(j-1)}(\cdot,\cdot)$ for $j = 2, \ldots, n$ and $t = 1, \ldots, j-1$. As already noted in Ref. 19, the likelihood $c(\mathbf{u}|\boldsymbol{\theta})$ of the D-vine copula for N n-dimensional realizations $\mathbf{u} := (\mathbf{u}_1, \ldots, \mathbf{u}_N)$ of $\mathbf{U} := (U_1, \ldots, U_n)^t$ can be calculated as

$$c(\mathbf{u}|\boldsymbol{\theta}) = \prod_{k=1}^{N}\left[\prod_{i=1}^{n-1} c(u_{i,k}, u_{i+1,k}|\rho_{i(i+1)}, \nu_{i(i+1)}) \prod_{j=2}^{n-1}\prod_{i=1}^{n-j}\right.$$

$$\left. \times c(v_{j-1,2i-1,k}, v_{j-1,2i,k}|\rho_{i(i+j)|(i+1):(i+j-1)}, \nu_{i(i+j)|(i+1):(i+j-1)})\right], \tag{12.7}$$

where for $k = 1, \ldots, N$

$$v_{1,1,k} := h(u_{1,k}|u_{2,k}, \rho_{12}, \nu_{12})$$

$$v_{1,2i,k} := h(u_{i+2,k}|u_{i+1,k}, \rho_{(i+1)(i+2)}, \nu_{(i+1)(i+2)}), \ i = 1, \ldots, n-3,$$

$$v_{1,2i+1,k} := h(u_{i+1,k}|u_{i+2,k}, \rho_{(i+1)(i+2)}, \nu_{(i+1)(i+2)}), \ i = 1, \ldots, n-3,$$

$$v_{1,2n-4,k} := h(u_{n,k}|u_{n-1,k}, \rho_{(n-1)n}, \nu_{(n-1)n}),$$

$$v_{j,1,k} := h(v_{j-1,1,k}|v_{j-1,2,k}, \rho_{1(1+j)|2:j}, \nu_{1(1+j)|2:j}), \ j = 2, \ldots, n-2,$$

$$v_{j,2i,k} := h(v_{j-1,2i+2,k}|v_{j-1,2i+1,k}, \rho_{i(i+j)|(i+1):(i+j-1)},$$
$$\nu_{i(i+j)|(i+1):(i+j-1)}) \quad \text{for } n > 4, \ j = 2, \ldots, n-3 \text{ and}$$
$$i = 1, \ldots, n-j-2$$

$$v_{j,2i+1,k} := h(v_{j-1,2i+1,k}|v_{j-1,2i+2,k}, \rho_{i(i+j)|(i+1):(i+j-1)},$$
$$\nu_{i(i+j)|(i+1):(i+j-1)}) \quad \text{for } n > 4, \ j = 2, \ldots, n-3 \text{ and}$$
$$i = 1, \ldots, n-j-2$$

$$v_{j,2n-2j-2,k} := h(v_{j-1,2n-2j,k}|v_{j-1,2n-2j-1,k}, \rho_{(n-j)n|(n-j+1):(n-1)},$$
$$\nu_{(n-j)n|(n-j+1):(n-1)}) \quad \text{for } j = 2, \ldots, n-2.$$

Note that $v_{j,2i,k}$ and $v_{j,2i+1,k}$ in (12.7) are jth-fold superpositions of the h-function (12.6).

12.4 Bayesian Inference for D-Vine PCC Based on t-Copulae

Estimation of D-vine PCCs in an MCMC framework is straightforward and similar to estimation of any multivariate distribution with many parameters. The nature of parameters defined by the specific choice of the copula family should be taken into account. In contrast to multivariate density functions, the arguments in a conditional D-vine density term is a complicated function of arguments and the parameters of earlier pair D-vine densities. This makes the evaluation of the log-likelihood time-consuming. Further, the parameter update of PCCs is usually performed by a Metropolis–Hastings (MH) algorithm (see Refs. 18 and 13).

Min and Czado[19] develop and implement one such MCMC algorithm for the estimation of parameters of PCCs. They use noninformative priors for ρ's and ν's. Since estimation of df ν is unstable for large true ν's, its support should be restricted to some finite interval $(1, U)$. A noninformative prior for each ρ results in a uniform distribution on $(-1, 1)$. There are several other possibilities for the choice of priors for ν. Thus we use a Cauchy distribution in Chapter 13 while Valle[6] utilizes a truncated Poisson distribution. Joe[16] developed a uniform prior on the space of positive definite correlation matrices, which imply different beta priors for corresponding partial correlations arising from a D-vine. This alternative prior choice has been used in Chapter 13.

There are also several choices of the proposal distributions needed for the MH algorithm. Min and Czado[19] use a modification of a random normal walk proposal, which is a normal distribution truncated to the support of parameters. Variances of normal distributions are tuned to achieve acceptance rates between 20% and 80% as suggested by Besag *et al.*[4] Another choice is an independence proposal distribution which is independent of the current value of the sampled parameter. A common independence proposal is a normal distribution with the same mode and inverse curvature at the mode as the target distribution described for example in Ref. 11. This has been used in Chapter 13 for the joint MCMC estimation of marginal AR(1) and D-vine copula parameters. Generalizations of the normal independence proposal using t-distribution with low degrees of freedom ν, say $\nu = 3$ or $\nu = 5$, are also often used.

The number of pair-copulae $n_c = n(n-1)/2$ in (12.7) increases quadratically with dimension n of the data. However, if independence or conditional independence is present in the data, then the number of factors in

(12.7), respectively, may reduce drastically. This (conditional) independence is characterized by a unit pair-copula density. Therefore the first task on model selection for D-vine PCCs is to determine its non-unit pair-copula terms. Min and Czado[20] derive and implement a reversible jump (RJ) MCMC by Green.[12] The algorithm by Green[12] allows a huge number of models to be explored since only visited models will be fitted. Therefore it is well-accepted by the Bayesian community though its derivation and implementation for a particular problem are not simple tasks. Another model selection approach by Congdon[5] is discussed and utilized in Ref. 19. The recent approach of Congdon's[5] is easy to implement but it compares only among prespecified models.

The key points of the RJ MCMC algorithm of Min and Czado[20] are the introduction of a model indicator vector of dimension n_c and the RJ mechanism for a model change. They associate models with subdecompositions of (12.7) consisting of k ($1 \leq k \leq n_c$) pair-copula terms. To specify the model indicator, pair-copulae in full decomposition (12.7) has to be ordered. Otherwise an identifiability problem occurs since PCCs are invariant with respect to the permutation of factors. According to the labeling in Ref. 20, for $n = 4$ the full decomposition of a multivariate copula density is given as follows.

$$c(u_1, u_2, u_3, u_4) = c_{12} c_{23} c_{34} c_{13|2} c_{24|3} c_{14|23},$$

where we omit arguments and parameters of pair-copulae for brevity. The model indicator $\mathbf{m}_f$ is given by a six-dimensional vector $(1,1,1,1,1,1)$, where 1 indicates the presence of the corresponding pair-copula term. If now some pair-copula terms are not present in the decomposition, then the corresponding ones are replaced by zeros. For example, a model indicator $\mathbf{m} = (1,1,1,1,1,0)$ corresponds to the subdecomposition $c(u_1, u_2, u_3, u_4) = c_{12} c_{23} c_{34} c_{13|2} c_{24|3}$ without the last pair-copula $c_{14|23}$.

Any RJ MCMC algorithm consists of so-called birth and death moves. For birth moves the dimension of the model parameter increases while for death moves the dimension decreases. Min and Czado[20] derive acceptance probabilities for both moves in detail. As a proposal distribution for the parameters of the sth pair-copula, they use a bivariate normal distribution $N_2(\hat{\boldsymbol{\theta}}_s^{\text{MLE}}, \Sigma_s)$ truncated to $(-1,1) \times (1, U)$. Here $\hat{\boldsymbol{\theta}}_s^{\text{MLE}} = (\rho_s^{\text{MLE}}, \nu_s^{\text{MLE}})'$ denotes the corresponding two-dimensional sub-vector of the maximum likelihood estimate (MLE) $\hat{\boldsymbol{\theta}}_{\mathbf{m}_f}^{\text{MLE}}$ in the full model $\mathbf{m}_f$. Note that there are n_c covariance matrices Σ_s's, which govern the reversible jump mechanism. They are taken of the form $\Sigma_s = diag(\sigma_{s,\rho}^2, \sigma_{s,\nu}^2)$, where $diag(a,b)$ denotes a diagonal matrix with a and b on the main diagonal.

12.5 Application: Australian Electricity Loads

In this section we illustrate the above discussed estimation algorithms for the Australian electricity loads from Chapter 13. We are solely interested in estimating the dependence structure and therefore marginal AR(1)s are first fitted to extract independent i.i.d. residuals. Now copula data for the Australian electricity loads can be obtained using probability integral transformations (PITs). One choice for the PIT is the empirical PIT while the other is the standard normal PIT given by $U = \Phi(Z)$, where Z is a normal $N(0, 1)$ random variable and $\Phi(\cdot)$ is the standard normal cdf. Here we consider both choices for producing copula data. The copula data produced by the empirical PIT we call nonparametric copula data, while the one produced by normal PIT we call parametric copula data. To facilitate comparison to the models considered in Chapter 13, we now investigate the following PCC here and in the sequel:

$$c(u_Q, u_N, u_V, u_S) = c_{QN} \cdot c_{NV} \cdot c_{VS} \cdot c_{QV|N} \cdot c_{NS|V} \cdot c_{QS|NV}, \quad (12.8)$$

where the parameter dependence of each bivariate t-copula and their arguments are dropped to keep the expression short. The subindexes Q, N, V and S correspond to Queensland, New South Wales, Victoria and South Australia, respectively.

For both copula data we run the MH algorithm specified in Ref. 19 for 10000 iterations using Cauchy priors truncated to $(1, 100)$ for the df parameter of each pair as in Ref. 9, namely $\pi(\nu) \propto 1/[1 + (\nu - 1)^2/4]$. The first 500 iterations are considered the burn-in. Proposal variances were determined in pilot runs and resulted in acceptance rates between 23%–77% for all parameters after 10000 iterations. Autocorrelations among the MCMC iterates suggested sub-sampling to reduce these correlations and every 10th iteration was recorded. Table 12.1 summarizes the estimated posterior distributions for all parameters based on the recorded iterations for both copula data. For comparison we also include the corresponding maximum likelihood estimates (MLE) in the last column of Table 12.1. For the convenience of the reader, results of the joint estimation of AR(1) margins and the D-vine PCC model (12.8) for copula parameters from Chapter 13 are displayed in Table 12.2. There, the joint MCMC estimation uses a slightly different prior for the ρ parameter.

For both copula data, the Bayesian estimates of $\rho_{QV|N}, \rho_{NS|V}$ and $\rho_{QS|NV}$ are not credible at the 5% level since the corresponding credible intervals contain 0. They are also not credible at the 10% level except for $\rho_{QV|N}$ when the parametric copula data are used. Posterior mode estimates of ν's are larger

Table 12.1. Estimated posterior mean, mode and quantiles of MCMC as well as MLE of copula parameters for the copula data obtained from the preprocessed Australian load data using Cauchy prior for degrees of freedom ν's truncated to $(1, 100)$.

Copula	2.5%	5%	50%	95%	97.5%	mean	mode	MLE
\multicolumn{9}{c}{Nonparametric copula data}							
ν_{QN}	3.27	3.39	4.46	6.18	6.75	4.59	4.30	4.46
ν_{NV}	2.61	2.68	3.31	4.32	4.58	3.37	3.27	3.26
ν_{VS}	4.27	4.53	6.04	8.65	9.34	6.24	5.79	6.24
$\nu_{QV|N}$	10.29	11.50	28.38	79.72	88.49	34.31	22.33	51.80
$\nu_{NS|V}$	4.96	5.27	7.56	12.63	14.64	8.13	6.92	7.73
$\nu_{QS|NV}$	8.74	10.01	21.62	73.13	83.50	28.41	17.61	32.18
ρ_{QN}	0.24	0.25	0.30	0.35	0.36	0.30	0.31	0.31
ρ_{NV}	0.28	0.29	0.35	0.40	0.41	0.35	0.35	0.35
ρ_{VS}	0.53	0.53	0.57	0.60	0.61	0.57	0.57	0.57
$\rho_{QV|N}$	−0.02	−0.01	0.03	0.08	0.09	0.03	0.03	0.03
$\rho_{NS|V}$	−0.04	−0.03	0.03	0.08	0.09	0.03	0.03	0.03
$\rho_{QS|NV}$	−0.03	−0.02	0.03	0.08	0.08	0.03	0.03	0.03
\multicolumn{9}{c}{Parametric copula data}							
ν_{QN}	5.04	5.31	7.24	10.48	11.43	7.48	6.92	7.32
ν_{NV}	4.08	4.24	5.50	7.62	7.96	5.66	5.33	5.52
ν_{VS}	6.79	7.16	10.72	18.03	20.22	11.43	9.99	11.17
$\nu_{QV|N}$	13.71	15.25	31.13	78.32	86.09	36.77	26.50	38.55
$\nu_{NS|V}$	5.73	6.15	8.41	13.34	15.55	8.97	7.92	8.34
$\nu_{QS|NV}$	12.49	14.07	30.11	79.18	87.96	35.91	24.58	58.07
ρ_{QN}	0.27	0.28	0.33	0.38	0.38	0.33	0.33	0.33
ρ_{NV}	0.32	0.33	0.38	0.43	0.44	0.38	0.38	0.39
ρ_{VS}	0.53	0.54	0.57	0.60	0.61	0.57	0.57	0.57
$\rho_{QV|N}$	−0.00	0.01	0.06	0.12	0.13	0.06	0.07	0.06
$\rho_{NS|V}$	−0.03	−0.02	0.03	0.09	0.10	0.03	0.03	0.04
$\rho_{QS|NV}$	−0.03	−0.02	0.03	0.08	0.09	0.03	0.03	0.03

than 10 only for $\rho_{QV|N}$ and $\rho_{QS|NV}$ while it is smaller than 10 for $\nu_{NS|V}$. At 95% credibility we conclude for both copula data that conditional independence between loads of Queensland and Victoria, given loads of New South Wales, as well as between loads of New South Wales and South Australia, given loads of Victoria, are present. Therefore the decomposition (12.8) can be reduced by pair-copulae $c_{QV|N}$ and $c_{QS|NV}$. However, it is difficult to decide from the above results whether loads of New South Wales and South Australia are conditionally independent given loads of Victoria. More sophisticated Bayesian model selection procedures discussed in the next section address this problem. For the parametric copula data, posterior mode estimates for df's are usually higher than the corresponding one for the nonparametric copula

Table 12.2. Estimated posterior mean, mode and quantiles of joint MCMC estimation for copula parameters using Cauchy prior for degrees of freedom ν's truncated to $(1, 100)$.

Copula	2.5%	5%	50%	95%	97.5%	Mean	Mode
	Joint estimation of marginal and copula parameters						
ν_{QN}	5.17	5.45	7.37	11.72	12.71	7.80	6.92
ν_{NV}	4.11	4.26	5.57	7.65	8.60	5.76	5.36
ν_{VS}	7.08	7.69	11.45	24.29	29.68	12.89	10.23
$\nu_{QV|N}$	15.01	16.28	36.89	84.56	91.77	41.44	29.80
$\nu_{NS|V}$	4.20	4.73	14.25	78.22	93.44	22.84	11.60
$\nu_{QS|NV}$	12.32	14.58	34.43	78.22	86.81	38.94	29.00
ρ_{QN}	0.27	0.28	0.34	0.38	0.39	0.34	0.34
ρ_{NV}	0.33	0.35	0.40	0.45	0.45	0.40	0.40
ρ_{VS}	0.54	0.55	0.59	0.62	0.63	0.59	0.59
$\rho_{QV|N}$	−0.01	−0.00	0.05	0.10	0.11	0.05	0.05
$\rho_{NS|V}$	−0.01	0.00	0.05	0.11	0.12	0.06	0.05
$\rho_{QS|NV}$	−0.04	−0.03	0.02	0.07	0.08	0.02	0.03

data. Difference in estimates of ρ's is negligible here. Further, we observe that the joint estimation of AR(1) margins and copula parameters gives results similar to ones for the parametric copula data.

We now compare the two-step estimation procedures (estimate margins first, then form standardized residuals and transform to copula data, either using nonparametric or parametric transformations) to the one-step estimation procedure using joint MCMC. For this comparison we see that the credible intervals are similar for the two-step parametric and joint estimation procedure except for $\nu_{NS|V}$. For the nonparametric two-step estimation procedure, the posterior means and modes for the df parameters are lower than for the parametric and joint estimation procedure, thus indicating more heavy-tailedness in the data than what is present. Here we consider the joint estimation method as the most appropriate estimation method, since the marginal residuals do not violate the marginal AR(1) model assumption. Overall we see that the loss in efficiency is not huge if one uses a two-step estimation procedure compared to a joint estimation procedure for this data set.

12.6 Bayesian Model Selection for Australian Electricity Loads

Based on simulation studies, Min and Czado[20] have advocated using $U = 20$ as the upper limit of the prior distribution for ν's. Then the model selection performance of RJ MCMC for PCCs based on t-copulae significantly increases. Here, we follow their approach.

We run the MH algorithm presented in Ref. 19 to tune proposal variances for the full PCC in (12.8). These tuned variances are used in the stay move to update the corresponding new parameter values. For the birth move we propose new values for $\boldsymbol{\theta_{m_s}}$, $s = 1, \ldots, 6$ according to the normal $N_2(\hat{\boldsymbol{\theta}}_{m_s}^{\text{MLE}}, \Sigma)$ distribution truncated to $(-1, 1) \times (1, U)$, where $\hat{\boldsymbol{\theta}}_{m_s^{(\text{new})}}^{\text{MLE}} = (\rho_s^{\text{MLE}}, \nu_s^{\text{MLE}})'$ denotes the corresponding two-dimensional sub-vector of the ML estimate $\hat{\boldsymbol{\theta}}_{\mathbf{m}_f}^{\text{MLE}}$ in the full model $\mathbf{m}_f$. The MLE $\hat{\boldsymbol{\theta}}_{\mathbf{m}_f}^{\text{MLE}}$ is determined under the constraints $-1 < \rho < 1$ for ρ's and $1 < \nu < 20$ for ν's. We consider two birth proposal covariance matrices Σ's, namely $\Sigma_1 = diag(10^2, 100^2)$ and $\Sigma_2 = diag(0.1^2, 3^2)$, to investigate robustness of the procedure. Further, we use $\hat{\boldsymbol{\theta}}_{\mathbf{m}_f}^{\text{MLE}}$ and $\mathbf{m}_f$ as initial values for $\boldsymbol{\theta}$ and $\mathbf{m}$, respectively.

Note that there are six copula terms in (12.8) which can be present or not in the model. Disregarding the model of complete independence, this gives that there are $63 = 2^6 - 1$ models to be explored by the RJ MCMC algorithm. We enumerate models in binary. Thus the full decomposition $\mathbf{m} = (1, 1, 1, 1, 1, 1)$ in (12.8) corresponds to 63. If the pair-copulae $c_{QV|N}$ and $c_{QS|NV}$ are set equal to 1, then the corresponding model vector is given by $\mathbf{m} = (1, 1, 1, 0, 1, 0)$. This binary representation corresponds to 58.

Table 12.3 displays estimated posterior probabilities for all possible 63 PCC models and for both copula data obtained from the preprocessed Australian electricity loads based on 100000 iterations with a burn-in of 10,000. For comparison, the last column of Table 12.3 gives approximations to posterior probabilities for the seven models based on the approach by Congdon.[5] The implementation of Congdon's algorithm is similar to that presented in Chapter 13. Thus our RJ MCMC algorithm with $U = 20$ shows that the PCC model without pairs $c_{MB|T}$ and $c_{ST|M}$ has the highest estimated posterior probability for both choices of Σ independent of the transformation used to obtain copula data. Congdon's approach also supports the above model though with less confidence. However, Robert and Marin[23] note that there might be considerable bias in Congdon's method.

12.7 Summary and Discussion

This chapter reviews methods on Bayesian inference for D-vine PCCs and illustrates their use for a specific data set. The methodology can easily be extended to cover any regular vine PCC model. Since the classical ML

Table 12.3. Estimated posterior model probabilities $\hat{P}_k = \hat{P}(M_k|\text{data})$ of all 63 models for the nonparametric copula data obtained from the preprocessed Australian load data using an empirical cdfs. The model probabilities in the third and fourth columns, and in the last fifth column are obtained using RJ MCMC and Congdon's approach, respectively. The corresponding model probability estimates for the parametric copula data obtained from the preprocessed Australian load data are given in parentheses. Further, $U = 20$, $\Sigma_1 = diag(10^2, 100^2)$ and $\Sigma_2 = diag(0.1^2, 3^2)$.

Model	Formula	$\hat{P}_k$ Σ_1	Σ_2	Cong.
M_{63}: $\mathbf{m} = (1,1,1,1,1,1)$	$c_{QN}c_{NV}c_{VS}c_{QV|N}$ $\times c_{NS|V}c_{QS|NV}$	0.001 (0.000)	0.000 (0.000)	0.002 (0.009)
M_{62}: $\mathbf{m} = (1,1,1,1,1,0)$	$c_{QN}c_{NV}c_{VS}c_{QV|N}$ $\times c_{NS|V}$	0.013 (0.054)	0.014 (0.065)	0.062 (0.185)
M_{60}: $\mathbf{m} = (1,1,1,1,0,0)$	$c_{QN}c_{NV}c_{VS}c_{QV|N}$	0.001 (0)	0.001 (0.000)	0.004 (0.001)
M_{59}: $\mathbf{m} = (1,1,1,0,1,1)$	$c_{QN}c_{NV}c_{VS}$ $\times c_{NS|V}c_{QS|NV}$	0.022 (0.012)	0.025 (0.011)	0.090 (0.070)
M_{58}: $\mathbf{m} = (1,1,1,0,1,0)$	$c_{QN}c_{NV}c_{VS}$ $\times c_{NS|V}$	0.923 (0.933)	0.933 (0.923)	0.710 (0.726)
M_{57}: $\mathbf{m} = (1,1,1,0,0,1)$	$c_{QN}c_{NV}c_{VS}c_{QS|NV}$	0.001 (0)	0.000 (0.000)	0.003 (0.000)
M_{56}: $\mathbf{m} = (1,1,1,0,0,0)$	$c_{QN}c_{NV}c_{VS}$	0.039 (0.000)	0.026 (0.001)	0.129 (0.009)
M_i: for $i \neq 63, 62, 60, \ldots, 56$		0 (0)	0 (0)	– (–)

approach to PCCs will give only reliable point estimates but not reliable standard error estimates, a Bayesian approach is followed here.

To assess the influence of prior distributions, we have run the original MH algorithm specified in Min and Czado[19] for the Australian load data for 10000 iterations using uniform priors on $(1, 100)$ for each df parameter. This means a median value of 50.5 for each df parameter while the truncated Cauchy prior in Chapter 13 has its median at 3. This results, as expected, in a considerable increase of posterior means for ν's which are >20, while the ρ parameters are not affected. We observe differences in Bayesian estimates only for df's if the corresponding MLE or posterior mode estimates are larger than 20. For small estimated ν's, the influence of the prior for ν is negligible. In contrast, Bayesian estimates for ρ's are robust with respect to prior distributions for ρ.

Min and Czado[19] report that numerically evaluated Hessian matrix and bootstrap methods are good alternatives for getting reliable standard errors if the dimension of data is $n < 4$. The estimated Hessian matrix can often fail to give reliable standard estimates since negative variance estimates might occur. Further, for high-dimensional data of $n > 4$ with thousands of multivariate observations, the bootstrap and Hessian approaches become much more time-consuming in contrast to MCMC methods as Min and Czado[19] find. In a simulation study, Min and Czado[20] show that for model selection purposes, the upper limit of ν should be set to 20. Then the model performance of their RJ MCMC algorithm is significantly improved. The results of the RJ MCMC analysis for the Australian load data are robust with regard to the choice of proposal distribution for the birth move as Min and Czado[20] notice. In the next step, we plan to derive and implement a RJ MCMC algorithm where the copula family of pair-copulae is not fixed anymore and it can vary within a catalogue of bivariate copulae, including the independence copula.

In some problems, joint estimation of marginal and copula parameters has recently been found to be important. Thus it is shown in Ref. 17 that a separate estimation of the marginal parameters may have an essential influence on the parameter estimation of multivariate copulae. Therefore inference based on joint estimates might lead to quite different results compared to the inference ignoring estimation errors in the marginal parameters. In financial applications, marginal time series usually follow ARMA or GARCH models. Here, our future research will concentrate on joint estimation of marginal (ARMA) GARCH and PCC parameters. Finally, this can all be generalized to PCC models with time-varying parameters since financial data usually shows that the dependence structure changes over time (see, for example, Ref. 22).

Acknowledgments

Claudia Czado acknowledges the support of the Deutsche Forschungsgemeinschaft (Cz 86/1-3).

References

1. Aas K. and Berg D. (2009). Models for construction of multivariate dependence: A comparison study. *European Journal of Finance*, 15:639–659.

2. Aas K., Czado C., Frigessi F. and Bakken H. (2009). Pair-copula constructions of multiple dependence. *Insurance: Mathematics and Economics*, 44(2):182–198.
3. Bedford T.J. and Cooke R.M. (2002). Vines: A new graphical model for dependent random variables. *Annals of Statistics*, 30(4):1031–1068.
4. Besag J., Green P., Higdon D. and Mengersen K. (1995). Bayesian computation and stochastic systems. *Statistical Science*, 10(1):3–66.
5. Congdon P. (2006). Bayesian model choice based on Monte Carlo estimates of posterior model probabilities. *Computational Statistics and Data Analysis*, 50:346–357.
6. Dalla Valle L. (2009). Bayesian copulae distributions with application to operational risk management. *Methodology and Computing in Applied Probability*, 11(1):95–115.
7. Embrechts P., Lindskog F. and McNeil A.J. (2003). Modelling dependence with copulae and applications to risk management. In: *Handbook of Heavy Tailed Distributions in Finance*. Elsevier/North-Holland, Amsterdam.
8. Fischer M., Köck C., Schlüter S. and Weigert F. (2009). Multivariate copula models at work. *Quantitative Finance*, 9(7): 839–854.
9. Gärtner F. (2008). Bayesian analysis of multivariate time series models based on pair-copula construction. Diploma thesis, Zentrum Mathematik, Technische Universität München.
10. Genest C., Ghoudi K. and Rivest L.-P. (1995). A semiparametric estimation procedure of dependence parameters in multivariate families of distributions. *Biometrika*, 82(3):543–552.
11. Gilks W.R., Richardson S. and Spiegelhalter D.J. (1996). Introducing Markov chain Monte Carlo. In: *Markov Chain Monte Carlo In Practice*, W.R. Gilks et al. (eds.), pp. 1–19. Chapman & Hall, London.
12. Green P.J. (1995). Reversible jump Markov chain Monte Carlo computation and Bayesian model determination. *Biometrika*, 82(4):711–732.
13. Hastings W. (1970). Monte Carlo sampling methods using Markov chains and their applications. *Biometrika*, 57:97–109.
14. Heinen A. and Valdesogo A. (2008). A canonical vine autoregressive model for large dimensions. Technical report, Department of Finance and Management Science, Université Catholique de Louvain.
15. Joe H. (1996). Families of m-variate distributions with given margins and m(m-1)/2 bivariate dependence parameters. In: *Distributions with Fixed Marginals and Related Topics*, L. Rüschendorf and B. Schweizer and M. D. Taylor (eds.). Institute of Mathematical Statistics, California.
16. Joe H. (2006). Generating random correlation matrices based on partial correlations. *Journal of Multivariate Analysis*, 97:2177–2189.
17. Kim G., Silvapulle M.J. and Silvapulle P. (2007). Comparison of semiparametric and parametric methods for estimating copulae. *Computational Statistics and Data Analysis*, 51(6):2836–2850.
18. Metropolis N., Rosenbluth A.W., Rosenbluth M.N., Teller A.H. and Teller E. (1953). Equations of state calculations by fast computing machines. *Journal of Chemical Physics*, 21:1087–1092.
19. Min A. and Czado C. (2010). Bayesian inference for multivariate copulas using pair-copula constructions. *Journal of Financial Econometrics*. In press.
20. Min A. and Czado C. (2009). Bayesian model selection for multivariate copulae using pair-copula constructions. Technical report, Zentrum Mathematik, Technische Univerität München.

21. Morales-Nápoles O., Cooke R.M. and Kurowicka D. (2008). The number of vines and regular vines on n nodes. Submitted to *Discrete Applied Mathematics*.
22. Patton A.J. (2004). On the out-of-sample importance of skewness and asymmetric dependence for asset allocation. *Journal of Financial Econometrics*, 2:130–168.
23. Robert C. and Marin J.-M. (2008). Some difficulties with some posterior probability approximations. *Bayesian Analysis*, 3:427–442.

CHAPTER 13

Analysis of Australian Electricity Loads Using Joint Bayesian Inference of D-Vines with Autoregressive Margins

Claudia Czado[*,†], Florian Gärtner[*,‡] and Aleksey Min[*,§]

Technische Universität München, Zentrum Mathematik
Boltzmannstr. 3, 85747 Garching, Germany
†*cczado@ma.tum.de,* ‡*flo-gaertner@web.de*
§*aleksmin@ma.tum.de*

Sklar's theorem allows the construction of models for dependent components using a multivariate copula together with marginal distributions. For estimation of the copula and marginal parameters, a two-step procedure is often used to avoid high-dimensional optimization. Here, marginal parameters are estimated first, then used to transform to uniform margins and in a second step, the copula parameters are estimated. This procedure is not efficient. Therefore, we follow a joint estimation approach in a Bayesian framework using Markov chain Monte Carlo (MCMC) methods. This allows also for the assessment of parameter uncertainty using credible intervals. D-vine copulae are utilized and as marginal models we allow for autoregressive models of first order. Finally, we apply these methods to Australian electricity loads, demonstrating the usefulness of this approach. Bayesian model selection is also discussed and applied using a method suggested by Congdon.[12]

13.1 Introduction 266
13.2 Multivariate Time Series with D-Vine Dependency and Marginal Autoregressive Structure 267
13.3 Bayesian Analysis of Multivariate Time Series with D-Vine Dependency and Marginal Autoregressive Structure 269

---

*Technische Universität München, Zentrum Mathematik, Boltzmannstr. 3, 85747 Garching, Germany.

13.4 Modeling Australian Electricity Loads 270
13.5 Bayesian Model Selection 274
13.6 Summary and Discussion 279
References 279

13.1 Introduction

The celebrated result of Sklar[25] shows that dependence among random variables can be separated from the marginal distributions. This forms the basis for the construction of many multivariate models in statistics and finance (see, for example, Refs. 16, 23 and 10). While there are many bivariate copulae available for modeling bivariate dependence, the catalogue of multivariate copulae is less rich. Joe[16] used a decomposition into pair-copulae to construct multivariate distributions. Bedford and Cooke[7] systemized these constructions using a graphical tree representation and Kurowicka and Cooke[19] give an overview.

Aas et al.[1] derived statistical inference methods based on these pair-copula constructions (PCC) using bivariate t-copulae. While the maximum likelihood estimation is feasible in small dimensions, the number of copula pairs increases quadratically with the number of dimensions. Therefore, assessing the variability of the estimates is difficult, requiring the numerical inversion of large dimensional Hessian matrices. Hence, we prefer to use a Bayesian approach for estimation and inference. Min and Czado[22] develop such a Bayesian approach using Markov chain Monte Carlo (MCMC) methods to estimate the posterior distribution for PCCs with bivariate t-copulae as building blocks for multivariate copula data, i.e., for an i.i.d. multivariate sample with uniform margins.

For financial applications, one usually starts with multivariate time series and, in the first step, one estimates for each marginal time series its structure as an ARMA or GARCH structure. In the second step, one determines standardized residuals, which are assumed to form an i.i.d marginal sample. Depending on whether the distribution of the residuals are known or unknown, one uses a parametric or empirical probability transform to transform to data with approximate uniform margins. This separates the marginal distribution from the dependence structure. In the final step, this dependency is modeled using a multivariate copula and copula parameters are estimated. The statistical properties of such two-step estimation procedures are investigated by Joe[17] for a known standardized residual distribution and by Chen[9] for unknown standardized residual distribution, respectively. It is known that such two-step procedures are not efficient. The loss in efficiency depends on the specific data structure and model.

The main contribution of this chapter is to provide a Bayesian analysis which jointly estimates the marginal and copula parameters. For this, we start with a simple marginal time series structure such as the first-order autoregressive structure. In Section 13.2, we introduce the model and in Section 13.3, we derive an MCMC algorithm which will be used to facilitate the Bayesian analysis. In Section 13.4, we apply our methods to Australian electricity load data. Here, we remove the trend and seasonal effects. In Section 13.5, we consider Bayesian model selection when one wants to choose among a small number of alternative model specifications based on the approach suggested by Congdon.[12] Section 13.6 closes this chapter with conclusions and a discussion on future research.

13.2 Multivariate Time Series with D-Vine Dependency and Marginal Autoregressive Structure

First we describe the marginal time series structure. For this, let $\boldsymbol{Y}^t := (Y_{1t}, \ldots, Y_{nt})$ for $t = 1, \ldots, N$ denote n-dimensional time series with marginal AR(1) time series structure and normal errors, i.e.,

$$Y_{it} = \gamma_i \cdot Y_{i(t-1)} + \epsilon_{it} \quad t = 1, \ldots, N, \text{ and}$$

$$\epsilon_{it} \sim \mathcal{N}(0, \sigma_i^2) \text{ i.i.d.}, \quad i = 1, \ldots, n$$

$$Y_{i0} \sim \mathcal{N}\left(0, \frac{\sigma_i^2}{1 - \gamma_i^2}\right) \quad \text{and} \quad |\gamma_i| < 1.$$

From this, it follows that $\boldsymbol{Y_i} := (Y_{i1}, \ldots, Y_{iN})' \sim \mathcal{N}_N(\boldsymbol{0}, \Sigma_i)$ with the (r,s)th element of Σ_i given by $\Sigma_{i,r,s} = \frac{\sigma_i^2}{1-\gamma_i^2} \gamma_i^{|r-s|}$ for $i = 1, \ldots, n$. We first transform each marginal time series $\boldsymbol{Y_i}$ to achieve i.i.d. margins by defining

$$\boldsymbol{Z_i} := \Sigma_i^{-1/2} \boldsymbol{Y_i} \sim \mathcal{N}_N(\boldsymbol{0}, I_N), \tag{13.1}$$

where I_N is the identity matrix of size N. Wise[28] shows that Σ_i^{-1} is a tridiagonal matrix and it is easy to determine the Cholesky factorization

$\Sigma_i^{-1/2}$ as

$$\Sigma_i^{-1/2} = \sigma_i^{-1} \begin{pmatrix} 1 & 0 & \cdots & \cdots & 0 \\ -\gamma_i & 1 & 0 & & \vdots \\ 0 & -\gamma_i & 1 & \ddots & \vdots \\ \vdots & \ddots & \ddots & \ddots & 0 \\ 0 & \cdots & 0 & -\gamma_i & \sqrt{1-\gamma_i^2} \end{pmatrix}$$

and

$$\det \Sigma_i^{-1/2} = \sigma_i^{-N} \sqrt{1-\gamma_i^2}.$$

For the dependency structure between the n marginal time series, we now impose an n-dimensional D-vine (see Refs. 6 and 19) structure on the i.i.d. random vectors $\boldsymbol{Z}^t := (Z_{1t}, \ldots, Z_{nt})$ for $t = 1, \ldots, N$. In particular, each random vector $\boldsymbol{Z}^t$ has a D-vine density $f(z_1, \ldots, z_n)$ (see, e.g., Chapter 12) given by

$$\prod_{k=1}^{n} f(z_k) \prod_{j=1}^{n-1} \prod_{i=1}^{n-j} c_{i,i+j|i+1,\ldots,i+j-1}$$
$$\times \{F(z_i|z_{i+1}, \ldots, z_{i+j-1}), F(z_{i+j}|z_{i+1}, \ldots, z_{i+j-1})\}, \quad (13.2)$$

where $c_{i,i+j|i+1,\ldots,i+j-1}(\cdot,\cdot)$ are arbitrary bivariate copula densities depending on parameters $\boldsymbol{\theta}_{i,i+j|i+1,\ldots,i+j-1}$. Here, $F(\cdot|\cdot)$ denotes a conditional cdf. Utilizing (13.1), we can take (13.2) together with the density transformation theorem to construct a multivariate density for $\boldsymbol{Y}^t$ for $t = 1, \ldots, N$. In the application, we will consider the case $n = 4$, where the D-vine density is given by

$$\begin{aligned} f(x_1, x_2, x_3, x_4) = & f_1(x_1) \cdot f_2(x_2) \cdot f_3(x_3) \cdot f_4(x_4) \\ & \cdot c_{12}(F_1(x_1), F_2(x_2)) \cdot c_{23}(F_2(x_2), F_3(x_3)) \\ & \cdot c_{34}(F_3(x_3), F_4(x_4)) \\ & \cdot c_{13|2}(F(x_1|x_2), F(x_3|x_2)) \cdot c_{24|3}(F(x_2|x_3), F(x_4|x_3)) \\ & \cdot c_{14|23}(F(x_1|x_2, x_3), F(x_4|x_2, x_3)). \end{aligned} \quad (13.3)$$

For details and exact density expressions for $\boldsymbol{Y}^t$ in the case $n = 4$, see Chapter 5 of Ref. 14. In our application, we use bivariate t-copulae $c(u, v) = c(u, v|\boldsymbol{\theta})$ with $\boldsymbol{\theta} = (\nu, \rho)$, where ν is the degree of freedom (df) parameter and ρ the correlation parameter. Correlation refers to the corresponding t-distribution and not to the copula.

13.3 Bayesian Analysis of Multivariate Time Series with D-Vine Dependency and Marginal Autoregressive Structure

The multivariate time series models introduced in the previous section have many parameters to estimate, for example, the model when $n = 4$ and bivariate t-copulae are used requires 20 parameters. We could follow a two-step approach by first estimating marginal parameters using the R function arima within the stats package. In a second step, we transform to the Z-level using these estimated parameters and apply the D-vine R package of Daniel Berg (private communication) to determine maximum likelihood estimates of the D-vine parameters. However, at the moment, there is no efficient estimation of the standard errors of such estimates, reflecting the uncertainty in the marginal parameters. Therefore, we follow a Bayesian approach where credible intervals can be determined naturally to assess the significance of the parameter.

First, we have to specify prior distributions to complete the model specification. In particular, we would like to use a uniform prior for the correlation matrix in the D-vine based on bivariate t-copulae. Using the results in Ref. 21 and the calculations of Joe,[18] we assume a $Beta((4-k)/2, (4-k)/2)$ distribution on $(-1, 1)$ for $\rho_{ij|i_1,\ldots,i_k}$, where k is the cardinality of the set of conditioning variables. This choice of conditional correlations results in an unconditional correlation matrix which is uniformly distributed over the space of correlation matrices. For the degree of freedom parameter, we choose a half Cauchy distribution as the prior, i.e., $\pi(\nu) \propto [1+(\frac{\nu-1}{2})^2]^{-1}, \nu \in (1, \infty)$, and for the marginal error variances σ_i^2 an inverse Gamma prior, given by $\pi(u) = \frac{0.001}{\Gamma(1)} u^{-2} \exp(-\frac{0.001}{u})$ as suggested on p. 15 of Congdon.[11] Finally, for the autoregressive parameters γ_i, we choose an $N(0, 10)$ prior on the transformed scale $a_i := \frac{1}{2} \log \frac{1+\gamma_i}{1-\gamma_i}$. In addition, we assume prior independence among all parameters.

The posterior distribution is however not analytically tractable, therefore we use Markov chain Monte Carlo (MCMC) methods (see, for example, Ref. 13). In particular, we utilize the Metropolis–Hastings (MH) algorithm with the independence proposal introduced by Ref. 26. As the proposal distribution, we take a normal distribution centered around the mode with a covariance matrix a multiple of the inverse Hessian matrix evaluated at the mode as suggested on p. 83 of Gamerman and Lopes.[13] For determining the mode and Hessian on the transformed scale of the parameters, the delta method is applied. To reduce computing time, we update the proposal distribution only every 20th iteration. The corresponding acceptance probabilities for the MH algorithm were developed in Chapter 4 of Ref. 14. In particular, each parameter is updated individually.

13.4 Modeling Australian Electricity Loads

Initially, the Australian electricity supply was organized as a vertically integrated monopoly with almost no trade or connection between different states. The liberalization started with the opening of the National Electricity Market (NEM) in December 1998. The first members were Victoria, Queensland, New South Wales and the Australian Capital Territory; South Australia and Tasmania joined in 2005. It is a wholesale electricity market that supplies retailers and end-users.

The connection between the electricity producers and electricity consumers is facilitated by the establishment of the National Electricity Market Management Company (NEMMCO). This company manages a pool where the output of all generators is aggregated and scheduled to meet the forecasted demand.

Wholesale trading is done as a real-time market where supply and demand are instantaneously matched through a centrally dispatched process. As the offers are submitted by the generators every five minutes, NEMMCO determines the necessary plants and they are dispatched into production. So the market clearing price is determined every five minutes and is averaged for each trading interval (30 minutes).

Since the Australian electricity market is an energy-only market, there are a lot of price spikes. But experience shows that these price spikes are incentive enough for the electricity companies to build new generation plants, for example, as seen in South Australia in the period 1998–2003 when generation capacities increased substantially after a series of price spikes to meet these peak demands. For more details on the Australian electricity market, see Ref. 27.

The load data used consist of four time series of daily observations dating from May 16, 2005, to June 30, 2008, in total 1142 data points per time series. It describes the average daily load demand in gigawatt (GW) for the regions Queensland, New South Wales, Victoria and South Australia, calculated by averaging the half-hourly observed data for one day. These data are available at www.nemmco.com.au.

Electricity demand clearly shows seasonal fluctuations, mostly due to changing climate conditions like temperature or the number of daylight hours. We follow the classical technique of seasonal decomposition by thinking of a trend component T_t, a seasonal component S_t and the remaining stochastic component Y_t, i.e., we can represent the observed daily load data $\{x_1, \ldots, x_N\}$ as

$$x_t = T_t + S_t + y_t; \quad t = 1, \ldots, N.$$

Using techniques presented in Refs. 27 and 8, we investigate the original time series to identify the trend and seasonal component. In a preprocessing step, these components will be removed before we analyze the marginal dependency and that across time series.

Looking at the original data in Fig. 13.1, we see that the trend component is negligible. Based on estimated (partial) autocorrelations and periodograms, we see evidence of a weekly and yearly cycle. Weron[27] uses a rolling volatility technique to remove the annual seasonality. We apply this to our data. In a second step, we remove the weekly cycle by fitting a moving average MA(7) model where only the seventh moving average coefficient is not equal to zero, i.e.,

$$X_t = \epsilon_t + \theta_7 \epsilon_{t-7}, \quad \theta_7 \neq 0, \quad \epsilon_t \sim WN(0, \sigma^2).$$

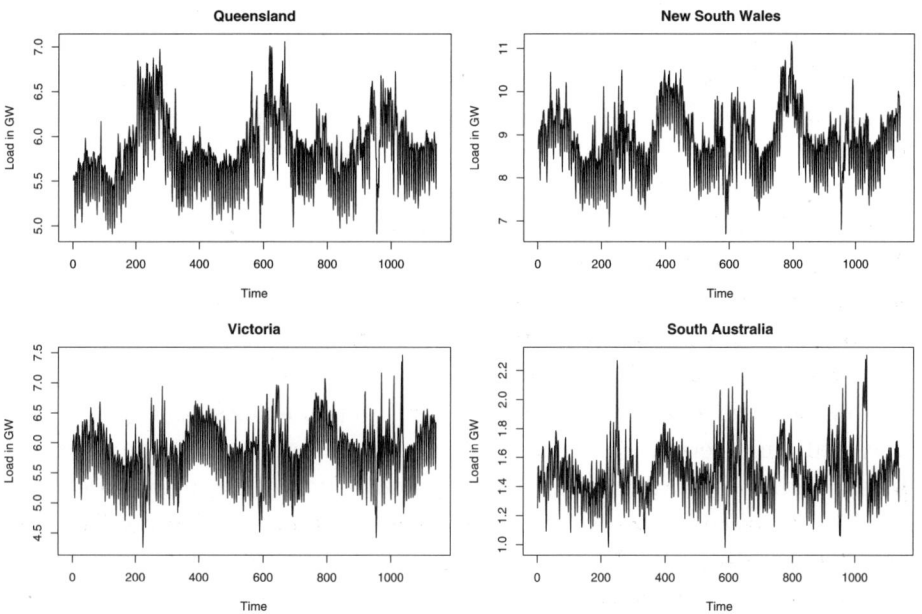

Figure 13.1. Time series plots of the observed data in the four different states.

Table 13.1. Tests for stationarity and unit root for the preprocessed time series.

	KPSS		ADF		PP	
	stat.	p-value	stat.	p-value	stat.	p-value
QLD	0.0671 accept K0	>0.1	−12.74 reject H0	<0.01	−334.31 reject H0	<0.01
NSW	0.126 accept K0	>0.1	−13.70 reject H0	<0.01	−347.94 reject H0	<0.01
VIC	0.099 accept K0	>0.1	−15.11 reject H0	<0.01	−413.37 reject H0	<0.01
SA	0.065 accept K0	>0.1	−15.04 reject H0	<0.01	−421.71 reject H0	<0.01

After the preprocessing, we want to test if these data is stationary or if there is a unit root. We use the augmented Dickey–Fuller test (ADF) with lag order 1, a Phillips–Perron test (PP) (both testing for a unit root) and the KPSS test for stationarity (for details, cf. Refs. 4 and 20). The KPSS test has the null hypothesis "K0: The time series is stationary" versus the alternative "K1: The time series is not stationary" and the ADF and PP tests have the null hypothesis "H0: The time series has a unit root, i.e., the autoregressive coefficient has an absolute value of 1" against the alternative "H1: The absolute value of the autoregressive coefficient is smaller than 1". The test results are given in Table 13.1. For each of the preprocessed time series, we cannot reject the KPSS test for stationarity. In addition, we have to reject the ADF and PP tests for unit roots. So, we assume stationarity for all four time series. Therefore, we will model them marginally with an AR(1) process. The preprocessed time series are given in Fig. 13.2. For more details on the preprocessing methods and alternative preprocessing, see Chapter 3 of Ref. 14.

Now, we are ready to investigate the dependencies among the preprocessed time series. For this, we have the preprocessed time series for Queensland, New South Wales, Victoria and South Australia, each consisting of 1134 available observations. We chose this order for geographical reasons. These states are adjacent and connected with all its infrastructure in this way along the Eastern Coast of Australia, beginning with Queensland in the North, then New South Wales, then Victoria and South Australia following in the South West.

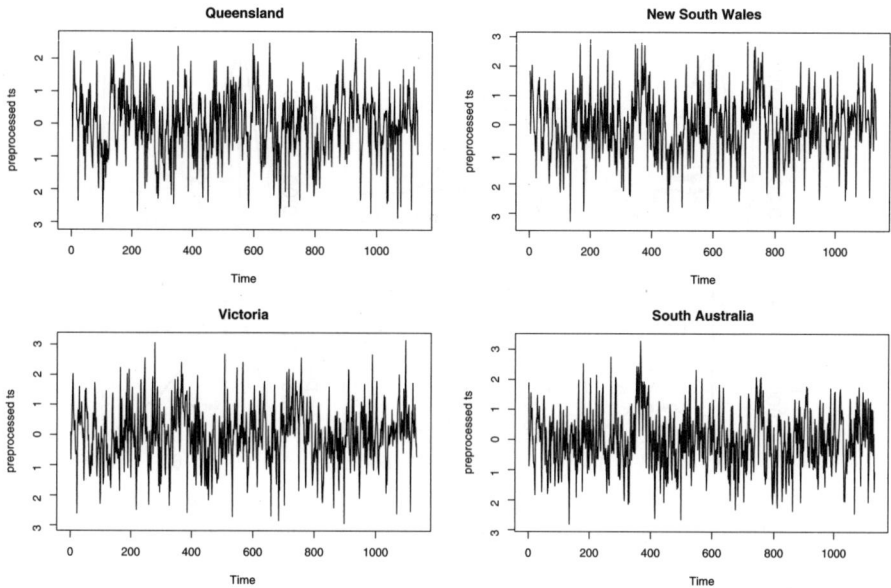

Figure 13.2. Time series plots of the preprocessed data in the four different states.

Applying the joint MCMC (JMCMC) approach developed in Section 13.2, we ran 10000 MCMC iterations. Using trace plots and estimated autocorrelation (see Section 5.3 in Ref. 14), we determined an appropriate burn-in and thinning parameters. A burn-in of 1000 iterations for all but one parameter and a thinning to every 20th iterations were sufficient.

For comparison, we also determined the two-step MLE estimates, i.e., first marginal parameters are estimated (marg. MLE) and used to transform to the Z-level. For the copula parameters, the Z-level time series are further transformed using the standard normal cdf to a time series with uniform margins. Finally, these data are used to determine the MLEs of the copula parameters (C-MLE). The resulting kernel density estimates of the posterior distribution for each parameter are given in Figs. 13.3 and 13.4, while summary statistics are presented in Table 13.2.

The MLEs of the marginal autoregressive parameters are higher than the corresponding JMCMC estimates, indicating the effect of ignoring the joint dependency. Some of these differences are credible since the marginal MLE values for the parameters γ_2, γ_3 and γ_4 lie outside of the estimated posterior 95% credibility interval. There is however good agreement for the remaining parameters. Some of the df parameters are quite high, especially those corresponding to conditional copula pairs. This leads to the question of

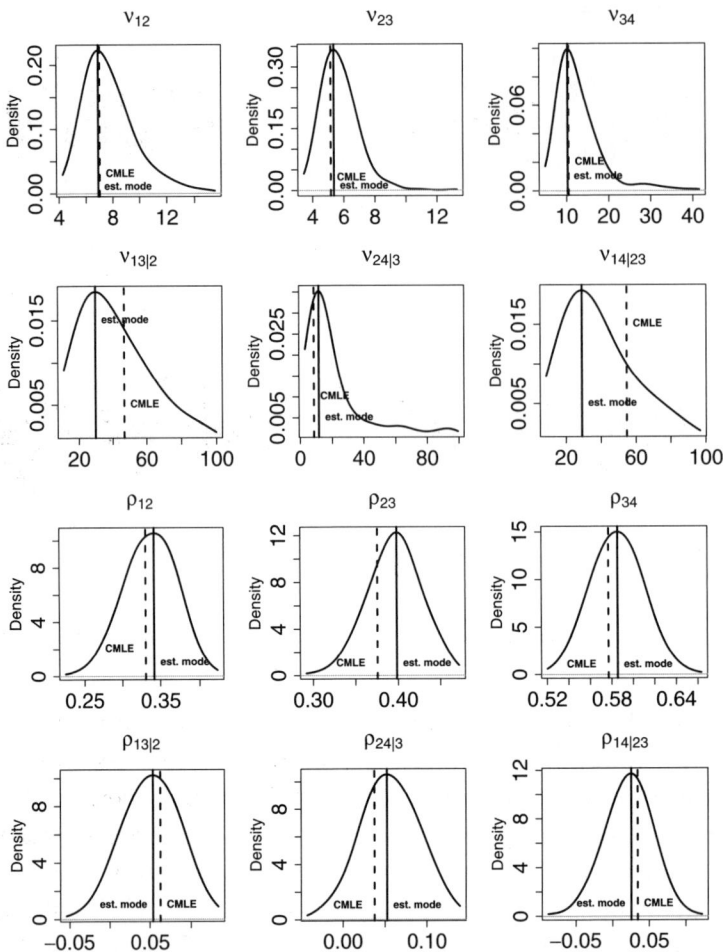

Figure 13.3. Plots of the estimated kernel density of ν and ρ of the observed real data based on the thinned out MCMC chain.

whether one can reduce the copula dependency model to Gaussian bivariate copulae for those conditional copula pairs. We will investigate this question in the following section.

13.5 Bayesian Model Selection

As we have seen in the previous section, we might want to compare several model specifications in a Bayesian setup. For this we want to compare posterior model probabilities for models of interest. Assume that we have

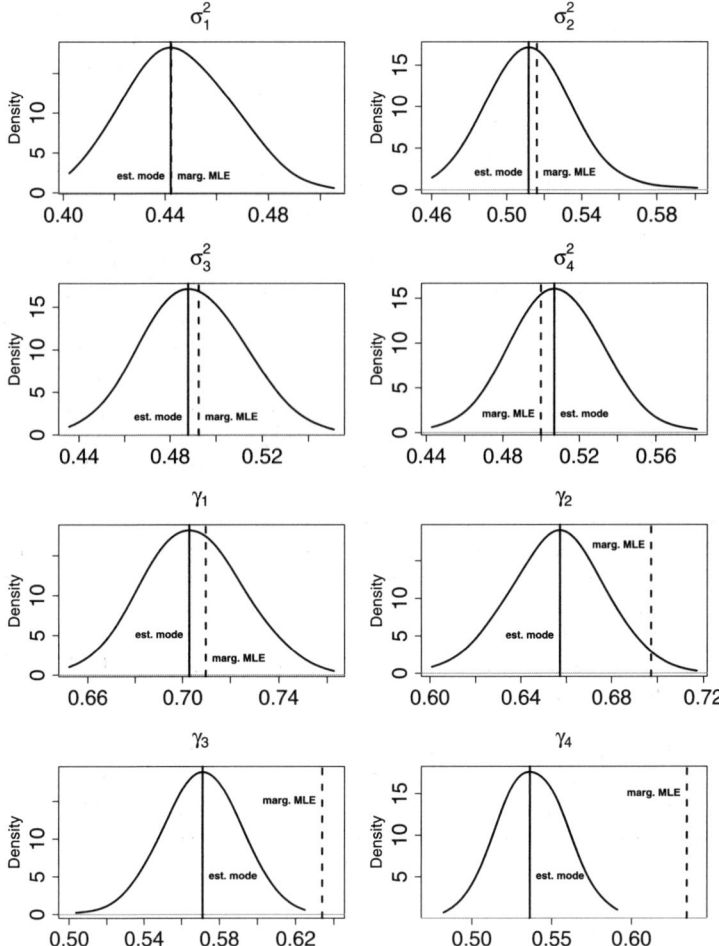

Figure 13.4. Plots of the estimated kernel density of σ^2 and γ of the observed real data based on the thinned out MCMC chain.

fitted models $M_1, \ldots, M_K$ with an MCMC method where model M_k has parameters $\boldsymbol{\theta}_k$ and we want to estimate

$$P(\text{Model } M_k | \text{data}), \quad k = 1, \ldots, K.$$

Congdon[12] gives the following estimation procedure under the assumption that the distribution of the data under Model M_k is independent of $\{\boldsymbol{\theta}_{j \neq k}\}$ and that there is independence among all $\boldsymbol{\theta}_k$ given Model M. Then he shows that the posterior distributions are independent and can be sampled

Table 13.2. Estimated posterior mean, mode and quantiles of JMCMC as well as marginal MLE, starting values and C-MLE for the preprocessed Australian load data.

Copula	2.5%	5%	50%	95%	97.5%	mean	mode	C-MLE
ν_{12}	5.17	5.45	7.37	11.72	12.71	7.80	6.92	7.32
ν_{23}	4.11	4.26	5.57	7.65	8.60	5.76	5.36	5.52
ν_{34}	7.08	7.69	11.45	24.29	29.68	12.89	10.23	11.17
$\nu_{13|2}$	15.01	16.28	36.89	84.56	91.77	41.44	29.80	38.55
$\nu_{24|3}$	4.20	4.73	14.25	78.22	93.44	22.84	11.60	8.34
$\nu_{14|23}$	12.32	14.58	34.43	78.22	86.81	38.94	29.00	58.07
ρ_{12}	0.27	0.28	0.34	0.38	0.39	0.34	0.34	0.33
ρ_{23}	0.33	0.35	0.40	0.45	0.45	0.40	0.40	0.39
ρ_{34}	0.54	0.55	0.59	0.62	0.63	0.59	0.59	0.57
$\rho_{13|2}$	−0.01	−0.00	0.05	0.10	0.11	0.05	0.05	0.06
$\rho_{24|3}$	−0.01	0.00	0.05	0.11	0.12	0.06	0.05	0.04
$\rho_{14|23}$	−0.04	−0.03	0.02	0.07	0.08	0.02	0.03	0.03
Marginal	2.5%	5%	50%	95%	97.5%	mean	mode	marg. MLE
σ_1^2	0.41	0.41	0.44	0.48	0.48	0.45	0.44	0.44
γ_1	0.67	0.67	0.70	0.74	0.74	0.70	0.70	0.71
σ_2^2	0.47	0.48	0.51	0.55	0.56	0.51	0.51	0.52
γ_2	0.62	0.62	0.66	0.69	0.69	0.66	0.66	0.70
σ_3^2	0.45	0.46	0.49	0.52	0.53	0.49	0.49	0.49
γ_3	0.53	0.54	0.57	0.60	0.61	0.57	0.57	0.63
σ_4^2	0.47	0.47	0.51	0.54	0.55	0.51	0.51	0.50
γ_4	0.50	0.51	0.54	0.57	0.58	0.54	0.54	0.63

individually. He uses the relationship

$$P(M = M_k | \text{data}, \boldsymbol{\theta}) \propto P(\text{data} | \boldsymbol{\theta}, M = M_k)$$
$$P(\boldsymbol{\theta} | M = M_k) P(M = M_k). \quad (13.4)$$

We assume now that the K independent MCMC runs result in

$$M_1 : \boldsymbol{\theta}_1^{(t)}, r = 1, \ldots R \qquad \qquad p(\boldsymbol{\theta}_1 | \text{data})$$
$$\vdots \qquad \text{which approximate} \qquad \vdots$$
$$M_K : \boldsymbol{\theta}_K^{(t)}, r = 1, \ldots R \qquad \qquad p(\boldsymbol{\theta}_K | \text{data}).$$

We use $\{\boldsymbol{\theta}^{(r)} := (\boldsymbol{\theta}_1^{(r)}, \ldots, \boldsymbol{\theta}_K^{(r)}), r = 1, \ldots R\}$ and hence, we can approximate

$$P(M | \text{data}) = \int P(M | \boldsymbol{\theta}, \text{data}) p(\boldsymbol{\theta} | \text{data}) d\boldsymbol{\theta}$$

by

$$\hat{P}(M|\text{data}) := \frac{1}{R}\sum_{r=1}^{R} P(M|\boldsymbol{\theta}^{(r)},\text{data}).$$

Using Eq. (13.4), we can estimate $P(M = M_k|\text{data}, \boldsymbol{\theta}^{(r)})$ by

$$w_k^{(r)} := \frac{G_k^{(r)}}{\sum_{j=1}^{K} G_j^{(r)}},$$

where

$$G_k^{(r)} := \exp(L_k^{(r)} - L_{\max}^{(r)})$$
$$L_k^{(r)} := \log(P(\text{data}|\boldsymbol{\theta}^{(r)}, M = M_k)P(\boldsymbol{\theta}^{(r)}|M = M_k)P(M = M_k))$$
$$L_{\max}^{(r)} := \max_{k=1,\ldots,K} L_k^{(r)}.$$

Therefore, we get

$$\hat{T}_k := \frac{1}{R}\sum_{r=1}^{R} w_k^{(r)}$$

as an estimator for $P(M = M_k|\text{data})$.

We investigated six models for the Australian load data which are described in Table 13.3 together with their estimated posterior model probability. These results indicate clearly that the model with a marginal AR(1) structure and a four-dimensional t-copula where the conditional correlation parameters are fixed to 0 gives the best fit to the observed data.

Finally, we present the parameter estimates for Model M6 in Table 13.4. This shows that there are strong marginal autocorrelations present in the four time series. The strongest is observed in Queensland, while the lowest is in South Australia. The dependence between the time series on the Z variable level has a first-order Markov structure determined by the unconditional bivariate t-copulae. Since conditional and partial correlations are the same for elliptical distributions (see Ref. 3) and there exists a one-to-one relationship between partial correlations and unconditional correlations, the remaining unconditional correlations in M6 can be determined. In particular, the posterior mode for ρ_{13}, ρ_{24} and ρ_{14} are estimated to be $0.13, 0.23$ and 0.075 respectively. This shows that the strongest dependence is between Victoria and South Australia, followed by New South Wales and Victoria. This is reasonable since South Australia and New South Wales are adjacent to Victoria which is the most populated region among the four regions.

Table 13.3. Estimated posterior model probabilities for six models for the Australian load data.

Model	Model	$\hat{T}_k$
M1	Joint Bayesian estimation with marginal AR(1) and D-vine of pair t-copulae	$4.2689 \cdot 10^{-06}$
M2	Joint Bayesian estimation of reduced model M1: marginal AR(1), unconditional pair-copulae as t-copulae, conditional pair-copulae as t-copulae with correlation 0 and df = 100	0.0351
M3	Joint Bayesian estimation with marginal AR(1) and D-vine of normal pair-copulae (approximated by a t-copula with df = 100)	$4.5999 \cdot 10^{-13}$
M4	Joint Bayesian estimation of reduced model M4: marginal AR(1), unconditional pair-copulae as Gauss copulae (approximated by a t-copula with df = 100), conditional pair-copulae as pair t-copulae with correlation 0	$3.2536 \cdot 10^{-14}$
M5	Marginal AR(1) and four-dimensional t-copula with common df	0.3245
M6	Marginal AR(1) and four-dimensional t-copula with common df and the conditional correlation parameters fixed to 0	0.6404

Table 13.4. Estimated posterior mean, mode and quantiles of the joint MCMC as well as marginal MLE and C-MLE for marginal AR(1) and four-dimensional t-copula with the conditional correlation parameters fixed to 0.

	2.5%	5%	50%	95%	97.5%	mean	mode	C-MLE
ν	5.80	6.08	7.44	9.30	9.79	7.54	7.32	9.59
ρ_{12}	0.27	0.28	0.33	0.38	0.39	0.33	0.33	0.30
ρ_{23}	0.34	0.35	0.40	0.45	0.46	0.40	0.40	0.34
ρ_{34}	0.54	0.55	0.59	0.62	0.63	0.59	0.59	0.58
	2.5%	5%	50%	95%	97.5%	mean	mode	marg. MLE
σ_1^2	0.40	0.41	0.44	0.48	0.48	0.44	0.44	0.44
γ_1	0.67	0.67	0.71	0.74	0.74	0.71	0.71	0.71
σ_2^2	0.46	0.47	0.50	0.54	0.54	0.50	0.50	0.52
γ_2	0.62	0.63	0.66	0.69	0.70	0.66	0.66	0.70
σ_3^2	0.45	0.46	0.49	0.52	0.53	0.49	0.49	0.49
γ_3	0.52	0.53	0.56	0.59	0.60	0.56	0.56	0.63
σ_4^2	0.47	0.48	0.51	0.55	0.56	0.51	0.51	0.50
γ_4	0.50	0.50	0.54	0.57	0.57	0.54	0.54	0.63

13.6 Summary and Discussion

This chapter developed a joint Bayesian analysis of a multivariate copula model with AR(1) time series margins. This avoids the efficiency loss introduced by the usual two-step estimation procedure. Model selection was facilitated by applying the approach by Congdon.[12] This approach has however been criticized by Robert and Marin[24] in general. Alternatively, one can use reversible jump MCMC (RJMCMC) developed by Green.[15] However, it is our experience that Congdon's method is a close approximation to RJMCMC for copula models based on vines.

Several extensions are of interest, for example, using higher order autoregressive models and especially using GARCH margins for financial applications. For Bayesian approaches to univariate GARCH models, see Refs. 5 and 2. The joint Bayesian estimation of such marginal models together with copula models based on vines is a subject of current research.

Acknowledgments

Claudia Czado acknowledges the support of the Deutsche Forschungsgemeinschaft (Cz 86/1-3).

References

1. Aas K., Czado C., Frigessi A. and Bakken H. (2009). Pair-copula constructions of multiple dependence. *Insurance: Mathematics and Economics*, 44(2):182–198.
2. Ardia D. (2008). *Financial Risk Management with Bayesian Estimation of GARCH Models*. Springer-Verlag, New York.
3. Baba K. and Sibuya M. (2005). Equivalence of partial and conditional correlation coefficients. *Journal of Japanese Statistical Society*, 35:1–19.
4. Banerjee A., Dolado J.J., Galbraith J.W. and Hendry D.F. (1993). *Cointegration, Error Correction and the Econometric Analysis of Non-Stationary Data*. Oxford University Press, Oxford.
5. Bauwens L. and Lubrano M. (1998). Bayesian inference on GARCH models using the Gibbs sampler. *Econometrics Journal*, 1(1):23–46.
6. Bedford T.J. and Cooke R.M. (2001). Probability density decomposition for conditionally dependent random variables modeled by vines. *Annals of Mathematics and Artificial Intelligence*, 32:245–268.
7. Bedford T.J. and Cooke R.M. (2002). Vines: A new graphical model for dependent random variables, *Annals of Statistics*, 30(4):1031–1068.
8. Brockwell P.J. and Davis R.A. (1991). *Time Series: Theory and Methods*, 2nd ed., Springer Series in Statistics. Springer, New York.
9. Chen X. and Fan Y. (2006). Estimation and model selection of semiparametric copula-based multivariate dynamic models under copula misspecification. *Journal of Econometrics*, 135:125–154.

10. Cherubini U., Luciano E. and Vecchiato W. (2004). *Copula Methods in Finance*. Wiley & Sons, New York.
11. Congdon P. (2003). *Applied Bayesian Modelling*. Wiley & Sons, Chichester.
12. Congdon P. (2006). Bayesian model choice based on Monte Carlo estimates of posterior model probabilities. *Computational Statistics and Data Analysis*, 50:346–357.
13. Gamerman D. and Lopes H.F. (2006). *Markov Chain Monte Carlo: Stochastic Simulation for Bayesian Inference*, 2nd ed. CRC Press, Boca Raton.
14. Gärtner F. (2008). Bayesian analysis of multivariate time series models based on pair-copula construction. Diploma thesis, Zentrum Mathematik, Technische Universität München.
15. Green P.J. (1995). Reversible jump Markov chain Monte Carlo computation and Bayesian model determination. *Biometrika*, 82(4):711–732.
16. Joe H. (1997). *Multivariate Models and Dependence Concepts*. Chapman & Hall, London.
17. Joe H. (2005). Asymptotic efficiency of the two-stage estimation method for copula-based models. *Journal of Multivariate Analysis*, 94:401–419.
18. Joe H. (2006). Generating random correlation matrices based on partial correlations. *Journal of Multivariate Analysis*, 97:2177–2189.
19. Kurowicka D. and Cooke R.M. (2006). *Uncertainty Analysis with High Dimensional Dependence Modelling*. Wiley & Sons, New York.
20. Kwiatkowski D., Phillips P.C., Schmidt P. and Shin Y. (1992). Testing the null hypothesis of stationarity against the alternative of a unit root. *Journal of Econometrics*, 54:159–178.
21. Lewandowski D., Kurowicka D. and Joe H. (2009). Generating random correlation matrices based on vines and extended onion method. *Journal of Multivariate Analysis*, 100:1989–2001.
22. Min A. and Czado C. (2008). Bayesian inference for multivariate copulas using pair-copula constructions. *Journal of Financial Econometrics*. In press.
23. Nelsen R.B. (2006). *An Introduction to Copulas*, 2nd ed., Springer Series in Statistics. Springer, New York.
24. Robert C. and Marin J.-M. (2008). Some difficulties with some posterior probability approximations. *Bayesian Analysis*, 3:427–442.
25. Sklar A. (1959). Fonctions de répartitions à n dimensions et leur marges. *Publications de l'Institut de Statistique de l'Université de Paris*, 8:229–231.
26. Tierney L. (1994). Markov chains for exploring posterior distributions. *Annals of Statistics*, 22(4):1701–1762.
27. Weron R. (2006). *Modeling and Forecasting Electricity Loads and Prices*. Wiley & Sons, Chichester.
28. Wise J. (1955). The autocorrelation function and the spectral density function. *Biometrika*, 42(1/2):151–159.

CHAPTER 14

Non-Parametric Bayesian Belief Nets versus Vines

Anca Hanea

Delft University of Technology, Institute of Applied Mathematics
Mekelweg 4, 2628 CD Delft, the Netherlands
A.Hanea@ewi.tudelft.nl

This chapter reviews aspects of non-parametric Bayesian belief nets (NPBBN). The theory behind NPBBNs is closely related to that of regular vines and it has benefited from developments in the latter. It also offers an alternative to undirected graphical models in general, and to regular vines in particular. The differences and similarities in modeling using directed versus undirected graphs are discussed in this chapter from the perspective of NPBBNs and vines. Until recently, Bayesian belief nets (BBNs) were either discrete or discrete-normal. Despite their popularity, both suffer from severe limitations. Discrete BBNs are limited by size and complexity, discrete-normal BBNs are limited by the assumption of joint normality. NPBBNs were introduced to overcome these limitations. Algorithms for specifying, sampling and analyzing high-dimensional distributions using NPBBNs have been developed and successfully applied in decision support systems.

14.1 Introduction or: How to Represent Information Burdened by Uncertainty 282
14.2 Non-Parametric Bayesian Belief Nets: Sampling and Conditionalizing 288
    14.2.1 Sampling an NPBBN 291
    14.2.2 Conditionalizing an NPBBN 293
14.3 Data Mining with NPBBNs 297
14.4 Applications of NPBBNs 299
14.5 Conclusions 300
References 302

14.1 Introduction or: How to Represent Information Burdened by Uncertainty

Understanding and representing multivariate distributions along with their dependence structure is a highly active area of research. A large body of scientific work treating multivariate models is available. This chapter in particular, and this book in general, advocates using graphical models to represent high-dimensional distributions with complex dependence structures.

Graphical models proved to be a flexible probabilistic framework and their use has increased substantially, hence the theory behind them has been constantly developed and extended.

There are two main types of graphical models: directed, based on directed acyclic graphs (DAGs), and undirected, generally referred to as Markov networks. The regular vines are a generalization of Markov trees, hence they fall into the former category, whereas the Bayesian belief nets (BBNs) belong to the latter. Why or when to use one graphical model or another is not a question with a straightforward answer. This chapter will provide some insights into the differences and similarities between the two types of models, and hopefully these will serve as guidelines for modelers.

Both directed and undirected models consist of a qualitative and a quantitative part. The qualitative part is represented by the graph itself together with the (in)dependence relationships entailed. Perhaps the most important difference between directed and undirected graphs, in general, is that they make different statements of conditional independence. We will first focus on the differences arising from the graphical structures, rather than the quantification of a joint multivariate distribution.

The absence of a link between two nodes means that any dependence between these two variables is mediated via some other variables, hence they encode conditional (in)dependence statements between variables. Given the nested tree structure of a regular vine, one can consider them as fully connected graphs. In this sense, in regular vines, the concept of conditional independence is weakened to allow for various forms of conditional dependence. Is this an advantage or a disadvantage of vines? The answer depends on many factors, which may lead to the conclusion that the question is ill-posed.

A number of examples will shed some light on the matter. Consider three random variables X_1, X_2 and X_3 represented as nodes in a graphical structure. The node that corresponds to variable X_i is denoted by i. Let node 3 have converging links. This is a configuration that yields conditional

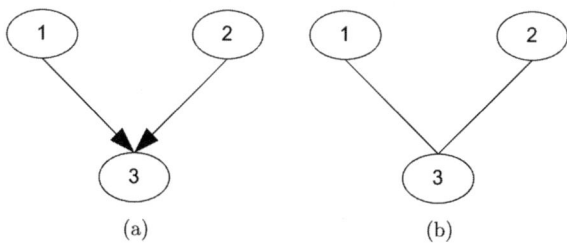

Figure 14.1. Nodes with converging links. (a) Node with converging arrows in a BBN. (b) Node with converging edges in a Markov network.

independence in Markov networks and conditional dependence in BBNs. The structure in Fig. 14.1(a) entails the *conditional dependence* of X_1 and X_2, given X_3,[a] whereas Fig. 14.1(b) entails the *conditional independence* of X_1 and X_2, given X_3.

The possibility of representing the combination of statements in Fig. 14.1(a) may be regarded as an advantage of BBNs over undirected structures, since it permits the display of induced and non-transitive dependencies. This configuration also represents the main difference between the separation properties in the directed and undirected graphs. In directed graphs, the direction-dependent criterion of connectivity called the *d-separation criterion* consists of the above rule for converging arrows, plus the usual cutset criterion of Markov networks, whenever the arrows are diverging or cascaded.[16] If two nodes of a BBN are d-separated by a set of nodes, then the corresponding variables are conditionally independent, given that set.

Remark 14.1. If two nodes are not d-separated it does not necessarily mean that the corresponding variables are not conditionally independent. In other words, whenever an arc or an unblocked path[b] exists between two nodes, it is not necessarily the case that the corresponding variables are dependent.

Regular vines, however, may also be used to represent the independence of X_1 and X_2, and the conditional dependence of X_1 and X_2, given X_3. Nevertheless, the graphical structure alone will not suffice in completing the task and this might be viewed as a disadvantage of regular vines. Given the

---

[a]The independence of X_1 and X_2.
[b]Intuitively, an unblocked path *may* carry information, or dependence between end nodes. For exact definitions we refer to Ref. 16.

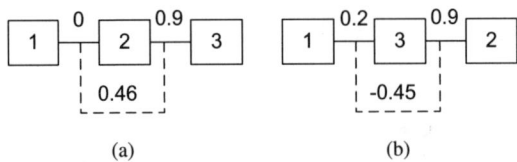

Figure 14.2. D-vines "representing" induced and non-transitive dependencies. (a) D-vine representing the independence of X_1 and X_2. (b) D-vine representing the conditional dependence of X_1 and X_2, given X_3.

full connectivity of vines, (conditional) dependencies and/or independencies can only be represented through quantification. Edges of a regular vine can be associated with (conditional) rank correlations. If these rank correlations are realized by copulae with the zero independence property, representing the independence of X_1 and X_2 reduces to associating the edge between them with a zero rank correlation. This is shown in Fig. 14.2(a). Yet, the conditional dependence of X_1 and X_2, given X_3, is not obvious. A few calculations are needed in order to verify that, and a different graph is needed to actually visualize it. Figure 14.2(b) shows a non-zero conditional rank correlation between X_1 and X_2, given X_3, but fails to represent the independence of X_1 and X_2.

It is worth remembering that the present discussion solely regards the representation of certain (conditional) (in)dependencies using different graphical structures, and not the full representation/quantification of joint distributions. The specification of (conditional) rank correlations on the edges of a regular vine serves here this purpose only.

Another feature of BBNs that can be regarded as an advantage over regular vines is that conditional independencies are represented by missing arcs, therefore certain conditional independencies become visible in the graph by deleting arcs. Consider the D-vine on four variables in Fig. 14.3. In this example and further in this chapter, the copulae used to realize the (conditional) rank correlations associated with the edges of a regular vine will possess the zero independence property.

Variables X_1 and X_3 are independent, given X_2. Independence is denoted by $\perp$, e.g. $X_1 \perp X_3|X_2$. The notation $X_2 \not\perp X_4|X_3$ means that X_2 and X_4 are not conditionally independent, given X_3. If a (conditional) rank correlation from the D-vine is not replaced by zero, the corresponding variables are considered to be (conditionally) dependent. The information represented by the D-vine in Fig. 14.3 can be represented using a saturated BBN, from which the arc between X_1 and X_3 is deleted. In this way, the dependence

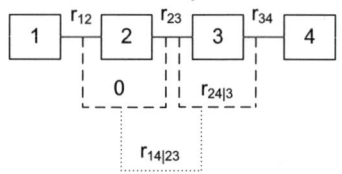

Figure 14.3. A D-vine on four variables representing the following: $X_1 \perp X_3|X_2$; $X_2 \not\perp X_4|X_3$; $X_1 \not\perp X_4|(X_2, X_3)$; $X_1 \not\perp X_2$, $X_2 \not\perp X_3$, $X_3 \not\perp X_4$.

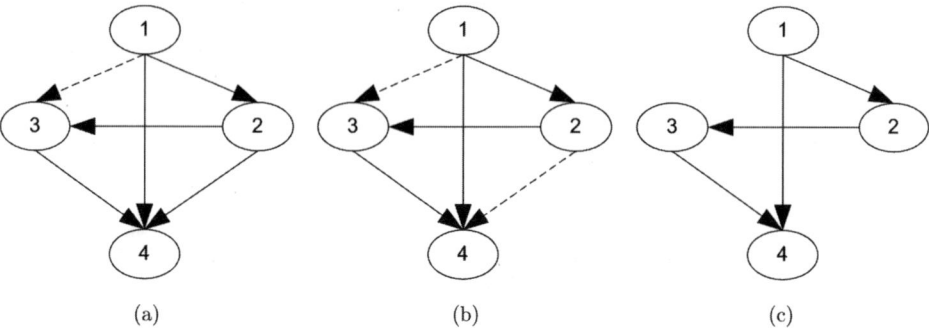

Figure 14.4. (a) A BBN with four nodes and five arcs representing $X_1 \perp X_3|X_2$. (b) A BBN with four nodes and four arcs representing $X_1 \perp X_3|X_2$. (c) The same BBN as in (b).

between the two variables is mediated only via X_2 (see Fig. 14.4(a)). Further, X_2 and X_4 are conditionally dependent, given X_3. Since the presence of arcs does not guarantee dependence between variables (see Remark 14.1), this statement cannot be represented with a BBN. The best one could do is to avoid representing the opposite (i.e., $X_2 \perp X_4|X_3$). The dependence between X_2 and X_4 is not mediated only through X_3, therefore the arc between them can be deleted (see Fig. 14.4(b)). This of course will introduce a new conditional independence statement, i.e., $X_2 \perp X_4|(X_1, X_3)$, but it will not necessarily violate the requirements imposed by the D-vine. The resultant structure is presented in Fig. 14.4(c).

Only four arcs are necessary in order to represent the same conditional independence statements as in the D-vine. The reduction in the number of arcs constitutes a major advantage since it results in a sparser, more readable structure. Another example of a set of conditional independence statements represented on a D-vine with 15 edges versus a BBN with six arcs is presented in Figs. 14.5 and 14.6.

Following the same strategy as before, i.e., starting with the saturated BBN and removing the arcs corresponding to the independence statements,

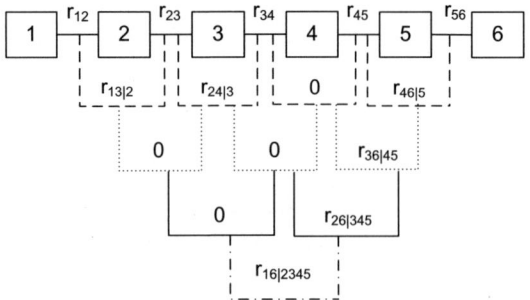

Figure 14.5. A D-vine on six variables representing four conditional independence statements.

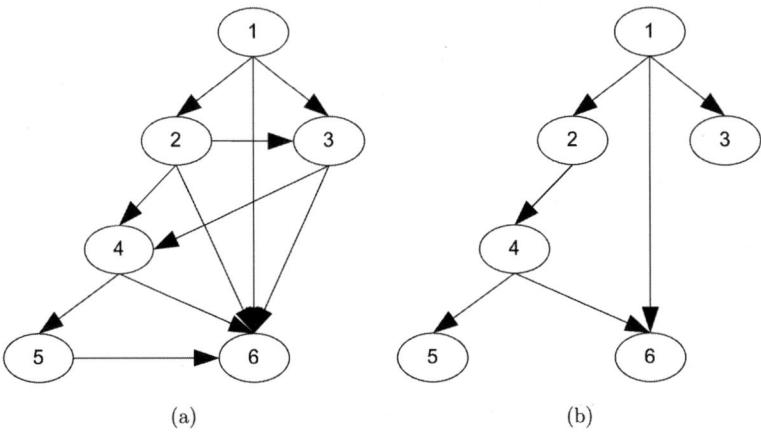

Figure 14.6. (a) BBN with six nodes and 11 arcs representing the same conditional independence statements as the D-vine in Fig. 14.5. (b) BBN with six nodes and six arcs representing the same conditional independence statements as the D-vine in Fig. 14.5.

results in the BBN in Fig. 14.6(a). As expected, the number of arcs is reduced to 11. Nevertheless, if one only wants to preserve the conditional independence statements shown in the regular vine and not violate the conditional dependencies, the structure can be reduced even further, e.g., Fig. 14.6(b).

In larger structures, with many conditional independence statements present, the reduction might be even more dramatic. Nevertheless, there are configurations in which deleting arcs from a saturated BBN (corresponding to a regular vine) does not result in a better "picture". Consider the D-vine in Fig. 14.7.

Starting with the saturated BBN and deleting the arcs between X_1, X_3 and X_2, X_4 will result in the BBN in Fig. 14.4(c). But in this structure

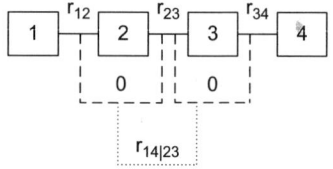

Figure 14.7. A D-vine on four variables representing the following: $X_1 \perp X_3|X_2$; $X_2 \perp X_4|X_3$; $X_1 \perp X_4|(X_2, X_3)$; $X_1 \not\perp X_2$, $X_2 \not\perp X_3$, $X_3 \not\perp X_4$.

X_2 and X_4 are not d-separated by X_3. This does not imply that they are not conditional independent, given X_3. They might be, but this conditional independence is not visible anymore, and the BBNs' advantage of being more visually intuitive vanishes. Any reorientation of the arcs will fail to represent — via d-separation — both conditional independence statements.

On the other hand, starting with the BBN structure in Fig. 14.4(c) (rearranged as in Fig. 14.8(a)) and trying to represent its conditional independencies with a vine might prove difficult. Figure 14.8(a) encodes $X_4 \perp X_2|X_1, X_3$ and $X_1 \perp X_3|X_2$. To represent the first statement on a D-vine, variable X_1 has to be before variable X_2 in the first tree (see Fig. 14.8(b)), whereas to represent the second statement the order of these variables has to change (see Fig. 14.8(c)).

The choice between representing a multivariate distribution using a regular vine or using a BBN depends on many factors. A few of them, related exclusively to the graphical representation of (in)dependence statements were discussed above. Other factors will be explored throughout this chapter.

The rest of the chapter is organized as follows. We first introduce non-parametric Bayesian belief nets (NPBBNs) and their connection with

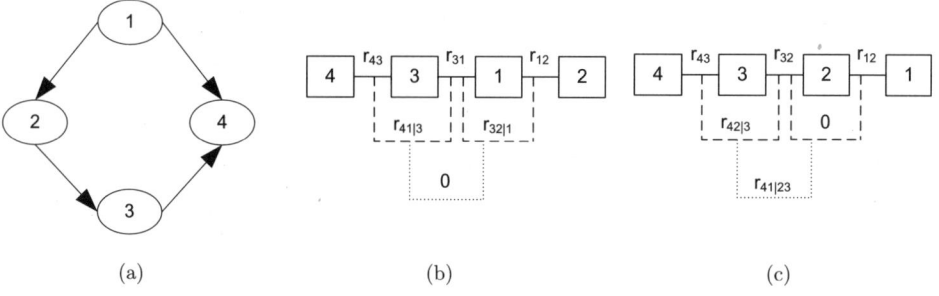

Figure 14.8. (a) A BBN representing $X_4 \perp X_2|X_1, X_3$ and $X_1 \perp X_3|X_2$. (b) A D-vine representing $X_4 \perp X_2|X_1, X_3$. (c) A D-vine representing $X_1 \perp X_3|X_2$.

regular vines. Differences in sampling and performing inference using an NPBBN versus using a regular vine are further discussed. The issues of model learning and validation are addressed and some applications of the NPBBNs methodology are finally presented. The last section gathers conclusions.

14.2 Non-Parametric Bayesian Belief Nets: Sampling and Conditionalizing

This chapter concentrates on BBNs. As already mentioned, BBNs are DAGs, whose nodes represent univariate random variables and whose arcs represent direct influences.[c]

The origin of BBNs can be traced back to the early decades of the 20th century when Sewell Wright,[19] in his pioneering work, developed path analysis to help the study of genetic inheritance.

In their most popular form, BBNs were introduced in the 1980s as a knowledge representation formalism to encode and use the information acquired from human experts in automated reasoning systems to perform diagnostic and prediction.[16]

BBNs provide a compact representation of high-dimensional distributions of a set of variables and encode their joint density/mass function by specifying a set of conditional independence statements and a set of probability functions. The graph itself and the (conditional) independence relations entailed form the qualitative part of a BBN model. The quantitative part of the model consists of the conditional probability functions associated with the variables. In Section 14.1 we concentrated on the qualitative part of BBNs. Subsequently we will mainly discuss their quantitative aspects and the techniques for building high-dimensional distributions.

Until recently, BBNs were discrete, normal or discrete-normal. In discrete BBNs, nodes represent discrete random variables. These models specify marginal distributions for source nodes, and conditional probability tables for child nodes. If the nodes of a BBN correspond to variables that follow a joint normal distribution, we talk of Gaussian BBNs (or normal BBNs).[16,18] Continuous BBNs developed for joint normal variables interpret the *influence* of the parents on a child as partial regression coefficients when the

---

[c]BBNs can also contain functional nodes, i.e., nodes which are functions of other nodes. The ensuing discussion refers to probabilistic nodes.

child is regressed on the parents. They require means, conditional variances and partial regression coefficients which can be specified in an algebraically independent manner.[18]

Despite their popularity, they suffer from severe limitations. Discrete BBNs are limited by size and complexity; normal and discrete-normal BBNs are limited by the assumption of joint normality.[d]

Uncertainty distributions may not be assumed to conform to any parametric form. Algorithms for specifying, sampling and analyzing high-dimensional distributions should therefore be non-parametric. Regular vines allow us to move beyond discrete BBNs without defaulting to the joint normal distribution. When no marginal distribution assumption is made, we talk of non-parametric BBNs, abbreviated NPBBNs. NPBBNs and their relationship with regular vines were introduced in Ref. 11 and extended in Ref. 5. The focus of this section is on quantifying and building a joint distribution using an NPBBN.

An *NPBBN* is a DAG, together with a set of (conditional) rank correlations, a copula class with the zero independence property, parameterized by rank correlation, and a set of marginal distributions. In NPBBNs nodes are associated with arbitrary distributions and arcs with (conditional) rank correlations that are realized by the chosen copula. In continuous NPBBNs nodes are associated with continuous invertible distribution functions. The nodes of an NPBBN will be assumed continuous unless otherwise specified. Further in this chapter, whenever we speak of NPBBNs, we mean the DAG together with the specification of rank correlations, copula and margins.

The DAG of an NPBBN induces a (non-unique) ordering and stipulates that each variable is conditionally independent of all predecessors in the ordering, given its direct predecessors. The direct predecessors of a node i, corresponding to variable X_i, are called *parents* and the set of all i's parents is denoted $Pa(i)$.

Each variable is associated with a conditional probability function of that variable, given its parents in the graph, $f_{i|Pa(i)}, i = 1, \ldots, n$. The conditional independence statements encoded in the graph allow us to write[e]:

$$f_{1,2,\ldots,n} = \prod_{i=1}^{n} f_{i|Pa(i)}. \tag{14.1}$$

---

[d]For a detailed discussion of the disadvantages of discrete and normal BBNs, we refer to Chapter 1 of Hanea.[4]
[e]This factorization is of course valid for BBNs in general and not only for NPBBNs.

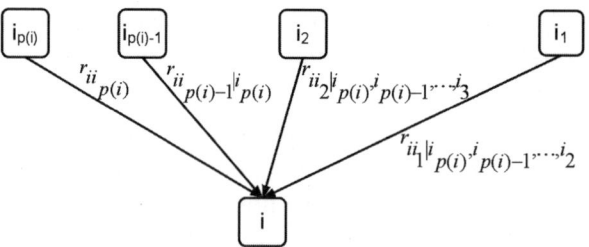

Figure 14.9. Node i of an NPBBN and the set of parent nodes for i.

For each variable i with parents $i_1 \ldots i_{p(i)}$, we associate the arc $i_{p(i)-k} \to i$ with the conditional rank correlation:

$$\begin{cases} r_{i,i_{p(i)}}, & k = 0 \\ r_{i,i_{p(i)-k}|i_{p(i)},\ldots,i_{p(i)-k+1}}, & 1 \leq k \leq p(i) - 1. \end{cases} \quad (14.2)$$

The assignment is vacuous if $\{i_1 \ldots i_{p(i)}\} = \emptyset$ (see Fig. 14.9).

Therefore, every arc in the NPBBN is assigned a (conditional) rank correlation between parent and child. These assignments are made according to a protocol presented in Ref. 11. The conditional rank correlations need not be constant, although they are taken to be constant in the following example.[f] We will illustrate the protocol for assigning (conditional) rank correlations to the arcs of an NPBBN with an example.

Example 14.1. Let us consider the undirected cycle on four variables in Fig. 14.10. This structure is similar to the structure presented in Fig. 14.8(a).

The DAG of this NPBBN induces two orderings[g] of the variables: 1, 2, 3, 4, or 1, 3, 2, 4. Let us choose 1, 2, 3, 4. The factorization of the joint distribution is:

$$P(1)P(2|1)P(3|1\underline{2})P(4|2\underline{31}). \quad (14.3)$$

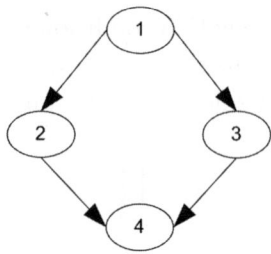

Figure 14.10. BBN with four nodes and four arcs.

---

[f]The conditional rank correlations must be constant when the normal copula is used.
[g]Such an ordering of the variables is referred to as *sampling order* or *topological order*.

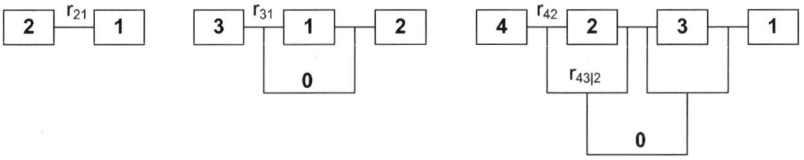

Figure 14.11. $\mathcal{D}^2, \mathcal{D}^3, \mathcal{D}^4$ for Example 14.1.

The underscored nodes in each conditioning set are the non-parents of the conditioned variable. Thus, they are not necessary in sampling the conditioned variable. This uses some of the conditional independence relations in the NPBBN. The correlation between the child and its first parent[h] will be an unconditional rank correlation, and the correlations between the child and its next parents (in the ordering) will be conditioned on the values of the previous parents. Hence, one set of (conditional) rank correlations that can be assigned to the edges of the NPBBN in Fig. 14.10 is: $\{r_{21}, r_{31}, r_{42}, r_{43|2}\}$. For each term i ($i = 1, \ldots, 4$) of the factorization (14.3), a D-vine on i variables is built. This D-vine is denoted by $\mathcal{D}^i$ and it contains: the variable i, the non-underscored variables, and the underscored ones, in this order. Figure 14.11 shows the D-vines built for variables 2, 3, 4.

Building the D-vines is not a necessary step in specifying the rank correlations,[i] but it is essential in proving a result that not only establishes the connection between NPBBNs and vines, but is also crucial for the development of NPBBNs. The result will be further formulated; for its proof we refer to Ref. 5:

Given a continuous NPBBN on n variables, the joint distribution of the variables is uniquely determined. This joint distribution satisfies the characteristic factorization (14.1) and the conditional rank correlations in (14.2) are algebraically independent.

The (conditional) rank correlations and the marginal distributions needed in order to specify the joint distributions represented by the NPBBN can be retrieved from data if available, or elicited from experts.[14]

14.2.1 Sampling an NPBBN

Since no analytical/parametric form of the joint distributions is available, the only way to stipulate it is by sampling it. In order to sample an NPBBN,

---

[h]The parents of each variable can be ordered in a non-unique way.
[i]These are assigned directly to the arcs of the BBN. Each arc is associated with a (conditional) parent–child rank correlation as in Fig. 14.9.

we will use the procedures for regular vines presented in Chapter 3. Variable X_i is sampled using the procedure for the vine $\mathcal{D}^i$. When using regular vines to sample a continuous NPBBN, it is not in general possible to keep the same order of variables in successive vines. In other words, we will have to re-order the variables before constructing $\mathcal{D}^{i+1}$ and sampling X_{i+1}, which will involve calculating some conditional distributions. If the order of variables does not change from one vine to another, the sampling procedure for the NPBBN coincides with the sampling procedure for the regular vine built for the last variable in the ordering (for details and examples, see Ref. 13). In Fig. 14.11, one can notice that the D-vine for the third variable is $\mathcal{D}^3 = D(3,1,2)$, and the order of the variables from $\mathcal{D}^4$ must be $D(4,3,2,1)$. Hence, this NPBBN cannot be represented as just one D-vine. This particularity of an undirected cycle was already noticed in Fig. 14.8 from Section 14.1. In order to sample X_4, we use the sampling procedure described in Chapter 3 of this book:

$$x_4 = F^{-1}_{r_{42};x_2}(F^{-1}_{r_{43|2};F_{r_{32};x_2}(x_3)}(F^{-1}_{r_{41|32};F_{r_{21|3};F_{r_{32};x_3}(x_2)}(F_{r_{31};x_3}(x_1))}(u_4))),$$

which, using the conditional independencies from the graph, reduces to:

$$x_4 = F^{-1}_{r_{42};x_2}(F^{-1}_{r_{43|2};F_{r_{32};x_2}(x_3)}(u_4)).$$

The conditional distribution $F_{r_{32};x_2}(x_3)$ is not given explicitly but it can be calculated as follows:

$$F_{3|2}(x_3) = \int_0^{x_3} \int_0^1 c_{21}(x_2, x_1) c_{31}(v, x_1) dx_1 dv,$$

where c_{i1} is the density of the chosen copula with correlation r_{i1}, $i \in \{2,3\}$.

For each sample, one needs to calculate the numerical value of the double integral. In this particular case, when only one double integral needs to be evaluated, it can be easily done without excessive computational burden. If the NPBBN contains an undirected cycle of five variables, and the same sampling procedure is applied, a triple integral will have to be calculated. The bigger the undirected cycle is, the larger the number of multiple integrals to be numerically evaluated.

If the multivariate distribution can be represented and assessed using one single regular vine, no extra calculations are needed in order to obtain samples from the joint distribution, hence the computational time is drastically reduced.

Nevertheless, the disadvantage mentioned above vanishes when the normal copula is used. A different sampling protocol based on the normal copula uses the properties of normal vines to realize the dependence structure specified via (conditional) rank correlations on the NPBBN. This sampling protocol is presented in Chapter 3. The main advantage of this method is that everything is calculated on the joint normal vine, hence we can reorder the variables (if necessary) and recompute all partial correlations needed. This results in a dramatic decrease in the computational time. For examples and comparisons, see Ref. 5.

It is worth mentioning that the approach to continuous NPBBNs using vines is extended to include ordinal discrete random variables. The dependence structure in the NPBBN is defined via (conditional) rank correlations, hence with respect to the underlying uniform variables. The rank correlation of two discrete variables and the rank correlation of their underlying uniforms are not equal. The relationship between them is established in Ref. 6. This relationship is based on a generalization of the population version of Spearman's rank correlation coefficient for the case of ordinal discrete random variables.

Since the sampling procedure for NPBBNs is based on the one for regular vines, we cannot talk about the advantages of the former compared to the latter.

14.2.2 *Conditionalizing an NPBBN*

Maybe one of the most important features of probabilistic graphical models is that they can be used for inference. One can calculate the distributions of unobserved nodes, given the values of the observed ones, i.e., conditional distributions.

For regular vines, if values of some variables are observed, the results of sampling the model — conditional on these values — can be obtained either by sampling again the structure (the cumulative approach) or by using the density approach, both of which are presented in Chapter 3. The new conditional distribution, although calculated, cannot be easily visualized and compared with the unconditional one. Even if this is merely an implementation issue for graphical software, NPBBNs still hold the advantage that conditionalization can be visualized and interpreted in terms of the directionality of arcs. In other words, if the reasoning is done "bottom-up" (in terms of the directionality), the NPBBN is used for diagnosis, whereas if it is done "top-down", the NPBBN serves for prediction. Following the principle that *a picture is worth a thousand words*, we will continue with an example. It is loosely based on an ongoing project undertaken by the European Union that uses the NPBBNs' methodology. The name of the project is Beneris (short for "Benefit and Risk") and it focuses on the analysis of health benefits and risks associated with food consumption.[j] The model introduced here is a highly simplified version of the NPBBN model used in the project.[10] The goal is to estimate the beneficial and harmful health effects in a specified population, as a result of exposure to various contaminants and nutrients through ingestion of fish.

Example 14.2. Figure 14.12(a) resembles the version of the model that we are considering for purely illustrative purposes.

The variables of interest for this model are the health endpoints resulting from exposure to fish constituents, namely cancer and cardiovascular risk. These risks are defined in terms of remaining lifetime risks. The three fish constituents considered are: dioxins/furans, polychlorinated biphenyls and fish oil. The first two are persistent and bio-accumulative toxins which cause cancer in humans. Fish oil is derived from the tissues of oily fish and has high levels of omega-3 fatty acids which regulate cholesterol and reduce inflammation throughout the human body. Personal factors such as smoking,

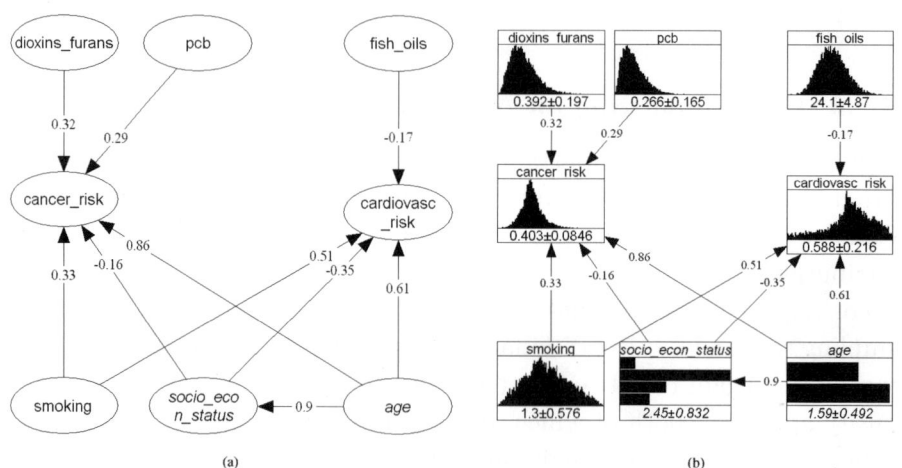

Figure 14.12. (a) Simplified fish consumption NPBBN. (b) Simplified fish consumption NPBBN with histograms.

---

[j]http://www.beneris.eu/

socioeconomic status and age may also influence cancer and cardiovascular risk. Smoking is measured as the yearly intake of nicotine during smoking and passive smoking, while socioeconomic status is measured by income and represented by a discrete variable with four income classes (from top to bottom in Fig. 14.12(b) unemployed, blue collar, white collar, and farmer and entrepreneur). Age is take as a discrete variable with two states, 15 to 34 years and 35 to 59 years.

The distributions of the variables are presented in Fig. 14.12(b) together with their means and standard deviations. They are chosen by the author for illustrative purposes only. So are the (conditional) rank correlations assigned to the arcs of the NPBBN.

We are interested in what-if? scenarios, in diagnosis and/or prediction, and moreover in visualizations and comparisons with the default situation. Examine the situation in which there is a very high risk of cancer. To do that, we conditionalize on the 0.9 value of cancer risk and study in what way the other variables in the graph are affected by this information. In this case, the NPBBN is used for diagnosis.

Figures 14.12 and 14.13 are obtained with UNINET, a software application where the approach to mixed non-parametric continuous and discrete

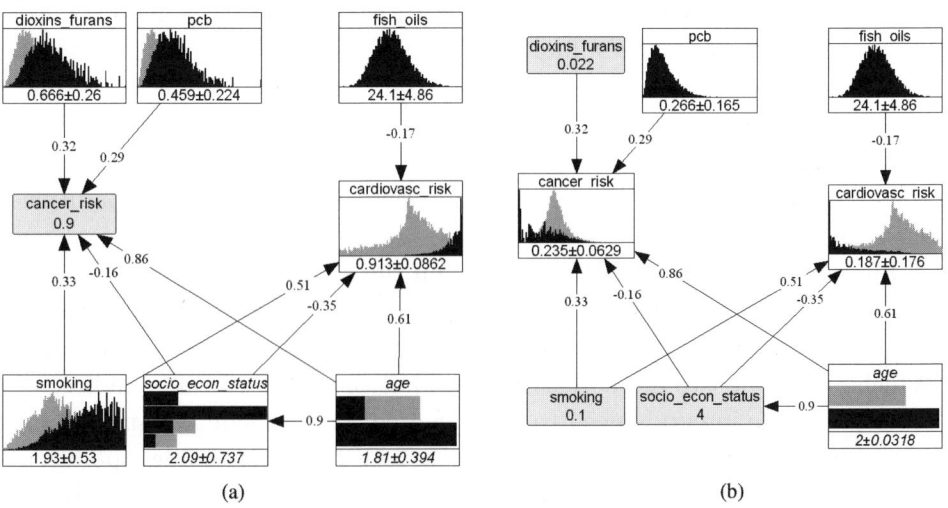

Figure 14.13. Diagnostic & predictive reasoning using the NPBBN. (a) Conditionalised NPBBN for cancer_risk = 0.9. (b) Conditionalised NPBBN for dioxins_furans = 0.022, smoking = 0.1, socio_econ_status = 4.

BBNs has been implemented.[k] In Fig. 14.13, the grey distributions in the background are the unconditional marginal distributions, provided for comparison. The conditional means and standard deviations are displayed under the histograms. In Fig. 14.13(a), we examine the situation of a very high cancer risk. We are interested in what can we infer about the factors influencing the cancer risk, when this risk is known to be 0.9. From the shift of the distributions, one can see that a person with this risk level is neither very young, nor very wealthy, smokes much and ingests a large amount of dioxins/furans and polychlorinated biphenyls. Because some of these factors also influence cardiovascular risk, the shift in their distributions causes an increase in cardiovascular risk as well.

The conditionalization in an NPBBN can also be used for prediction. For example, one can be interested in the cancer risk of a person that inhales a very small amount of nicotine, has a high socioeconomic status and ingests very little dioxins/furans. Figure 14.13(b) presents the flow of this information through the graph. The expected value of the cancer risk decreases from 0.4 to 0.23. A substantial decrease can also be seen in the cardiovascular risk. Because socioeconomic status and age are positively correlated, a high socioeconomic status results in a reduction of the population to the segment older than 35 years.

All the results and computations performed in this section are also possible if the model used is a regular vine rather than an NPBBN. Nevertheless the visualization of such results is not yet available and the interpretations, in terms of the flow of influences, might be somewhat cumbersome when using regular vines.

One might wonder how we actually calculated the conditional distributions presented in Fig. 14.13. There are several ways to perform conditionalization in NPBBNs.

Since sampling an NPBBN is based on the sampling procedure for regular vines, the cumulative or density approach for vines, mentioned in the beginning of this section, can be used to perform inference in NPBBNs. Whichever of the two methods is preferred, if the DAG contains undirected cycles, multiple integrals need to be evaluated for each sample, and for any new conditionalization. This might be a very time-consuming operation. Nevertheless, the problem owner might not be prepared to wait days or

---

[k]The software is available on http://dutiosc.twi.tudelft.nl/~risk/, together with supporting scientific documentation.

even hours for the results of new scenarios and policies. In these cases, the advantages of fast updating algorithms for discrete BBNs[3,16] are decisive. The reduced assessment burden and modeling flexibility of the NPBBNs are combined with the fast updating algorithms of discrete BBNs in the hybrid method presented in Ref. 5. Sampling a large NPBBN structure once and then discretizing it so as to enable fast updating provides an elegant solution to the above problem. This method is not applicable when working with regular vines, since no fast algorithms for vines on discrete variables are available.

The last and fastest way of conditionalizing in an NPBBN is in the particular case in which the normal copula is used to realize the rank correlations. Since all the calculations are performed on a joint normal vine, any conditional distribution will also be normal, so in this case conditioning can be performed analytically. This last method is implemented in UNINET, hence it was used to produce Fig. 14.13.

The advantages of the normal copula are also used in the next section where the model learning problem is discussed.

14.3 Data Mining with NPBBNs

In situations where data does not exist or is very limited, expert judgment must be used to define the graphical structure and assess the required parameters. However, if the data are available, we would like to extract a fitting model from the data. In the process of learning a model from data, two aspects are of interest: learning the parameters of the model, given the structure, and learning the structure itself. Both learning the parameters of a regular vine, given the structure, and learning the vine structure together with its parameters are discussed in Chapter 3. The ensuing discussion concentrates on learning the DAG of an NPBBN together with its parameters from an ordinal data set.

The idea behind model inference for NPBBNs coincides with the one for regular vines, and it is based on the factorization of the determinant of the correlation matrix on the arcs of the NPBBN. This factorization is similar to the one for regular vines and the proof of this is available in Ref. 7. Once again, the directed nature of an NPBBN and the possibility of excluding arcs that correspond to zero rank correlations make learning an NPBBN a more intuitive task than learning a regular vine.

An NPBBN induced from data can be used to investigate distant relationships between variables, as well as making predictions, by computing the conditional probability distribution of one variable given the values of some others (see the previous section).

The distinctive feature of learning an NPBBN from a data set is that the one-dimensional marginal distributions are taken directly from data, and the model assumes only that the joint distribution has a normal copula. That is to say, the variables' rank dependence structure is that of a joint normal distribution. The NPBBN methodology is based on representing (conditional) dependencies on the arcs of a DAG, hence our strategy for inferring an NPBBN from data searches conditional dependencies in the data and associates arcs with them. A detailed discussion is found in Ref. 7; here we only sketch the ideas.

The concepts of learning and validation are closely connected, as indeed the goal is to learn an NPBBN that is valid. Validation involves two steps: validating that the joint normal copula adequately represents the multivariate data, and validating that the NPBBN is an adequate model of the saturated graph. Validation requires an overall measure of multivariate dependence on which statistical tests can be based. A suitable measure in this case is the determinant of the rank correlation matrix.[7] The determinant is 1 if all variables are independent, and 0 if there is linear dependence between the normal versions of the variables. We distinguish three determinants: DER is the determinant of the empirical rank correlation matrix. DNR is the determinant of the rank correlation matrix obtained by transforming the marginals to standard normals, and then transforming the product moment correlations to rank correlations using Pearson's transformation.[1] Finally, DBBN is the determinant of the rank correlation matrix of an NPBBN using the normal copula. DNR will generally differ from DER because DNR assumes the normal copula, which may differ from the empirical copula. A statistical test for the suitability of DNR for representing DER is to obtain the sampling distribution of DNR and check whether DER is within the 90% central confidence band of DNR. If DNR is not rejected on the basis of this test, we shall attempt to build an NPBBN which represents the DNR parsimoniously. The saturated NPBBN will induce a joint distribution whose rank determinant is equal to DNR, since the NPBBN uses the normal copula. However, many of the influences only reflect sample jitter and we will eliminate them from the model. Moreover, for a large number of variables, the saturated graph is dense and unintuitive.

---

[1]Pearson's transformation[17] is characteristic of the normal distribution. The normal copula assumption implies that the variables are assumed to have the distribution of transforms of a joint normal vector.

Once the normal copula is validated, we will build the NPBBN by adding arcs between variables only if the rank correlation between those two variables is among the largest. The second validation step is similar to the first. The general procedure can then be represented thus:

(1) Verify that DER is not outside the plausible central confidence band for DNR. If so, the normal copula hypothesis is not rejected;
(2) Construct a skeletal NPBBN by adding arcs to capture known causal or temporal relations;
(3) If DNR is within the 90% central confidence band of the determinant of the skeletal NPBBN, then stop, else continue with the following steps;
(4) Find the pair of variables such that the arc between them is not in the DAG and their rank correlation is greater than the rank correlation of any other pair not in the DAG. Add an arc between them and recompute DBBN together with its 90% central confidence band;
(5) If DNR is within the 90% central confidence band of DBBN, then stop, else repeat step 4.

The procedure for building an NPBBN to represent a given data set is not fully automated, as it is impossible to infer directionality of influence from multivariate data. Insight into the causal processes generating the data should be used, whenever possible, in constructing an NPBBN. Because of this fact, there are different NPBBN structures that are wholly equivalent, and many non-equivalent NPBBNs may provide statistically acceptable models of a given multivariate ordinal data set.

This approach is already used in several studies that try to link $PM_{2.5}$ concentrations to stationary source emissions.[7,8,15] Other applications of the NPBBN methodology are briefly mentioned in the next section.

14.4 Applications of NPBBNs

In Example 14.2.2, we have already mentioned one of the ongoing applications that uses NPBBNs, namely Beneris, a project undertaken by the European Union. The name of the project is short for "Benefit and Risk" and it focuses on the analysis of health benefits and risks associated with food consumption.[10]

Another project that uses NPBBNs is CATS, which stands for "Causal Model for Air Transport Safety". It is a large-scale application on risks in the aviation industry, and is currently under development. The project is commissioned by the Netherlands Ministry of Transport and Water Management.[1,2]

It is worth mentioning that both Beneris and CATS models use NPBBNs with hundreds of nodes and arcs. Models involving hundreds of variables benefit greatly from the advantages of the directed structure of an NPBBN. The use of regular vines in such situations would be somewhat cumbersome if not impossible.

A third application employs NPBBNs as a tool to estimate the extent of a fire in a building, given any combination of possible conditions and any unexpected course of events during an emergency.[9]

The latest attempt to use an NPBBN-based approach is in the field of reservoir engineering, namely in the estimation of surface characteristics (see www.data-assimilation.com/ssda).

All of the above projects use UniNet, the software application mentioned in Section 14.2.2. UniNet was initially developed to support the CATS project, and it is under constant development. The main program features are presented in the Appendix of Ref. 4.

14.5 Conclusions

In this book, graphical models have been chosen to represent multivariate distributions with complex dependence structures. More specifically, regular vines were advocated for this purpose. This chapter proposes NPBBNs as an alternative to regular vines and discusses the differences and similarities between the two.

The most important difference between NPBBNs and regular vines turned out to be the different statements of conditional (in)dependence that they make through their undirected and directed nature, respectively. In the DAG of an NPBBN, the absence of an arc encodes (conditional) independence statements. Regular vines, on the other hand, can be viewed as fully connected graphs that represent (conditional) dependence statements. Accordingly, the *absence* of edges in a regular vine is only possible for very special structures.[m] Nevertheless, the presence of arcs in NPBBNs does not guarantee dependence between variables (see Remark 14.1). Consequently, if one graph fails to represent dependencies, the other fails to represent independencies.

---

[m]If all conditional rank correlations in the higher-order trees of a vine are zero, then the edges of these trees can be removed.[12]

The possibility of excluding arcs from an NPBBN, whenever a (conditional) independence statement is known, produces a more perspicuous graphical structure. In order to visualize (conditional) independence statements on a regular vine, one has to assign zero (conditional) rank correlations to the edges; the edges cannot be removed. In this way, similar independence statements can be represented using both structures, and comparisons can be made. After such an analysis, no definite conclusion emerged. Some combinations of statements are better represented using a regular vine, whereas others benefit from the representation in a DAG form. This is only true for small structures. When hundreds of variables are involved, the saturated nature of regular vines constitutes a disadvantage in modelling and visualizing. Moreover the directed structure of NPBBNs holds the advantage of a more intuitive representation in terms of the flow of influences between variables.

When it comes to the quantitative part of the models, both NPBBNs and regular vines require marginal distributions and (conditional) rank correlations. Once these are obtained, the joint distribution is stipulated through a sampling procedure. The sampling procedure for NPBBNs uses the one for regular vines, hence we cannot talk about the advantages of the former compared to the latter. Moreover, in DAG structures that contain large undirected cycles, sampling an NPBBN involves extra numerical calculations that might be time consuming. These calculations are not necessary if the multivariate distribution can be represented and assessed using a regular vine. However, this disadvantage of NPBBNs vanishes when the normal copula is used.

Possessing a joint distribution allows us to perform inference. We can calculate the conditional distributions of unobserved variables, given the values of the observed ones. To achieve this, similar calculations are performed in both graphical models. Numerical complications that might arise for DAGs containing undirected cycles are circumvented by using a hybrid method that combines the flexibility of NPBBNs with the fast updating algorithms of discrete BBNs. When regular vines are used, the new conditional distributions, although calculated, cannot be easily visualized and compared with the unconditional one. This is purely an implementation issue for graphical software, hence it might be viewed as a recommendation for future development. Nonetheless, NPBBNs hold the advantage that conditionalization can be interpreted in terms of the directionality of arcs. In other words, if the reasoning is done "bottom-up" (in terms of the directionality), then the NPBBN is used for diagnosis, whereas if it is done "top-down", the NPBBN serves for prediction.

When data are available we are interested in learning a fitting model from data. In this process we could either learn the parameters of the model, given the structure, or learn the structure itself. The subject of learning the parameters of an NPBBN, given the structure, has not yet been addressed. Future research could investigate the methodology presented in Chapter 3 and its applicability to NPBBNs.

The idea behind learning the DAG of an NPBBN together with its parameters from an ordinal data set coincides with the one for learning regular vines. Still, the directed nature of an NPBBN and the possibility of including only arcs that correspond to the highest rank correlations make learning an NPBBN a more intuitive task than learning a regular vine.

References

1. Ale B., Bellamy L., Cooke R., Goossens L., Hale A., Roelen A. and Smith E. (2006). Towards a causal model for air transport safety: An ongoing research project. *Safety Science*, 44(8):657–673.
2. Ale B., Bellamy L.J., van der Boom, R., Cooper J., Cooke R., Goossens L.H.J., Hale A.R., Kurowicka D., Morales O., Roelen A. and Spouge J. (2009) Further development of a causal model for air transport safety (CATS). Building the mathematical heart. *Reliability Engineering and System Safety*, 94(4):1433–1441.
3. Cowell R., Dawid A., Lauritzen S. and Spiegelhalter D. (1999). *Probabilistic Networks and Expert Systems*, Statistics for Engineering and Information Sciences. Springer-Verlag, New York.
4. Hanea A. (2008). Algorithms for non-parametric Bayesian belief nets. PhD Dissertation, Delft Institute of Applied Mathematics, Wöhrmann Print Service.
5. Hanea A., Kurowicka D. and Cooke R. (2006). Hybrid method for quantifying and analyzing Bayesian belief nets. *Quality and Reliability Engineering International*, 22(6):613–729.
6. Hanea A., Kurowicka D. and Cooke R. (2007). The population version of Spearman's rank correlation coefficient in the case of ordinal discrete random variables. *Proceedings of the Third Brazilian Conference on Statistical Modelling in Insurance and Finance*.
7. Hanea A.M., Kurowicka D., Cooke R.M. and Ababei D.A. (2010). Mining and visualising ordinal data with non-parametric continuous BBNs. *Computational Statistics and Data Analysis*, 54:668–687.
8. Hanea A.M. and Harrington W. (2009). Ordinal $PM_{2.5}$ data mining with non-parametric continuous Bayesian belief nets.
9. Hanea D. and Ale B. (2009). Risk of human fatality in building fires: A decision using Bayesian networks. *Fire Safety Journal*, doi:10.1016/j.firesaf.2009.01.006.
10. Jesionek P. and Cooke R. (2007). Generalized method for modeling dose-response relations-application to BENERIS project. Technical Report, European Union project.
11. Kurowicka D. and Cooke R. (2004). Distribution-free continuous Bayesian belief nets. *Proceedings of the Fourth International Conference on Mathematical Methods in Reliability Methodology and Practice*, Santa Fe, New Mexico.

12. Kurowicka D. and Cooke R. (2006). Completion problem with partial correlation vines. *Linear Algebra and Its Applications*, 418(1):188–200.
13. Kurowicka D. and Cooke R. (2006). *Uncertainty Analysis with High Dimensional Dependence Modelling*. Wiley, Chichester.
14. Morales-Nápoles O., Kurowicka D. and Roelen A. (2007). Eliciting conditional and unconditional rank correlations from conditional probabilities. *Reliability Engineering and System Safety*, 93(5):699–710.
15. Morgenstern R., Harrington W., Shis J., Cooke R., Krupnick A. and Bell M. (2008). Accountabilty analysis of title IV of the 1990 Clean Air Act Amendments: An approach using Bayesian belief nets. In *Annual Conference*, Health Effects Institute, Philadelphia.
16. Pearl J. (1988). *Probabilistic Reasoning in Intelligent Systems: Networks of Plausible Inference*. Morgan Kaufman Publishers, San Mateo.
17. Pearson K. (1907). Mathematical contributions to the theory of evolution. *Biometric*, Series VI.
18. Shachter R. and Kenley C. (1989). Gaussian influence diagrams. *Management Science*, 35(5):527–550.
19. Wright S. (1921). Correlation and causation. *Journal of Agricultural Research*, 20: 557–585.

CHAPTER 15

Modeling Dependence Between Financial Returns Using Pair-Copula Constructions

Kjersti Aas* and Daniel Berg[†]

*Norwegian Computing Center
P.O. Box 114 Blindern, N-0314 Oslo, Norway
Kjersti.Aas@nr.no

[†]University of Oslo and Norwegian Computing Center
daniel@danielberg.no

In this chapter, we compare three constructions for modeling higher-dimensional dependence: the Student copula, the partially nested Archimedean construction (PNAC) and the pair-copula construction. For the latter two, a multivariate data set is modeled using a cascade of lower-dimensional copulae. They differ, however, in their construction of the dependence structure. The PNAC is more restrictive than the PCC in two respects. There are strong limitations on the degree of dependence in each level of the PNAC, and all the bivariate copulae in this construction has to be Archimedean. The PCC, on the other hand, can be built using copulae from any class and there are no constraints on the parameters. We show through two applications that the PCC provides a better fit to financial data than the two other structures.

15.1 Introduction 306
15.2 Constructions of Higher-Dimensional Dependence 307
    15.2.1 Student copula 307
    15.2.2 Partially nested Archimedean construction (PNAC) 307
    15.2.3 Pair-copula construction (PCC) 309
15.3 Parameter Estimation 311
    15.3.1 Student copula 311
    15.3.2 PNAC 311
    15.3.3 PCC 312

15.4 Portfolio 1 313
    15.4.1 Data set 313
    15.4.2 Results 315
    15.4.3 Validation 317
15.5 Portfolio 2 319
    15.5.1 Data set 319
    15.5.2 Results 320
    15.5.3 Tail dependence 323
    15.5.4 Pair-copula decomposition with copulae from different families 324
15.6 Summary and Conclusions 326
References 327

15.1 Introduction

A copula is a multivariate distribution function with standard uniform marginal distributions. While the literature on copulae is substantial, most of the research is still limited to the bivariate case. Building higher-dimensional copulae is a natural next step. However, this is not an easy task. Apart from the multivariate Gaussian and Student copulae, the selection of higher-dimensional parametric copulae is still rather limited.[12]

Recent developments in this area tend toward hierarchical, copula-based structures. Perhaps the most promising of these is the pair-copula construction (PCC). Originally proposed in Ref. 18, it has been further discussed and explored by Refs. 3, 4, 22 (simulation) and 1 (inference). Lately, some publications on applications of PCCs have appeared in the literature, especially in finance. However, the Student copula, in particular, and an alternative structure for building higher-dimensional copulae — the nested Archimedean construction (NAC)[19] — are still more commonly used.

In this chapter, we compare the PCC to the Student copula and the NAC. The rest of the chapter is organized as follows. In Sections 15.2 and 15.3, we give short reviews of the three constructions to be compared and how to estimate their parameters, respectively. In Section 15.4, we fit the PCC and the NAC to a four-dimensional equity portfolio, while the PCC is compared to the Student copula in the context of a four-dimensional portfolio comprised of Norwegian and international stock and bond indices in Section 15.5. Finally, Section 15.6 provides some summarizing comments and conclusions.

15.2 Constructions of Higher-Dimensional Dependence

In this section, we give a short review of the three constructions to be compared.

15.2.1 *Student copula*

The n-dimensional Student copula has been used repeatedly for modeling multivariate financial return data. A number of papers, such as Ref. 23, have shown that the fit of this copula is generally superior to that of other n-dimensional copulae for such data. The density of the n-dimensional Student copula is given by Ref. 8:

$$c(\mathbf{u}) = \frac{\Gamma(\frac{\nu+n}{2})\Gamma(\frac{\nu}{2})^{n-1}\left(1+\frac{\mathbf{x}'\mathbf{R}^{-1}\mathbf{x}}{\nu}\right)^{-\frac{(\nu+n)}{2}}}{|\mathbf{R}|^{1/2}\Gamma(\frac{\nu+1}{2})^n \prod_{j=1}^n (1+\frac{x_j^2}{\nu})^{-\frac{\nu+1}{2}}}, \quad (15.1)$$

where $\mathbf{x} = (t_\nu^{-1}(u_1), \ldots, t_\nu^{-1}(u_n))$ and $\mathbf{R}$ and ν are the copula parameters.

The Student copula has only one parameter, i.e. ν, for modeling tail dependence, independent of dimension. Hence, if the tail dependence of different pairs of risk factors in a portfolio are very different, it might not be flexible enough.

15.2.2 *Partially nested Archimedean construction (PNAC)*

The Archimedean copula family (see, e.g., Ref. 19, for a review) is a class that has attracted particular interest due to numerous properties which make them simple to analyze. The most common way of defining a multivariate Archimedean copula is the exchangeable Archimedean copula (EAC), defined as

$$C(u_1, u_2, \ldots, u_n) = \varphi^{-1}\{\varphi(u_1) + \cdots + \varphi(u_n)\}, \quad (15.2)$$

where the function φ is a decreasing function known as the generator of the copula and φ^{-1} denotes its inverse (see, e.g., Ref. 27). Note that some authors define φ and φ^{-1} oppositely to what we have done here.

For $C(u_1, u_2, \ldots, u_n)$ to be a valid n-dimensional Archimedean copula, φ^{-1} should be defined in the range of zero to one, be monotonically decreasing and $\varphi(1) = 0$. Furthermore, if $\varphi(0) = \infty$ the generator is said to be strict. Archimedean copulae arise naturally in the context of Laplace transforms of distribution functions.[19] If φ, in addition, equals the inverse of the Laplace transform of a distribution function G on $\mathcal{R}^+$ satisfying $G(0) = 0$,[a] the copula in (15.2) is guaranteed to be a proper distribution.

The EAC is extremely restrictive, allowing the specification of only one generator, regardless of dimension. Hence, all k-dimensional marginal distributions ($k < d$) are identical. For several applications, one would like to have multivariate copulae which allow for more flexibility. There have been some attempts at constructing more flexible multivariate Archimedean copula extensions (see, e.g., Refs. 10, 17, 19, 26, 27 and 31). In this chapter, we use one class of such extensions, the partially nested Archimedean construction (PNAC). It was originally proposed in Ref. 19 and is also discussed in Refs. 17, 24, 25 where it is denoted partially exchangeable and 36. The lowest dimension for which there is a distinct construction of this class is four, when we have the following copula:

$$C(u_1, u_2, u_3, u_4) = C_{21}(C_{11}(u_1, u_2), C_{21}(u_3, u_4))$$
$$= \varphi_{21}^{-1}\{\varphi_{21}(\varphi_{11}^{-1}\{\varphi_{11}(u_1) + \varphi_{11}(u_2)\})$$
$$+ \varphi_{21}(\varphi_{12}^{-1}\{\varphi_{12}(u_3) + \varphi_{12}(u_4)\})\}. \quad (15.3)$$

The construction, which is shown in Fig. 15.1 for the four-dimensional case, is quite simple but notationally cumbersome. We first couple the two pairs (u_1, u_2) and (u_3, u_4) with copulae C_{11} and C_{12}, having generator functions

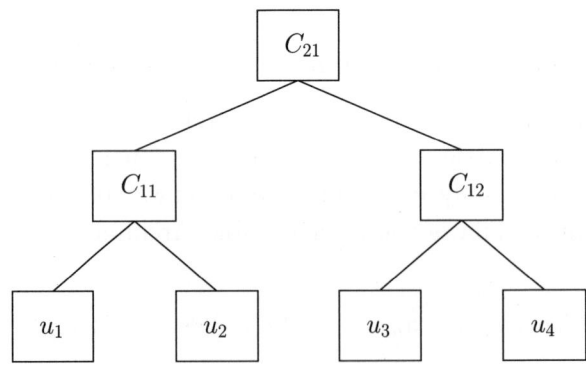

Figure 15.1. Partially nested Archimedean construction.

---

[a]The Laplace transform of a distribution function G on $\mathcal{R}^+$, satisfying $G(0) = 0$, is

$$\hat{G}(t) = \int_0^\infty e^{-tx} dG(x), t \geq 0.$$

φ_{11} and φ_{12}, respectively. We then couple these two copulae using a third copula C_{21}. Hence, in Fig. 15.1, the four pairs (u_1, u_3), (u_1, u_4) (u_2, u_3) and (u_2, u_4) will all have copula C_{21}, with dependence parameter θ_{21}. This means that in the PNAC, $d-1$ copulae and corresponding distributional parameters are freely specified, while the remaining copulae and parameters are implicitly given through the construction.

The PNAC is a construction of partial exchangeability and there are some technical conditions that need to be satisfied for (15.3) to be a proper n-dimensional copula. First, all the generators have to be strict with completely monotone inverses. Second, $\varphi_{2,1} \circ \varphi_{1,1}^{-1}$ and $\varphi_{2,1} \circ \varphi_{1,2}^{-1}$ must have completely monotone derivatives (see, e.g., Refs. 24 and 17). These conditions put restrictions on the parameters of the copulae involved. For instance, if all the generators are of the same type, e.g., Clayton, Ali-Mikhail-Haq, Gumbel, Frank or Joe type, the degree of dependence, as expressed by the copula parameter, must decrease with the level of nesting in order for the resulting n-dimensional distribution to be a proper copula. If generators belonging to different families are involved in a nested Archimedean construction, the parameter restrictions are even stronger. For example, if $\varphi_{2,1}$ and $\varphi_{1,1}$ are the generators for the Clayton and Gumbel copulae, respectively, $\varphi_{2,1} \circ \varphi_{1,1}^{-1}$ does not have a completely monotonic derivative for any parameter choice. For more information on this, see Refs. 24 and 17.

15.2.3 *Pair-copula construction (PCC)*

While the PNAC constitutes a large improvement compared to the EAC, it still only allows for the specification of up to $n-1$ copulae. An even more flexible construction, the PCC, allows for the free specification of $n(n-1)/2$ copulae. This construction was orginally proposed in Ref. 18, and it has been discussed in detail by Refs. 3, 4, 22 (simulation) and 1 (inference). Similar to the NAC, the PCC is hierarchical in nature. The modeling scheme is based on a decomposition of a multivariate density into $n(n-1)/2$ bivariate copula densities, of which the first $n-1$ are dependency structures of unconditional bivariate distributions, and the rest are dependency structures of conditional bivariate distributions.

While the PNAC is defined through its distribution functions, the PCC is usually represented in terms of the density. Two main types of PCCs have been proposed in the literature; canonical vines and D-vines.[21] Here, we concentrate on the D-vine representation, for which the density is[1]:

$$f(x_1,\ldots x_n) = \prod_{k=1}^{n} f(x_k) \prod_{j=1}^{n-1} \prod_{i=1}^{n-j} c\{F(x_i|x_{i+1},\ldots,x_{i+j-1}),$$
$$F(x_{i+j}|x_{i+1},\ldots,x_{i+j-1})\}. \quad (15.4)$$

In (15.4), $c(\cdot,\cdot)$ is a bivariate copula density and the conditional distribution functions are computed using[18]

$$F(x|\boldsymbol{v}) = \frac{\partial C_{x,v_j|\boldsymbol{v}_{-j}}\{F(x|\boldsymbol{v}_{-j}), F(v_j|\boldsymbol{v}_{-j})\}}{\partial F(v_j|\boldsymbol{v}_{-j})}. \quad (15.5)$$

In (15.5), $C_{x,v_j|\boldsymbol{v}_{-j}}$ is the dependency structure of the bivariate conditional distribution of x and v_j conditioned on $\boldsymbol{v}_{-j}$, where the vector $\boldsymbol{v}_{-j}$ is the vector $\boldsymbol{v}$ excluding the component v_j.

To use the D-vine construction to represent a dependency structure through copulae, we assume that the univariate margins are uniform in [0,1]. One four-dimensional case of (15.4) is then

$$c(u_1, u_2, u_3, u_4) = c_{11}(u_1, u_2) \cdot c_{12}(u_2, u_3) \cdot c_{13}(u_3, u_4)$$
$$\cdot c_{21}(F(u_1|u_2), F(u_3|u_2)) \cdot c_{22}(F(u_2|u_3), F(u_4|u_3))$$
$$\cdot c_{31}(F(u_1|u_2, u_3), F(u_4|u_2, u_3)),$$

where

$$F(u_1|u_2) = \partial C_{11}(u_1, u_2)/\partial u_2,$$
$$F(u_3|u_2) = \partial C_{12}(u_2, u_3)/\partial u_2,$$
$$F(u_2|u_3) = \partial C_{12}(u_2, u_3)/\partial u_3,$$
$$F(u_4|u_3) = \partial C_{13}(u_3, u_4)/\partial u_3,$$
$$F(u_1|u_2, u_3) = \partial C_{21}(F(u_1|u_2), F(u_3|u_2))/\partial F(u_3|u_2),$$
$$F(u_4|u_2, u_3) = \partial C_{22}(F(u_4|u_3), F(u_2|u_3))/\partial F(u_2|u_3).$$

Hence, the conditional distributions involved at one level of the construction are always computed as partial derivatives of the bivariate copulae at the previous level. Since only bivariate copulae are involved, the partial derivatives may be obtained relatively easily for most parametric copula families. Figure 15.2 illustrates this construction.

The copulae involved in (15.4) do not have to belong to the same family. In contrast to the PNAC, they do not even have to belong to the same class. The resulting multivariate distribution will be valid even if we choose, for each pair of variables, the parametric copula that best fits the data. As seen from (15.4), the PCC consists of $n(n-1)/2$ bivariate copulae of

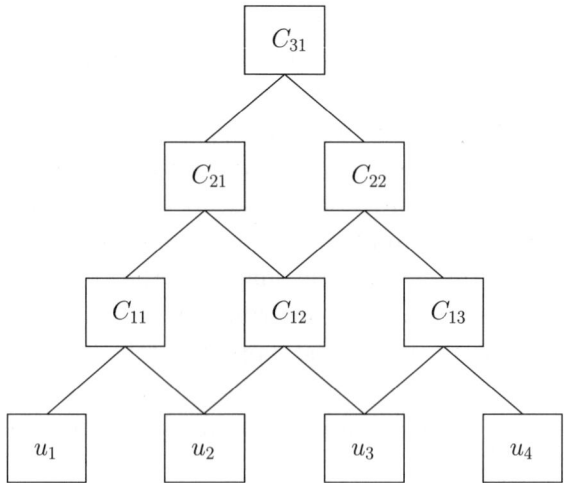

Figure 15.2. Pair-copula construction.

known parametric families, of which $n-1$ are copulae of pairs of the original variables, while the remaining $(n-1)(n-2)/2$ are copulae of pairs of variables constructed using (15.5) recursively. This means that in contrast to the NAC, the unspecified bivariate margins will not belong to a known parametric family in general.

15.3 Parameter Estimation

15.3.1 *Student copula*

To estimate the parameters of the Student copula, we use the two-step maximum likelihood method described in broad terms by Ref. 28 and later formalized and studied in Refs. 13 and 32. The estimation of the Student copula parameters requires numerical optimization of the log-likelihood function; see, for instance, Refs. 23 or 9.

15.3.2 *PNAC*

Full estimation of a PNAC should in principle consider the following three steps:

(1) the selection of a specific factorization
(2) the choice of pair-copula types,
(3) the estimation of the copula parameters.

Hence, before estimating the parameters of the construction, one has to choose which variables to join at each level of the PNAC as well as the parametric shapes of all pair-copulae involved. Due to the restrictions on the dependency parameters of the involved copulae described in Section 15.2.2, it is usually appropriate to join the variables that have the strongest tail dependence first. Recently, there has been an attempt at formalizing the procedure of determining the optimal structure of an NAC (see Ref. 29). Concerning the parametric shapes of the copulae, one may use a goodness-of-fit test, e.g., the one described in Section 15.4.2.1, for determining the copula family that most appropriately fits the data. A problem with the PNAC, however, is that many pairs have the same copula by construction. Hence, the choice of copula family for these pairs is not obvious.

Having determined the appropriate parametric shapes for each copula, all the parameters of the PNAC may be estimated by maximum likelihood. However, it is not straightforward to derive the density. Due to the complex structure of this construction, one has to use a recursive approach. One differentiates the n-dimensional top level copula with respect to its arguments using the chain rule. Hence, the number of computational steps needed to evaluate the density increases rapidly with the complexity of the copula, and parameter estimation becomes very time-consuming in high dimensions. See Ref. 31 for more details.

15.3.3 *PCC*

Full inference for a pair-copula construction should in principle consider the same three steps as described for the PNAC in Section 15.3.2. First, one has to choose which variables to join at the first level of the PCC. We then usually join the variables that have the strongest tail dependence. Having chosen the order of the variables at the first level, one has also determined which factorization to use.

Given the data and the chosen factorization, one must then specify the parametric shape of each pair-copula involved. The parametric shapes may, for instance, be determined using the following procedure:

(1) Determine which copula families to use at level 1 by plotting the observations, and/or applying a goodness-of-fit (GoF) test (see, e.g., Section 15.4 for a powerful GoF test).
(2) Estimate the parameters of the selected copulae.

(3) Determine the observations required for level 2 as the partial derivatives of the copulae from level 1.
(4) Determine which copula families to use at level 2 in the same way as at level 1.
(5) Repeat (1)–(3) for all levels of the construction.

This selection mechanism does not guarantee a globally optimal fit. Having determined the appropriate parametric shapes for each copulae, all parameters of the PCC are estimated by numerical optimization of the full likelihood. In contrast to the PNAC, the density is explicitly given. However, also for this construction, a recursive approach is used (see Algorithm 4 in Ref. 1). Hence, the number of computational steps to evaluate the density increases with the complexity of the copula, and parameter estimation becomes time-consuming in high dimensions.

15.4 Portfolio 1

In this section, the PCC is compared to the NAC in the context of a four-dimensional equity portfolio. We first describe the data set in Section 15.4.1. In Section 15.4.2, we show the results of fitting the PCC and NAC to this data set. Finally, in Section 15.4.3, the PCC is validated out-of-sample with respect to one-day Value-at-Risk.

15.4.1 *Data set*

The equity portfolio studied in this example is comprised of four time series of daily log-return data from the period August 14, 2003, to December 29, 2006 ($N = 852$ observations for each firm). The data set was downloaded from http://finance.yahoo.com. The firms are British Petroleum (BP), Exxon Mobil Corp (XOM), Deutsche Telekom AG (DT) and France Telecom (FTE). Financial log-returns are usually not independent over time. Hence, the original vectors of log-returns are processed by a GARCH filter before further modeling. We use the GARCH(1,1)-model[6]:

$$r_t = c + \epsilon_t$$
$$\mathrm{E}[\epsilon_t] = 0 \quad \text{and} \quad \mathrm{Var}[\epsilon_t] = \sigma_t^2 \qquad (15.6)$$
$$\sigma_t^2 = a_0 + a\,\epsilon_{t-1}^2 + b\,\sigma_{t-1}^2.$$

It has been known for a long time that GARCH models, coupled with the assumption of conditionally normally distributed errors, are unable to fully account for the tails of the distributions of daily returns.[7] In a study performed in Ref. 35, the NIG distribution outperforms a skewed Student's t-distribution and a non-parametric kernel approximation as the conditional distribution of a one-dimensional GARCH process. Hence, we follow Ref. 34 and use the normal inverse Gaussian (NIG) distribution[2] as the conditional distribution. After filtering the original returns with the GARCH model (15.6), the standardized residual vectors are converted to uniform pseudo-observations. Figures 15.3 and 15.4 show the filtered daily log-returns and pseudo-observations for each pair of assets, respectively.

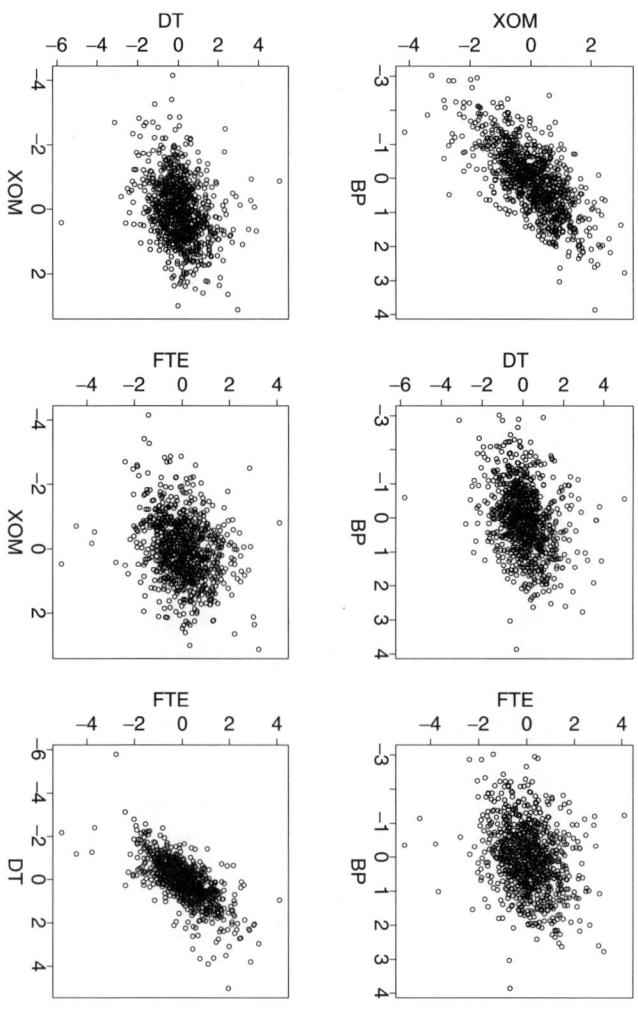

Figure 15.3. GARCH-filtered daily log-returns for our four stocks for the period from August 14, 2003 to December 29, 2006.

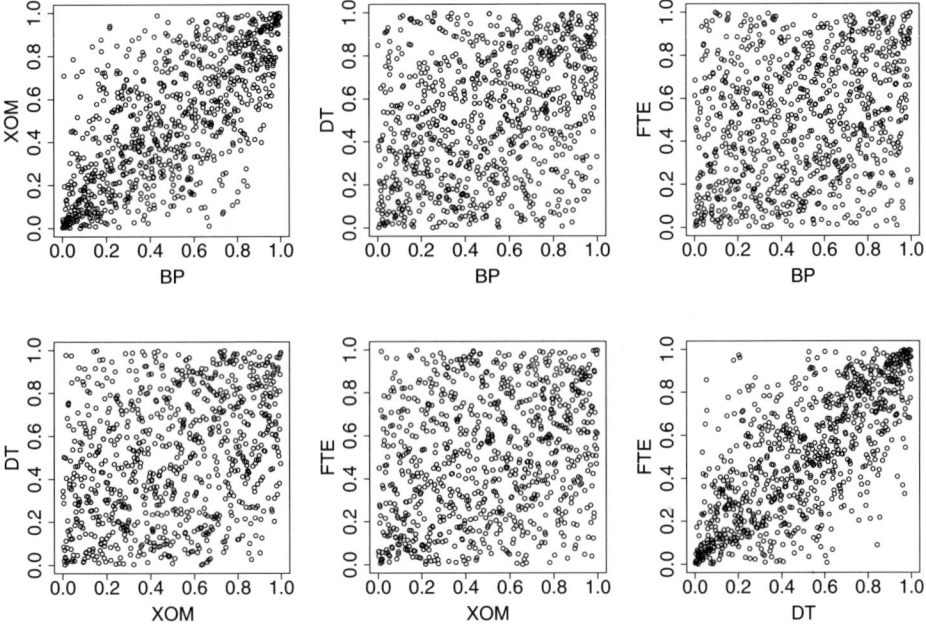

Figure 15.4. Pseudo-observations corresponding to Fig. 15.3.

Based on visual inspection and preliminary goodness-of-fit tests for bivariate pairs (the copulae taken into consideration were the Student, Clayton, survival Clayton, Gumbel and Frank copulae), we decided to examine a Frank NAC and Frank and Student PCCs.

15.4.2 Results

15.4.2.1 PNAC

We use the following PNAC:

$$C(u_1, u_2, u_3, u_4) = C_{21}(C_{11}(u_1, u_2), C_{21}(u_3, u_4)).$$

The most appropriate ordering of the variates in the construction is found by comparing Kendall's tau values for all bivariate pairs. The Kendall's tau values are shown in Table 15.1. As expected, stocks within one industrial sector are more dependent than stocks from different sectors. Hence, we choose C_{11} as the copula of BP and XOM, C_{12} as the copula of DT and FTE, and C_{21} as the copula of the remaining pairs. The leftmost column of Table 15.2 shows the estimated parameter values, resulting log-likelihood and p-values for the Frank PNAC. We use the Cramér–von-Mises statistic,

Table 15.1. Estimated Kendall's tau for pairs of variables for our four stocks.

Firm	XOM	DT	FTE
BP	0.45	0.19	0.20
XOM		0.23	0.17
DT			0.48

Table 15.2. Estimated parameters, log-likelihood and p-values for the Student copula, the NAC and the PCCs fitted to the filtered equity data.

	NAC	PCC	
Parameter	Frank	Frank	Student
$\theta_{11}\backslash\nu_{11}$	5.57	5.56	0.70\13.8
$\theta_{12}\backslash\nu_{12}$	6.34	1.89	0.32\134.5
$\theta_{13}\backslash\nu_{13}$	—	6.32	0.73\6.4
$\theta_{21}\backslash\nu_{21}$	1.78	0.91	0.14\12.0
$\theta_{22}\backslash\nu_{22}$	—	0.30	0.06\20.6
$\theta_{31}\backslash\nu_{31}$	—	0.33	0.07\17.8
Log-likelihood	616.45	618.63	668.49
p-value of S_n	0.006	0.008	0.410

defined by:

$$S_N = N \int_{[0,1]^n} \{C_N(\mathbf{u}) - C_{\theta_N}(\mathbf{u})\}^2 \mathrm{d}C_N(\mathbf{u})$$

$$= \sum_{j=1}^{N} \{C_N(\mathbf{U}_j) - C_{\theta_N}(\mathbf{U}_j)\}^2 \qquad (15.7)$$

for testing the goodness-of-fit. It is been verified that this test has the necessary asymptotic properties.[14,30] Further, Refs. 15 and 5 have shown, by bootstrapping p-values, that it is a very powerful procedure in most cases. Large values of S_N mean a poor fit, and lead to the rejection of the null hypothesis copula. In practice, the limiting distribution of S_N depends on θ. Hence, approximate p-values for the test must be obtained through a parametric bootstrap procedure. We adopt the procedure given in Appendix A of Ref. 15, setting the bootstrap parameters m and N to 5000 and 1000, respectively. The low p-value for the Frank NAC indicates that the fit is not very good.

15.4.2.2 PCC

We use the following PCC:

$$c(u_1, u_2, u_3, u_4) = c_{11}(u_1, u_2) \cdot c_{12}(u_2, u_3) \cdot c_{13}(u_3, u_4)$$
$$\cdot c_{21}(F(u_1|u_2), F(u_3|u_2)) \cdot c_{22}(F(u_2|u_3), F(u_4|u_3))$$
$$\cdot c_{31}(F(u_1|u_2, u_3), F(u_4|u_2, u_3)).$$

Like for the PNAC, the most appropriate ordering of the variates in the construction is determined by the size of the Kendall's tau values. Hence, we choose c_{11} as the copula density of BP and XOM, c_{12} as the copula density of XOM and DT, and c_{13} as the copula density of DT and FTE. The parameters of the PCC are estimated by the procedure described in Section 15.3.3. The two rightmost columns of Table 15.2 show the estimated parameter values, resulting log-likelihood and p-values for the Frank and Student PCCs. We see that the Frank PCC, like the Frank PNAC, is rejected. The Student PCC, however, provides a very good fit.

15.4.3 Validation

With the increasing complexity of models, there is always the risk of overfitting the data. To examine whether this is the case for the PCC, we validate it out-of-sample. More specifically, we use the fitted PCC from Section 15.4.2.2 to determine the risk of the return distribution for an equally weighted portfolio of BP, XOM, DT and FTE over a one-day horizon. The equally weighted portfolio is only meant as an example. In practice, the weights will fluctuate unless the portfolio is rebalanced every day.

The model estimated from the period August 14, 2003, to December 29, 2006, is used to forecast one-day VaR at different significance levels for each day in the period from December 30, 2006, to June 11, 2007 (110 days). The test procedure is as follows. For each day t in the test set:

(1) For each variable $j = 1, \ldots, 4$, compute the one-step ahead forecast of $\sigma_{j,t}$, given information up to time t.
(2) For each simulation $k = 1, \ldots$

- Generate a sample $u_1, \ldots, u_4$ from the estimated Student PCC.[b]

---

[b]The simulation algorithm for a D-vine is straightforward and simple to implement (see Algorithm 2 in Aas et al.[1]).

- Convert $u_1, \ldots, u_4$ to NIG(0,1)-distributed samples $z_1, \ldots, z_4$ using the inverses of the corresponding NIG distribution functions.
- For each variable $j = 1, \ldots, 4$, determine the log-return $r_{j,t} = c_{j,t} + \sigma_{j,t} z_j$. (Here $c_{j,t}$ is computed as the mean of the last 100 observed log-returns.)
- Compute the return of the portfolio as $r_{p,t} = \sum_{j=1}^{4} \frac{1}{4} r_{j,t}$.

(3) For significance levels $q \in \{0.005, 0.01, 0.05\}$
- Compute the one-day VaR_t^q as the qth quantile of the distribution of $r_{p,t}$.
- If VaR_t^q is greater than the observed value of $r_{p,t}$ this day, a violation is said to occur.

Figure 15.5 shows the actual log-returns for the portfolio in the period December 30, 2006, to June 11, 2007, and the corresponding VaR levels obtained from the procedure described above. Further, the two upper rows of Table 15.3 gives the number of violations, x, of VaR for each significance level and with the expected values, respectively. To test the significance of the differences between the observed and the expected values, we use the likelihood ratio statistic by Ref. 20. The null hypothesis is that the expected proportion of violations is equal to α. Under the null hypothesis, the likelihood ratio statistic given by

$$2\ln\left(\left(\frac{x}{N}\right)^x \left(1 - \frac{x}{N}\right)^{N-x}\right) - 2\ln(\alpha^x (1-\alpha)^{N-x}),$$

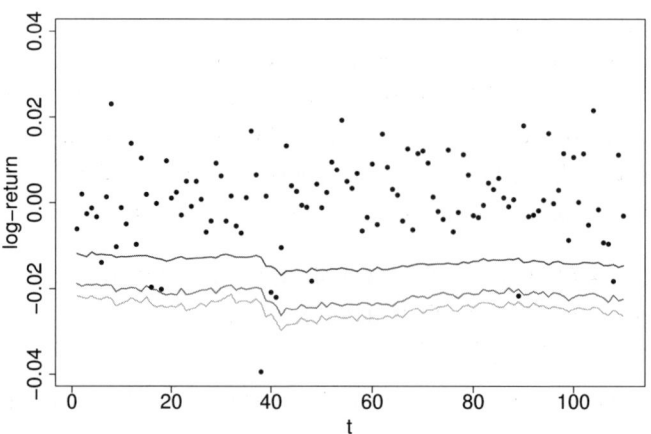

Figure 15.5. Log-returns for the equity portfolio for the period December 30, 2006 to June 11, 2007, along with 0.5%, 1%, 5% VaR simulated from the estimated GARCH–NIG–Student PCC.

Table 15.3. Number of violations of VaR, expected number of violations and p-values for the Kupiec test.

Significance level	0.005	0.01	0.05
Observed	1	2	9
Expected	0.55	1.1	5.5
p-value	0.13	0.44	0.16

where N is the length of the sample, is asymptotically distributed as $\chi^2(1)$. We have computed p-values of the null hypothesis for each quantile. The results are shown in the lower row of Table 15.3. If we use a 5% level for the Kupiec LR statistic, the null hypothesis is not rejected for any of the three quantiles. Hence, the GARCH–NIG–Student PCC seems to work very well out-of-sample.

15.5 Portfolio 2

In this section, we compare a four-dimensional PCC with Student copulae for all pairs with the four-dimensional Student copula. The n-dimensional Student copula has been used repeatedly for modeling multivariate financial return data. However, the Student copula has only one parameter for modeling tail dependence, independent of dimension. Hence, if the tail dependence of different pairs of risk factors in a portfolio are very different, we believe that a better description of the dependence structure can be achieved with the pair-copula decomposition with Student copulae for each pair.

The rest of this section is organized as follows. We first describe the data set in Section 15.5.1. In Section 15.5.2 we show the results of fitting the PCC and the Student copula to this data set, while Section 15.5.3 discusses the difference in tail dependence properties between the two structures. Finally, in Section 15.5.4, we investigate whether we would get an even better fit for our data set if we allowed the pair-copulae in the PCC to come from different parametric families.

15.5.1 *Data set*

The portfolio studied in this example is comprised of four time series of daily data: the Norwegian stock index (TOTX), the MSCI world stock index, the

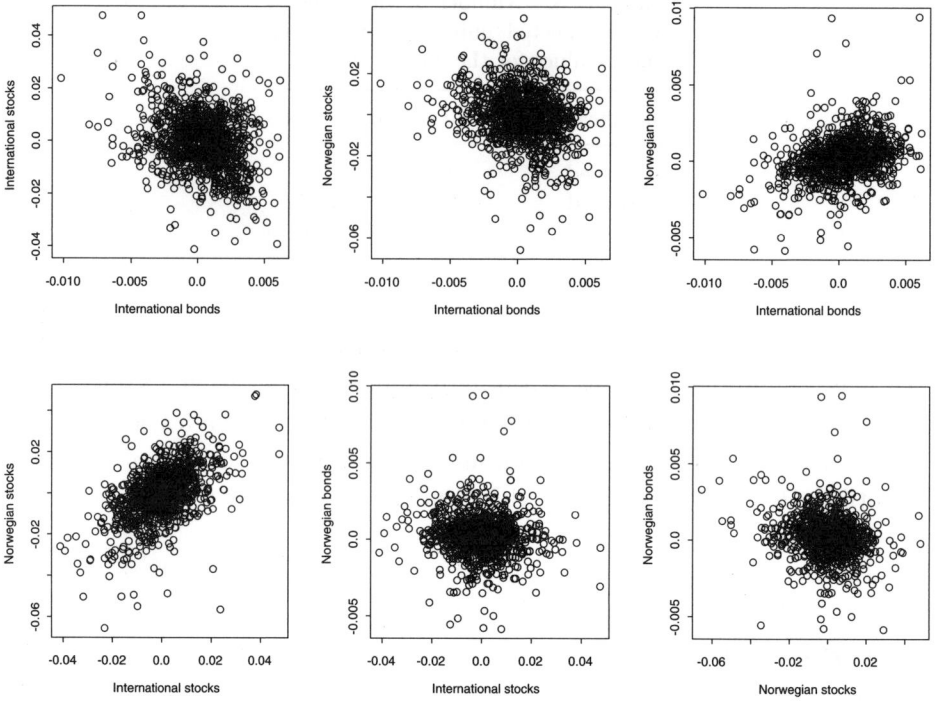

Figure 15.6. Log-returns for pairs of assets during the period from January 4, 1999 to July 8, 2003.

Norwegian bond index (BRIX) and the SSBWG hedged bond index, for the period from January 4, 1999 to July 8, 2003. Figure 15.6 shows the log-returns of each pair of assets. The four variables are denoted T, M, B and S.

As stated in Section 15.4, the observations of each variable must be independent over time. Hence, also for this data set, the original vectors of log-returns are processed by a GARCH(1,1)-filter before further modeling. Since we are mainly interested in estimating the dependence structure of the risk factors, the standardized residual vectors are converted to uniform variables using the empirical distribution functions before further modeling.

15.5.2 Results

15.5.2.1 PCC

The first step is to choose the most appropriate ordering of the risk factors. This is done as follows. First, we fit a bivariate Student copula to each pair

Table 15.4. Estimated numbers of degrees of freedom for bivariate Student copula for pairs of variables.

Between	M	T	B
S	4.21	34.16	14.47
M		8.03	15.48
T			12.60

of risk factors, obtaining estimated degrees of freedom for each pair. For this, we use the procedure described in Section 15.3.1. Having fitted a bivariate Student copula to each pair, the risk factors are ordered such that the three copulae to be fitted at level 1 of the PCC are those corresponding to the three smallest numbers of degrees of freedom. A low number of degrees of freedom indicates strong dependence. The numbers of degrees of freedom for the different pairs are shown in Table 15.4. The dependence is strongest between international bonds and stocks (S and M), international and Norwegian stocks (M and T), and Norwegian stocks and bonds (T and B). Hence, we want to fit the copulae $C_{S,M}$, $C_{M,T}$ and $C_{T,B}$ at level 1 of the PCC. Using a D-vine specification with the nodes S, M, T and B in the listed order gives the three above-mentioned copulae at level 1. See Fig. 15.7 for the whole D-vine structure.

The parameters of the D-vine are estimated using the approach described in Section 15.3.3. Table 15.5 shows the starting values obtained using the sequential estimation procedure (left column) and the final parameter values

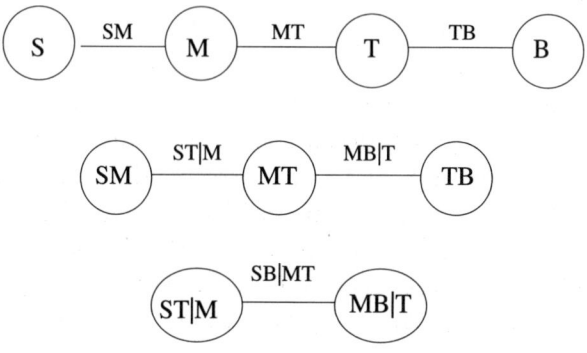

Figure 15.7. Selected D-vine structure for the data set in Section 15.5.1.

Table 15.5. Estimated parameters for four-dimensional pair–copula decomposition.

Param	Start	Final
ρ_{SM}	−0.25	−0.25
ρ_{MT}	0.47	0.47
ρ_{TB}	−0.17	−0.17
$\rho_{ST\|M}$	−0.11	−0.11
$\rho_{MB\|T}$	0.02	0.03
$\rho_{SB\|MT}$	0.29	0.28
ν_{SM}	4.21	4.34
ν_{MT}	16.65	16.26
ν_{TB}	12.60	13.17
$\nu_{ST\|M}$	300.00	300.00
$\nu_{MB\|T}$	130.33	45.59
$\nu_{SB\|MT}$	15.58	15.04
log.likelih.	267.86	268.17

(right column), together with the corresponding log-likelihood values. In the numerical search for the degrees of freedom parameter, we have used 300 as the maximum value. As can be seen from the table, the likelihood slightly increases when estimating all parameters simultaneously. The Akaike's information criterion (AIC) for the final model is −512.33.

15.5.2.2 Four-dimensional Student copula

In this section, we compare the results obtained with the PCC from Section 15.5.2.1 with those obtained with a four-dimensional Student copula. The parameters of the Student copula are shown in Table 15.6. The AIC for this model is −487.42, i.e., higher than that for the pair-copulae decomposition. All conditional distributions of a multivariate Student distribution are Student distributions. Hence, the n-dimensional Student copula is a special case of an n-dimensional D-vine with the needed pairwise copulae in the D-vine structure set to the corresponding conditional bivariate distributions of the multivariate Student distribution. Therefore, the four-dimensional Student copula is nested within the considered D-vine structure and the likelihood ratio test statistic is $2(268.17 − 250.71) = 34.92$ with $12 − 7 = 5$ degrees of freedom. This yields a p-value of 1.56e-006 and shows that the four-dimensional Student copula can be rejected in favor of the PCC.

Table 15.6. Estimated parameters for four-dimensional Student copula.

Param	Value
ρ_{SM}	−0.25
ρ_{ST}	−0.20
ρ_{SB}	0.30
ρ_{MT}	0.47
ρ_{MB}	−0.06
ρ_{TB}	−0.17
ν_{STMB}	14.56
log.likelih.	250.71

15.5.3 Tail dependence

Tail dependence properties are particularly important in many applications that rely on non-normal multivariate families.[18] This is especially the case for financial applications. Tail dependence in a bivariate distribution can be represented by the probability that the first variable exceeds its q-quantile, given that the other exceeds its own q-quantile. The limiting probability, as q goes to infinity, is called the *upper tail dependence coefficient*,[33] and a copula is said to be upper tail dependent if this limit is not zero. The lower tail dependence coefficient is analogously defined.

To illustrate the difference between the four-dimensional Student copula and the four-dimensional pair-copula decomposition, we computed the upper and lower tail dependence coefficients for the three bivariate margins SM, MT and TB for both structures. For the Student copula, the two coefficients are equal and given by Ref. 11:

$$\lambda_l(X,Y) = \lambda_u(X,Y) = 2\, t_{\nu+1}\left(-\sqrt{\nu+1}\sqrt{\frac{1-\rho}{1+\rho}}\right),$$

where $t_{\nu+1}$ denotes the distribution function of a univariate Student's t-distribution with $\nu+1$ degrees of freedom. Table 15.7 shows the tail dependency coefficients for the three margins and both structures. For the bivariate margin SM, the value for the pair-copula distribution is 279 times higher than the corresponding one for the Student copula. For a trader holding a portfolio of international stocks and bonds, the practical implication of this difference in tail dependence is that the probability of observing a large portfolio loss is much higher for the four-dimensional pair-copula decomposition than it is for the four-dimensional Student copula.

Table 15.7. Tail dependence coefficients.

Margin	Pair-copula decomp.	Student copula
SM	0.0279	0.0001
MT	0.0229	0.0317
TB	0.0005	0.0003

15.5.4 *Pair-copula decomposition with copulae from different families*

In this section, we investigate whether we would get an even better fit for our data set if we allowed the pair-copulae in the decomposition defined by Fig. 15.7 to come from different families. Figure 15.8 shows the data sets used to estimate the six pair-copulae in the decomposition described in Section 15.5.2.1. The three scatter plots in the upper row correspond to the three bivariate margins SM, MT and TB. The data clustering in the two opposite corners of these plots is a strong indication of both upper and lower tail dependence, meaning that the Student copula is an appropriate choice. In the two leftmost scatter plots in the lower row, the data seem to have no

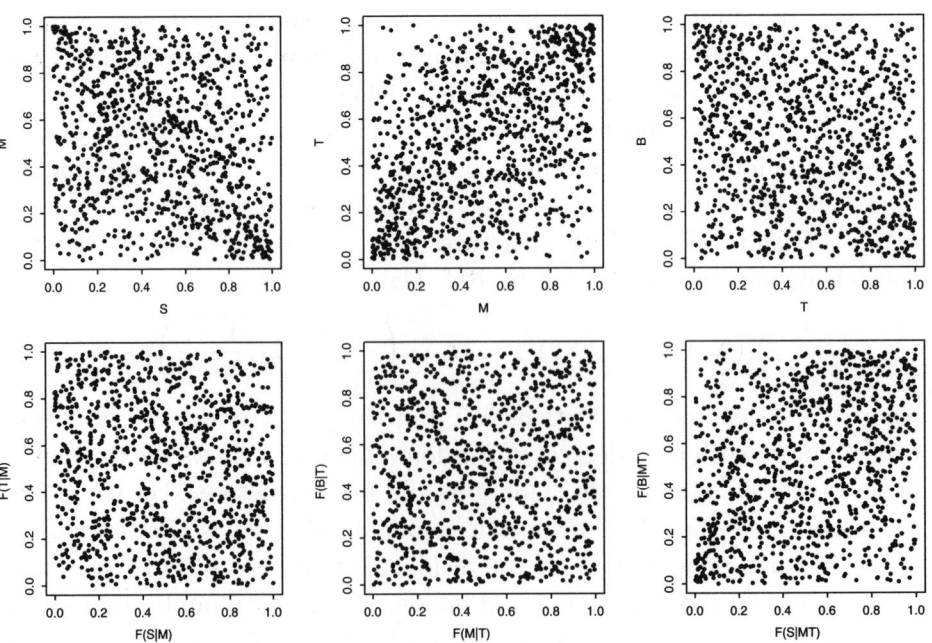

Figure 15.8. The data sets used to estimate the six pair-copulae in the decomposition described in Section 15.5.2.1.

tail dependence and the two margins also appear to be uncorrelated. This is in accordance with the parameters estimated for these data sets, $\rho_{ST|M}$, $\rho_{MB|T}$, $\nu_{ST|M}$, $\nu_{MB|T}$, shown in Table 15.5. The correlation parameters are close to 0 and the degrees of freedom parameters are very large, meaning that the two variables constituting each pair are close to being independent. If so, $c_{ST|M}(\cdot)$ and $c_{MB|T}(\cdot)$ are both 1, which means that the pair-copula construction defined by Fig. 15.7 may be simplified to

$$c_{SM}(x_S, x_M) \, c_{MT}(x_M, x_T) \, c_{TB}(x_T, x_B) \, c_{SB|MT}\{F(x_S|x_M), F(x_B|x_T)\}.$$

If we estimate this model instead, the parameters of copula $c_{SB|MT}(\cdot, \cdot)$ are slightly altered to $\rho_{SB|MT} = 0.28$ and $\nu_{SB|MT} = 15.22$. The log-likelihood for this reduced structure is 261.6 compared to 268.17 for the full one. The corresponding AIC values are -507.20 and -512.33, meaning that the full model is slightly better than the reduced one. This is also verified by the likelihood ratio statistic, which is $2(261.6-268.17) = 13.14$. With $12-8 = 4$ degrees of freedom this gives a p-value of 0.01 and shows that the reduced structure is rejected in favor of the full one.

Turning to the pair-copula $c_{SB|MT}(\cdot)$, there seems to be data clustering in the lower left corner of the scatter plot to the bottom right of Fig. 15.8, but not in the upper right. This indicates that the Clayton copula might be a better choice than the Student copula, since it has lower tail dependence but not upper. Hence, we have fitted the Clayton copula to this data set. The parameter was estimated to $\delta = 0.34$. The likelihood of the Clayton copula is lower than that of the Student copula (39.72 vs. 47.81). However, since the two copulae are non-nested, we cannot really compare the likelihoods. Instead we have used the procedure suggested by Ref. 16 for identifying the appropriate copula. According to this procedure, we examine the degree of closeness of the function $\lambda(z)$, given by

$$\lambda(z) = z - K(z).$$

Here $K(z)$ is the copula distribution function $K(z)$, defined by

$$K(z) = P(C(u_1, u_2) \le z).$$

For Archimedean copulae, $K(z)$ is given by an explicit expression, while for the Student copula it has to be numerically derived. In Fig. 15.9, the empirical lambda function and its confidence bands, computed as described in Ref. 16, are presented together with the fitted lambda functions for the Clayton copula and the Student copula. As can be seen from this figure, the Student copula fits the empirical data remarkably well.

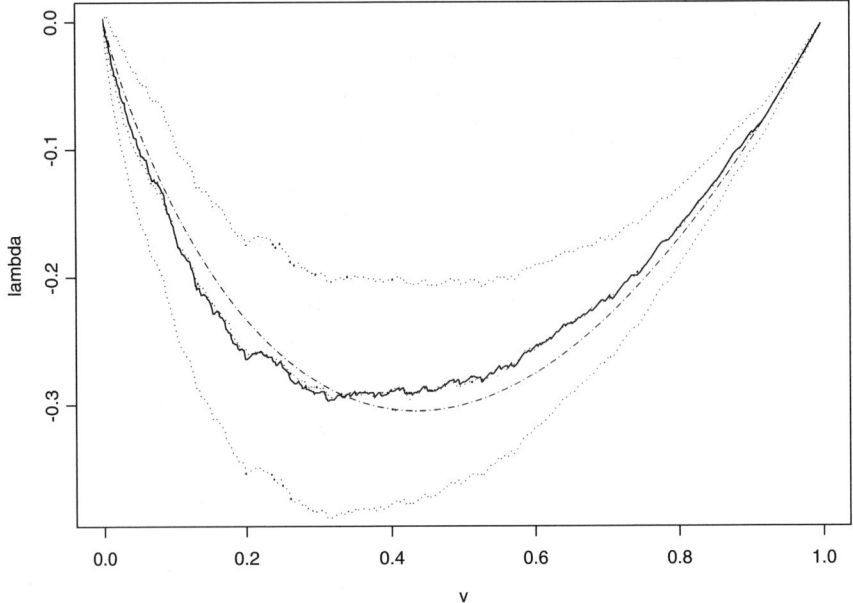

Figure 15.9. The empirical lambda function (solid line) and its confidence bands (dotted lines) are presented together with the fitted lambda functions for the Clayton copula (dashed line) and the Student copula (dotted line which can hardly be distinguished from the solid line).

15.6 Summary and Conclusions

In this chapter, we have compared three constructions for modeling higher-dimensional dependence: the Student copula, the partially nested Archimedean construction (PNAC) and the pair-copula construction (PCC). For the latter two constructions, a multivariate data set is modeled using a cascade of lower-dimensional copulae. The PNAC and PCC differ, however, in their construction of the dependence structure, the PCC being more flexible in that it allows for the free specification of $n(n-1)/2$ copulae, while the PNAC only allows for $n-1$. In addition, the PNAC has two important limitations. First, there are strong restrictions on the parameters of the construction. Second, the NAC is restricted to the Archimedean class, and there are even restrictions on which Archimedean copulae can be mixed. The PCC, on the other hand, can be built using copulae from any class and there are no constraints on the parameters of the construction.

We have shown through two examples that the PCC provides a better fit to financial data than the two other constructions. Moreover, through

VaR calculations we have shown that the PCC has not been overfitted to the training data, but also works very well out-of-sample.

Acknowledgments

This work is funded by Statistics for Innovation (sfi).[2]

References

1. Aas K., Czado C., Frigessi A. and Bakken H. (2009). Pair-copula constructions of multiple dependence. *Insurance: Mathematics and Economics*, 44:182–198.
2. Barndorff-Nielsen O.E. (1997). Normal inverse Gaussian distributions and stochastic volatility modelling. *Scandinavian Journal of Statistics*, 24:1–13.
3. Bedford T.J. and Cooke R.M. (2001). Probability density decomposition for conditionally dependent random variables modeled by vines. *Annals of Mathematics and Artificial Intelligence*, 32:245–268.
4. Bedford T.J. and Cooke R.M. (2002). Vines: A new graphical model for dependent random variables. *Annals of Statistics*, 30:1031–1068.
5. Berg D. (2007). Copula goodness-of-fit testing: An overview and power comparison. *European Journal of Finance*, 15:675–701.
6. Bollerslev T. (1986). Generalized autoregressive conditional heteroskedasticity. *Journal of Econometrics*, 31:307–327.
7. Bollerslev T. (1987). A conditionally heteroscedastic time series model for speculative prices and rates of return. *Review of Economics and Statistics*, 69:542–547.
8. Bouyé E., Durrleman V., Nikeghbali A., Riboulet G. and Roncalli T. (2000). Copulas for finance: A reading guide and some applications. Technical report, Crédit Lyonnais.
9. Demarta S. and McNeil A.J. (2005). The t-copula and related copulas. *International Statistical Review*, 73:111–129.
10. Embrechts P., Lindskog F. and McNeil A. (2003). Modelling dependence with copulas and applications to risk management. In S.T. Rachev (ed.), *Handbook of Heavy Tailed Distributions in Finance*. Elsevier, North-Holland.
11. Embrechts P., McNeil A.J. and Straumann D. (2001). Correlation and dependency in risk management: Properties and pitfalls. In: M.A.H. Dempster (ed.), *Risk Management: Value at Risk and Beyond*. Cambridge University Press, Cambridge.
12. Genest C., Gerber H.U., Goovaerts M.J. and Laeven R.J.A. (2009). Editorial to the special issue on modeling and measurement of risk in insurance and finance. *Insurance: Mathematics and Economics*, 44:143–145.
13. Genest C., Ghoudi K. and Rivest L.-P. (1995). A semi-parametric estimation procedure of dependence parameters in multivariate families of distributions. *Biometrika*, 82:543–552.
14. Genest C. and Rémillard B. (2005). Validity of the parametric bootstrap for goodness-of-fit testing in semiparametric models. Technical Report G-2005-51, GERAD, Montreal, Canada.
15. Genest C., Rémillard B. and Beaudoin D. (2009). Omnibus goodness-of-fit tests for copulas: A review and a power study. *Insurance: Mathematics and Economics*, 44: 199–213.

16. Genest C. and Rivest L.-P. (1993). Statistical inference procedures for bivariate Archimedean copulas. *Journal of the American Statistical Association*, 88:1034–1043.
17. Hofert M. (2008). Sampling Archimedean copulas. *Computational Statistics and Data Analysis*, 52:5163–5174.
18. Joe H. (1996). Families of m-variate distributions with given margins and $m(m-1)/2$ bivariate dependence parameters. In: L. Rüschendorf, B. Schweizer and M.D. Taylor (eds.), *Distributions with Fixed Marginals and Related Topics*. Institute of Mathematical Statistics, California.
19. Joe H. (1997). *Multivariate Models and Dependence Concepts*. Chapman & Hall, London.
20. Kupiec P. (1995). Techniques for verifying the accuracy of risk measurement models. *Journal of Derivatives*, 2:173–184.
21. Kurowicka D. and Cooke R.M. (2004). Distribution-free continuous Bayesian belief nets. *Fourth International Conference on Mathematical Methods in Reliability Methodology and Practice*, Santa Fe, New Mexico.
22. Kurowicka D. and Cooke R.M. (2006). *Uncertainty Analysis with High Dimensional Dependence Modelling*. Wiley, New York.
23. Mashal R. and Zeevi A. (2002). Beyond correlation: Extreme co-movements between financial assets. Technical report, Columbia University.
24. McNeil A.J. (2008). Sampling nested Archimedean copulas. *Journal of Statistical Computation and Simulation*, 78:567–581.
25. McNeil A.J., Frey R. and Embrechts P. (2006). *Quantitative Risk Management: Concepts, Techniques and Tools*. Princeton University Press, Princeton.
26. Morillas P.M. (2005). A method to obtain new copulas from a given one. *Metrika*, 61:169–184.
27. Nelsen R.B. (2006). *An Introduction to Copulas*, 2nd ed., Springer Series in Statistics. Springer, New York.
28. Oakes D. (1994). Multivariate survival distributions. *Journal of Nonparametric Statistics*, 3:343–354.
29. Okhrin O., Okhrin Y. and Schmid W. (2007). Radon Workshop on Financial and Actuarial Mathematics for Young Researchers, Linz, Austria.
30. Quessy J.-F. (2005). Théorie et application des copules: Tests d'adéquation, tests d'indépendance et bornes pour la valeur-à-risque. PhD thesis, Université Laval.
31. Savu C. and Trede M. (2006). Hierarchical Archimedean copulas. In *International Conference on High Frequency Finance*, Konstanz, Germany, May.
32. Shih J.H. and Louis T.A. (1995). Inferences on the association parameter in copula models for bivariate survival data. *Biometrics*, 51:1384–1399.
33. Sibuya M. (1960). Bivariate extreme statistics. *Annals of the Institute of Statistical Mathematics*, 11:195–210.
34. Venter J.H. and de Jongh P.J. (2002). Risk estimation using the normal inverse Gaussian distribution. *Journal of Risk*, 4(2):1–24.
35. Venter J.H. and de Jongh P.J. (2004). Selecting an innovation distribution for GARCH models to improve efficiency of risk and volatility estimation. *Journal of Risk*, 6(3): 27–53.
36. Whelan N. (2004). Sampling from Archimedean copulas. *Quantitative Finance*, 4: 339–352.

CHAPTER 16

Dynamic D-Vine Model

Andréas Heinen[*] and Alfonso Valdesogo[†]

[*]*Departamento de Estadística*
Universidad Carlos III de Madrid
126 Calle de Madrid, 28903 Getafe (Madrid) Spain
aheinen@est-econ.uc3m.es

[†]*CORE. Voie du Roman Pays 34*
B-1348 Louvain-la-Neuve, Belgium
alfonso.valdesogo@uclouvain.be

We model the dependence structure of multivariate financial returns with a time-varying D-vine copula. Vine copulae are flexible multivariate copulae that are obtained by a hierarchical construction, with bivariate copulae as building blocks. We focus on D-vines, which are a subclass of vine copulae. In order to take into account the fact that the dependence structure between financial returns is not constant over time, we allow each of the possible bivariate copulae to be time-varying. We use two different data sets, six exchange rates and five Asian equity indices. We find that most of the time variation is found in the first tree of the D-vine. Moreover, while currencies can be adequately modeled with symmetric copulae, Asian equity indices require asymmetric copulae.

16.1 Introduction 330
16.2 The Model 332
    16.2.1 Copulae 332
    16.2.2 D-vine copula 335
    16.2.3 Dynamic D-vine model 337
16.3 Components of the Model 338
    16.3.1 Marginal model 338
    16.3.2 Bivariate copulae 339
16.4 Empirical Results 341
    16.4.1 Data 341

            16.4.2   Marginal models 341
            16.4.3   Copula structure 345
    16.5   Conclusion 351
    References 352

16.1 Introduction

Financial returns are characterized by time-varying means and volatilities, as well as asymmetry and excess kurtosis. Traditionally, the financial literature uses correlation as a measure of dependence. Unfortunately, correlation is a good measure of dependence only in the elliptical world, for instance, when returns are multivariate Gaussian or Student t. In that case, returns are dependent whenever linear relations exist between them. In other situations, for instance, when there exist non-linear relations between variables, correlation can be misleading, as has been widely documented (see, for instance, Ref. 7). Given the stylized facts about financial returns, it is clear that elliptical distributions can be viewed only as a very crude approximation, and the same holds for correlation.

Copulae are a statistical tool that captures the dependence structure of a joint distribution independently from the features of the marginal distributions (see Refs. 16 and 11). The great advantage of copulae is that they permit flexible modeling of the dependence structure. There exists a large collection of bivariate copulae with different features that can match stylized facts of dependence. One such feature is asymmetric dependence, the fact that negative returns tend to be more dependent than positive ones. Another one is tail dependence, the fact that the dependence between returns does not vanish when the returns become extreme. This has important consequences, for instance, when calculating the risk of a portfolio. It is often of interest in finance to consider jointly a multivariate set of returns.

While there is a very large catalogue of bivariate copulae, the choice is much more restricted in the multivariate case. There exist some multivariate copulae, like multivariate Archimedean copulae, that impose strong restrictions like equal dependence on all pairs of variables. This is not general enough for most purposes. This explains why in applied work the two most widely used multivariate copulae that allow dependence to be modeled with a non-restricted dependence matrix are the Gaussian and the Student t. However, these copulae have some limitations in that they do not allow asymmetric dependence, one of the desirable features of a copula, and they are restrictive in the tail behavior, since the Gaussian does not allow tail dependence, while the Student t implies that the upper and lower tail dependence are equal, in contradiction with the stylized facts.

Recently, Bedford and Cooke[3] have introduced vine copulae, multivariate copulae based on graphical methods. These very flexible multivariate copulae are obtained by a hierarchical construction. The main idea is that a multivariate copula can be decomposed into a cascade of iteratively conditioned bivariate copulae. Vine copulae were introduced in the financial literature by Ref. 1, who also provided an estimation method. The great advantage of vine copulae is that they make the large choice of bivariate copulae available in the multivariate situation, therefore providing an incredible amount of flexibility.

A recent literature in empirical finance initiated by Ref. 14 has shown that correlations between financial returns are not constant over time and some models have been proposed to take this into account. The type of model that has achieved the highest level of popularity in this literature is certainly the DCC model of Engle[8] and the model of Tse and Tsui.[21] In the copula context, some models have been proposed by, e.g., Refs. 17 and 19. However, these models are limited to the bivariate case and they introduce dynamics by specifying a law of motion directly for the copula parameter. This makes comparison across models difficult since different copulae have different parameters with different support and interpretations. For instance, the Clayton copula parameter is defined on $[-1, \infty]\setminus\{0\}$, the Gumbel is defined on $[1, \infty]$, while the correlation coefficients of the Gaussian and Student t-copula are defined on $[-1, 1]$. Jondeau and Rokinger[12] use a time-varying multivariate copula model based on the Gaussian and the Student t. We follow the method suggested in Ref. 10, who propose a dynamic model for multivariate copulae, where the dynamics are comparable across copulae. They use a canonical vine decomposition where each of the possible bivariate copulae is allowed to be time varying. In this chapter we use a different vine structure, the D-vine, and allow each of the possible bivariate copulae to be time-varying as in Ref. 10.

We apply the dynamic D-vine to two different data sets. This allows us to answer two questions that are of empirical relevance: are the multivariate joint distributions of financial returns symmetric? Is the dependence constant over time? First we look at daily returns of six exchange rates, while the second data set consists of weekly returns on five Asian equity indices. In order to take into account possible dynamics in the conditional mean and volatility, we model the margins with an ARMA(p, q)-GARCH(1,1). The innovations are assumed to be skewed Student t-distributed. We use this distribution to take into account any possible asymmetry in the margins. With the marginals appropriately modeled, we specify the dependence structure. In both data sets, time-varying copulae are mainly found in the first tree of the D-vine structure. However, there is one marked difference between the bivariate copulae that compose the D-vine in the two data sets: exchange rates seem to be well described by symmetric copulae, whereas asymmetric copulae do a better job for the returns on the Asian stock exchanges.

The remainder of this chapter is organized as follows. In Section 16.2 we present the model: we briefly discuss copulae, copula-based dependence measures, D-vine copulae as well as the way in which we make the copulae time-varying. In Section 16.3 we discuss the components of the model, both in terms of the marginals, as well as the bivariate copulae that we use as building blocks of the D-vine. Section 16.4 presents the data and the results, and Section 16.5 concludes.

16.2 The Model

16.2.1 *Copulae*

Modeling the dependence between different risk factors is one of the key issues in most applications in finance such as Value-at-Risk and portfolio selection. Even though the notion of dependence has been traditionally linked to Pearson's correlation, it has some limitations. Consider, for example, two variables X and Y, where $X \sim \mathcal{N}(0,1)$ and $Y = X^2$. In this setup, $Cov(X,Y) = Cov(X, X^2) = Skewness(X)$. Therefore X and Y are uncorrelated, since their covariance is equal to the skewness of X, which is 0, by normality of X. Yet, clearly these variables are perfectly dependent. This simple example shows that correlation is not a good measure of dependence in all cases.[a] Pearson's correlation is only a good measure of dependence in the elliptical distributions, e.g., multivariate Gaussian or Student t-distributions. Given the stylized facts of financial returns, it is clear that elliptical distributions can be viewed only as a very crude approximation, and the same holds for correlation. In order to create more appropriate multivariate models, one can use the notion of copula.[b]

---

[a] For further examples, see Ref. 7.
[b] For related work on copulae as a modeling tool for returns, see Refs. 6 and 5.

Copula theory goes back to the work of Ref. 20, who showed that a joint distribution can be decomposed into its n marginal distributions and a copula, which fully characterizes dependence between the variables. This theorem provides an easy way to form valid multivariate distributions from known marginals that need not be of the same class. For example, it is possible to use a normal, Student or any other marginal, combine them with a copula and get a suitable joint distribution, which reflects the kind of dependence present in the series.[c] Specifically, let $H(y_1, \ldots, y_n)$ be a continuous n-variate cumulative distribution function with univariate margins $F_i(y_i)$, $i = 1, \ldots, n$, where $F_i(y_i) = H(\infty, \ldots, y_i, \ldots, \infty)$. According to Ref. 20, there exists a function C, called a copula, mapping $[0, 1]^n$ into $[0, 1]$, such that:

$$H(y_1, \ldots, y_n) = C(F_1(y_1), \ldots, F_K(y_n)). \tag{16.1}$$

The joint density function is given by the product of the marginals and the copula density:

$$\frac{\partial H(y_1, \ldots, y_n)}{\partial y_1 \cdots \partial y_n} = \prod_{i=1}^{n} f_i(y_i) \frac{\partial C(F_1(y_1), \ldots, F_n(y_n))}{\partial F_1(y_1) \cdots \partial F_n(y_n)}. \tag{16.2}$$

This allows us to define the copula as a multivariate distribution with uniform $[0, 1]$ margins:

$$C(z_1, \ldots, z_n) = H(F_1^{-1}(z_1), \ldots, F_n^{-1}(z_n)), \tag{16.3}$$

where $z_i = F_i(y_i)$, $i = 1, \ldots, n$ are the probability integral transformations (PIT) of the marginal models.

Evidently, with the use of copulae, we can map the univariate marginal distributions of n random variables, each supported in the $[0, 1]$ interval, to their n-variate distribution, supported on $[0, 1]^n$. This method applies, regardless of the type and degree of dependence among the variables.

16.2.1.1 Copula-based dependence measures

In order to describe the dependence that exists amongst variables that are not in the class of elliptical distributions, there exist several measures based on ranks of the variables. These measures are invariant with respect to any strictly increasing transformation of the data. Rank correlations are popular distribution-free measures of the association between variables. Unlike the

---

[c] A more detailed account of copulae can be found in Refs. 11 and 16 and in Ref. 4 who provide a more finance-oriented presentation.

traditional Pearson correlation, they work outside the range of the spherical and elliptical distributions and can detect certain types of non-linear dependence. The two most commonly used coefficients of rank correlation are Kendall's tau and Spearman's ρ. Both rely on the notion of concordance. Intuitively, a pair of random variables is concordant whenever large values of one variable are associated with large values of the other variable. More formally, if (y_i, x_i) and (y_j, x_j) are two observations of random variables (Y, X), we say that the pairs are *concordant* whenever $(y_i - y_j)(x_i - x_j) > 0$, and *discordant* whenever $(y_i - y_j)(x_i - x_j) < 0$.

Kendall's tau is defined as the difference between the probability of concordance and the probability of discordance. In general, it can be shown that the Kendall's tau between variables X and Y can be obtained as $\tau_{X,Y} = \frac{n_c - n_d}{n_c + n_d}$, where n_c is the number of concordant pairs and n_d is the number of discordant pairs. By definition, we then have that the total number of pairs is equal to the number of possible pairs with a sample of N bivariate observations $n_c + n_d = \frac{N!}{(N-2)!2!}$. Kendall's tau can also be expressed as a function of the copula:

$$\tau = 4 \int_{[0,1]^2} C(u,v) dC(u,v) - 1. \qquad (16.4)$$

16.2.1.2 *Asymmetric dependence, exceedance correlation and tail dependence*

An important feature of financial data is asymmetric dependence. There exist several measures that quantify this feature. In finance, it is of interest to measure both the usual sort of dependence between returns in the center of the distribution, and dependence amongst extreme events. The normal distribution captures the former, but risk theory deals mostly with the latter, as it is the negative extreme values in the distribution of asset returns that are crucial. There is a fairly large recent literature that studies this sort of extremal dependence. For example, Refs. 2, 15 and 18, among others, use exceedance correlation, which is defined as the correlation between two variables y_1 and y_2, conditional on both variables being above or below certain thresholds θ_1 and θ_2, respectively. Formally, lower exceedance correlation is defined as:

$$Corr(y_1, y_2 | y_1 \leq \theta_1, y_2 \leq \theta_2).$$

The main findings of these studies is that financial returns tend to exhibit excess correlation in bear markets, but not in bull markets. A Gaussian distribution cannot reproduce this feature. Therefore, while a Gaussian copula with Gaussian margins is unable to generate any exceedance correlation, an asymmetric copula with the same Gaussian marginals can produce this phenomenon. A weakness of this measure is that, like the Pearson correlation, it is not independent of the marginal distributions. Moreover, it is computed only from those observations that are below (above) the threshold, which means that as we move further out into the tails, the exceedance correlation is measured less and less precisely.

Quantile dependence is a somewhat different measure of the dependence in the tails of the distribution. If X and Y are random variables with distribution functions F_X and F_Y, then there is quantile dependence in the lower tail at threshold α, whenever $P[Y < F_Y^{-1}(\alpha)|X < F_X^{-1}(\alpha)]$ is different from zero. Finally, tail dependence obtains as the limit of this probability, as we go arbitrarily far out into the tails. The coefficient of the lower tail dependence of X and Y is:

$$\lim_{\alpha \to 0^+} P[Y < F_Y^{-1}(\alpha)|X < F_X^{-1}(\alpha)] = \lambda_L,$$

provided a limit $\lambda_L \in [0,1]$ exists. If $\lambda_L \in (0,1]$, X and Y are said to be asymptotically dependent in the lower tail; if $\lambda_L = 0$, they are asymptotically independent. If the marginal distributions of random variables X and Y are continuous, then the tail dependence of these random variables is a function only of their copula, and hence the amount of tail dependence is invariant under strictly increasing transformations. If a bivariate copula C is such that the limit

$$\lim_{u \to 0^+} C(u,u)/u = \lambda_L$$

exists, then C has lower tail dependence if $\lambda_L \in (0,1]$ and no lower tail dependence if $\lambda_L = 0$. Similarly, if a bivariate copula C is such that

$$\lim_{u \to 1^-} \bar{C}(u,u)/(1-u) = \lambda_U$$

exists, then C has upper tail dependence if $\lambda_U \in (0,1]$ and no upper tail dependence if $\lambda_U = 0$. $\bar{C}(u,v) = 1 - u - v + C(1-u, 1-v)$ denotes the survivor function of copula C.

16.2.2 D-vine copula

Until recently, an important limitation to the widespread use of copulae was the difficulty of constructing flexible families of multivariate copulae. In most applied work that dealt with a truly multivariate setting (more than two variables), researchers used either the Gaussian or the Student

t-copula. A recent advance in the statistics literature has provided such a construction. Vine copulae were introduced by Refs. 3 and 1. They are very flexible multivariate copulae, obtained by a hierarchical method. The main idea is that a multivariate copula can be decomposed into a cascade of iteratively conditioned bivariate copulae. Vine decompositions are very flexible (see Ref. 13). Two special cases are canonical vines and D-vines. Canonical vines can be viewed as factor models with one variable playing the role of the pivot (factor) in every tree of the dependence structure. In this chapter we limit our attention to the D-vine structure, as there is no economic reason to think that a factor structure should be relevant in our data. Formally, an n-dimensional D-vine consists of $n-1$ trees. Each tree j, for $j = 1,\ldots,n-1$, has $n+1-j$ nodes and $n-j$ edges. Each edge corresponds to a bivariate copula density. Figure 16.1 represents the dependence structure of a five-dimensional D-vine copula graphically.

The D-vine copula density corresponding to $f(y_1,\ldots,y_n)$ may be written as

$$\prod_{j=1}^{n-1}\prod_{i=1}^{n-j} c_{i,i+j|i+1,\ldots,i+j-1}(F(y_i|y_{i+1},\ldots,y_{i+j-1}),$$
$$F(y_{i+j}|y_{i+1},\ldots,y_{i+j-1})),$$

where index j identifies the trees, while i indicates the edges in each tree.

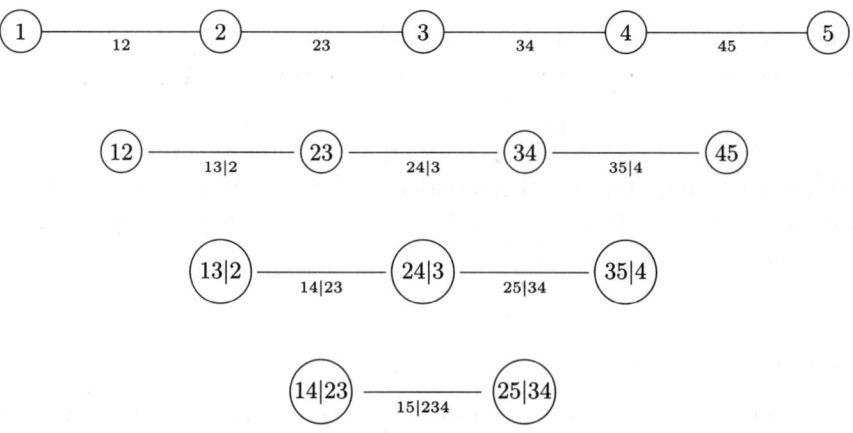

Figure 16.1. Dependence structure of a D-vine. This figure shows the structure of a D-vine copula with five variables. When there are five variables, there are ten bivariate copulae.

16.2.3 Dynamic D-vine model

In the previous section we defined the D-vine copula used in this chapter. With such a copula we can depart from linear dependence and capture asymmetric dependence and tail dependence in a multivariate setting. However, this model assumes that the dependence structure is constant over time. This goes against the stylized facts uncovered in the financial literature, for example, by Ref. 14, who show that correlations amongst returns are typically time-varying. Therefore we propose to introduce some dynamics into the dependence model. Recently, Ref. 10 introduced a dynamic canonical vine model in which each bivariate copula can be time-varying. In order to make the D-vine copula dynamic, we use the same methodology which we now briefly present.

The method consists of using the D-vine structure to construct multivariate copulae where the building blocks, the bivariate copulae, are allowed to be time-varying. The dynamics of the bivariate copula parameters are an extension of the DCC equations. As the inputs for the dynamic correlation in the DCC model are standardized residuals, we first apply the inverse CDF of the normal to the uniform random variables, the inputs of the copula, in order to transform them to standard normals[d]:

$$\epsilon_{i,t} = \Phi^{-1}(u_{i,t}).$$

$$\epsilon_t = [\epsilon_{1,t}, \epsilon_{2,t}].$$

We then use the dynamic equations of the DCC model:

$$Q_t = \Omega(1 - \alpha^C - \beta^C) + \alpha^C \epsilon_{t-1}\epsilon'_{t-1} + \beta^C Q_{t-1},$$

$$R_t = \{diag(Q_t)\}^{-1/2} Q_t \{diag(Q_t)\}^{-1/2},$$

where Ω is a symmetric 2×2 matrix, with ones on the diagonal, and the off-diagonal element equal to ω^C, while α^C and β^C are the autoregressive parameters. We define the dynamic Kendall's tau as:

$$\tau_t = 2\arcsin(\rho_t)/\pi. \tag{16.5}$$

ρ_t is the time-varying off-diagonal element of the matrix R_t and it is responsible for making Kendall's tau also time-varying, through Eq. (16.5). We know from Eq. (16.4) that there is a one-to-one relationship between Kendall's

---

[d]With the exception of the Student t-copula, where we apply the inverse CDF of the Student t-distribution with the corresponding degrees of freedom.

tau and the parameter of each copula. So for each period t, we can transform the corresponding Kendall's tau, τ_t, into the coefficient θ_t of each one of the copulae that we estimate. The model specification implies that ρ_t is the parameter of the Gaussian copula that would prevail, if the copula were indeed Gaussian. For those copulae that only allow positive dependence, we replace q_t, the off-diagonal element of Q_t, with $\max(q_t, 0)$, which ensures a well-defined model. Moreover, when there is negative dependence for prolonged periods, we usually do not end up selecting copulae that restrict the dependence to be positive, since they imply lower values of the likelihood.

16.3 Components of the Model

Although the components of the model are not nested, we use the BIC criterion to choose between them. We first describe the marginal model for each index or exchange rate. Second, we list the bivariate copulae that we use as candidates for the building blocks of the D-vine, along with their densities, as well as their tail dependence and Kendall's tau. Each copula is estimated in a constant and time-varying version and, again, we use the BIC to choose the better alternative.

16.3.1 *Marginal model*

In order to take into account the possible dynamics in the conditional mean and conditional variance, we model the marginal distributions of each one of our returns using an ARMA(p,q)-GARCH(1,1). After estimating up to two lags in each model, we select the best model by BIC criterion. The type of innovations that we use in the GARCH specification is a skewed Student t. This model can be viewed as the univariate skewed Student t GARCH model of Ref. 9, for the demeaned returns. Specifically, our system is expressed as

$$y_{i,t} = \mu_i + \eta_{i,t} + \sum_{j=1}^{p} \phi_{i,j} y_{i,t-j} + \sum_{k=1}^{q} \psi_{i,j} \eta_{i,t-k},$$

$$\eta_{i,t} = \sqrt{h_{i,t}} \cdot \epsilon_{i,t},$$

$$h_{i,t} = \omega_i + \alpha_i \eta_{i,t-1}^2 + \beta_i h_{i,t-1},$$

$$\epsilon_{i,t} \sim \text{skewed Student } t(\nu_i, \lambda_i),$$

where the skewed Student t-density is given by

$$g(z|\nu,\lambda) = \begin{cases} bc\left(1 + \dfrac{1}{\nu-2}\left(\dfrac{bz+a}{1-\lambda}\right)^2\right)^{-(\nu+1)/2} & z < -a/b \\ bc\left(1 + \dfrac{1}{\nu-2}\left(\dfrac{bz+a}{1+\lambda}\right)^2\right)^{-(\nu+1)/2} & z \geq -a/b \end{cases}$$

The constants a, b and c are defined as:

$$a = 4\lambda c\left(\frac{\nu-2}{\nu-1}\right), \quad b^2 = 1 + 3\lambda^2 - a^2, \quad c = \frac{\Gamma\left(\frac{\nu+1}{2}\right)}{\sqrt{\pi(\nu-2)}\Gamma\left(\frac{\nu}{2}\right)}.$$

A negative (positive) λ corresponds to a left(right)-skewed density, which means that there is more probability of observing large negative (positive) than large positive (negative) returns. This is what we expect to observe when we work with equity returns, since it captures the large negative returns associated with market crashes that are the cause of the skewness.

16.3.2 Bivariate copulae

We present the bivariate copulae that we use as components of the D-vine. For each of these copulae we estimate both a static and a dynamic version and, again, for each pair of returns, we use the BIC to choose the best of all static and dynamic copulae.

16.3.2.1 Gaussian copula

The Gaussian copula is related to the Gaussian distribution. The bivariate density is:

$$c_N(u_1, u_2; \rho) = \frac{1}{\sqrt{1-\rho^2}} \exp\left[\frac{-(q_1^2 + q_2^2 - 2q_1 q_2)}{2(1-\rho^2)} + \frac{q_1^2 + q_2^2}{2}\right],$$

where $q_i = \Phi^{-1}(u_i)$, Φ^{-1} denotes the inverse cumulative density of the standard normal and ρ is a correlation coefficient that lies between -1 and 1. The Gaussian copula has zero upper and lower tail dependence, $\lambda_U = \lambda_L = 0$, except in the case of perfect correlation, $\rho = 1$. The relation between the Kendall's tau and the parameter of the Gaussian copula is given by:

$$\rho = \sin(\tau\pi/2),$$

where ρ is the parameter of the Gaussian copula and τ is the Kendall's tau.

16.3.2.2 Student t-copula

The Student t-copula can be obtained from the Student t-distribution. The bivariate density is:

$$c_T(u_1, u_2; \rho, \nu) = \frac{\Gamma\left(\frac{\nu+2}{2}\right)}{\Gamma\left(\frac{\nu}{2}\right)\nu\pi\sqrt{1-\rho^2}}$$

$$\cdot \frac{\left(1 + \frac{T_\nu^{-1}(u_1)^2 + T_\nu^{-1}(u_2)^2 - 2\rho T_\nu^{-1}(u_1) T_\nu^{-1}(u_2)}{\nu(1-\rho^2)}\right)^{-\left(\frac{\nu+2}{2}\right)}}{f_\nu(T_\nu^{-1}(u_1)) f_\nu(T_\nu^{-1}(u_2))},$$

where $T_\nu^{-1}(v)$ is the inverse of the cumulative distribution function of the univariate Student t with ν degrees of freedom, $f_\nu(.)$ is the density of the Student t-distribution with ν degrees of freedom and $\rho \in (-1, 1)$ is the correlation parameter. The relation between the Kendall's tau and the parameter of the Student t-copula does not depend on the degrees of freedom and is the same as the Gaussian copula. The Student t-copula has the same lower and upper tail dependence for every pair of variables: $\lambda_U = \lambda_L = 2T_{\nu+1}\left(-\sqrt{\nu+1}\sqrt{\frac{1-\rho}{1+\rho}}\right)$.

16.3.2.3 Frank copula

The Frank copula has the following density:

$$c_F(u_1, u_2; \theta) = \frac{\theta(1 - e^{-\theta})e^{-\theta(u_1+u_2)}}{(1 - e^{-\theta}) - (1 - e^{-\theta u_1})(1 - e^{-\theta u_2})},$$

where $\theta \in (-\infty, \infty) \backslash 0$. The Frank copula has zero upper and lower tail dependence, $\lambda_U = \lambda_L = 0$, except in the limit when $\theta \to \infty$. Mapping the Kendall's tau to the parameter of the Frank copula should be done numerically.

16.3.2.4 Gumbel and rotated Gumbel copula

Unlike the previous copulae, the Gumbel copula is not symmetric. It has the following density:

$$c_G(u_1, u_2, \theta) = \frac{C_G(u_1, u_2, \theta)(\log u_1 \cdot \log u_2)^{\theta-1}}{u_1 u_2((-\log u_1)^\theta + (-\log u_2)^\theta)^{2-1/\theta}}$$

$$\cdot (((-\log u_1)^\theta + (-\log u_2)^\theta)^{1/\theta} + \theta - 1),$$

where $\theta \in [1, \infty)$. The Gumbel copula has only upper tail dependence, $\lambda_U = 2 - 2^{1/\theta}$ and the connection with the Kendall's tau is $\theta = 1/(1-\tau)$.

We use also the rotated version of the Gumbel defined as $c_{RG}(u_1, u_2, \theta) = c_G(1 - u_1, 1 - u_2, \theta)$. The rotated Gumbel has only lower tail dependence, $\lambda_L = 2 - 2^{1/\theta}$, and the relation with Kendall's tau is the same as for the Gumbel copula.

16.3.2.5 Clayton copula

Like the Gumbel, the Clayton copula is asymmetric. Its density is

$$c_C(u_1, u_2; \theta) = (1 + \theta)(u_1 v_1)^{-\theta-1}(u_1^{-\theta} + u_2^{-\theta} - 1)^{-2-1/\theta},$$

where $\theta \in [-1, \infty) \backslash 0$.

The Clayton copula has lower but not upper tail dependence: $\lambda_L = 2^{-1/\theta}$. The relation between the Kendall's tau and the parameter of the Clayton copula is given by $\theta = 2\tau/(1 - \tau)$.

16.4 Empirical Results

16.4.1 Data

We use two data sets of returns. The first sample comprises six exchange rates downloaded from Datastream (all against the euro): Australian dollar (AUD), Canadian dollar (CAD), Japanese yen (JPY), US dollar (USD), Swiss franc (CHF) and British pound (GBP). We use daily data from January 1, 1999, to December 31, 2007, which gives us 2,346 returns. The second data set consists of returns on five Asian equity indices: Hong Kong (HK), Korea (KR), Singapore (SG), Taiwan (TW) and Thailand (TH). The Asian equity indices are weekly MSCI price series from October 10, 1989, to May 30, 2006, where all prices are in US dollars. This gives us a sample of 868 weekly returns. We analyze the log-differences of each series multiplied by 100 in both data sets.

16.4.2 Marginal models

Table 16.1 presents summary statistics of the two data sets. For the currencies, the annualized average returns range from 2.336% for the US dollar to −2.378% for the Canadian dollar, while for Asian indexes the annualized average returns range from 8.193% for Hong Kong to −3.003% for Taiwan. The annualized standard deviations are quite different for both data sets. They are around 9% for the currencies, ranging from 10.743% for the Japanese yen to only 3.362% for the Swiss franc, which follows the euro

Table 16.1. Summary statistics. This table contains descriptive statistics for the returns of the six daily currencies and the weekly indices for the five Asian countries. The first data set spans the period from January 1, 1999, to December 31, 2007, with 2,346 daily returns, while for the Asian equity indices, the sample ranges from October 10, 1989, to May 30, 2006, with 868 weekly returns. The means and the standard deviations are annualized. All returns are log-difference of prices multiplied by 100.

	Mean	Stand. Deviation	Skewness	Kurtosis	Min	Max
AUD	−1.487	10.015	0.593	6.260	−2.623	4.599
CAD	−2.378	9.711	0.257	4.414	−2.262	3.601
JPY	2.233	10.743	−0.106	5.447	−3.116	4.480
USD	2.336	9.183	0.190	4.125	−2.266	3.321
CHF	0.276	3.362	−0.583	10.185	−2.145	1.342
GBP	0.424	6.689	0.191	4.440	−1.840	2.167
HK	8.193	27.402	−0.568	9.339	−29.439	18.465
KR	2.674	38.487	−0.470	10.089	−44.118	24.264
SG	4.218	27.235	−0.273	7.171	−18.776	18.303
TW	−3.003	32.970	−0.494	5.077	−23.710	17.966
TH	−1.230	40.235	0.839	10.839	−21.129	45.035

closely. The annualized standard deviation of the five Asian equity indices is around three times higher than the standard deviation of the currencies, and they range from 40.235% for Taiwan to 27.402% for Hong Kong. All series present clear signs of non-normality. This can be seen from their skewness and kurtosis. For the currencies, the skewness is positive in four cases and negative in only two cases. It ranges from 0.593 for the Canadian dollar to −0.583 for the Swiss franc. When we consider the Asian equity indices, the skewness is negative in all cases except for Thailand and ranges from 0.839 for Thailand to −0.568 for Hong Kong. All returns of currencies and Asian equity indices have kurtosis above 3. In general the kurtosis of the Asian equity indices is higher than the one observed in the currencies.

In order to take into account any possible dynamics in the conditional mean, we determine the appropriate ARMA(p,q) model. We use the BIC criterion in order to select the most parsimonious model, and we consider all possible models with $p, q \leq 2$. We reject the constant mean in only two cases. For the Australian dollar, we select an AR(1) with an autoregressive parameter of 0.066 and for the Swiss franc we select an MA(2), with moving average parameters of 0.055 and −0.054. None of the Asian equity indices present any dynamics in the conditional mean. The results of each of the univariate skewed Student t GARCH are presented in columns two to five of Table 16.2. The coefficients of the lagged conditional variance, β, are

Table 16.2. GARCH estimates, goodness-of-fit statistics. Columns two to six are parameter estimates of univariate skewed Student t GARCH(1,1) models of Ref. 9. Standard deviations of the parameters are in brackets. Columns seven to eleven report p-values of goodness-of-fit (GoF) statistics of the Probability Integral Transformation (PIT) of the marginal models. We present the p-values for the following tests. The Kolmogorov–Smirnov (KS) test evaluates the null hypothesis that the population cdf is uniform [0, 1]. $KS+$ tests the null against the alternative hypothesis that the population cdf is below a uniform [0, 1], while $KS-$ tests the null against the alternative hypothesis that the population cdf is above a uniform [0, 1]. AD refers to the Anderson–Darling test for uniformity, while K stands for the Kuiper test for uniformity, which puts more weight on the tails of the distribution than the other tests.

	ω	α	β	ν	λ	KS	KS+	KS−	AD	K
AUD	0.002 (0.001)	0.023 (0.006)	0.973 (0.007)	5.769 (0.666)	0.106 (0.027)	0.831	0.611	0.459	0.994	0.654
CAD	0.003 (0.002)	0.020 (0.006)	0.973 (0.009)	7.986 (1.297)	0.042 (0.027)	0.665	0.685	0.347	0.994	0.590
JPY	0.002 (0.001)	0.025 (0.006)	0.971 (0.007)	8.106 (1.255)	−0.109 (0.028)	0.879	0.639	0.500	0.993	0.740
USD	0.001 (0.001)	0.020 (0.005)	0.978 (0.005)	10.497 (2.233)	0.045 (0.027)	0.319	0.388	0.160	0.996	0.087
CHF	0.001 (0.000)	0.073 (0.014)	0.899 (0.019)	6.166 (0.753)	−0.055 (0.026)	0.941	0.609	0.570	0.994	0.785
GBP	0.001 (0.000)	0.025 (0.006)	0.972 (0.007)	10.163 (2.022)	0.099 (0.027)	0.220	0.110	0.404	0.997	0.057
HK	0.181 (0.108)	0.087 (0.024)	0.901 (0.028)	9.288 (2.665)	−0.156 (0.049)	0.979	0.639	0.754	0.993	0.957
KR	0.351 (0.180)	0.078 (0.023)	0.908 (0.026)	11.268 (3.580)	−0.008 (0.033)	0.896	0.516	0.788	0.994	0.905
SG	0.207 (0.124)	0.098 (0.029)	0.890 (0.032)	8.119 (1.879)	−0.075 (0.050)	0.985	0.764	0.658	0.993	0.968
TW	0.945 (0.376)	0.130 (0.032)	0.825 (0.041)	13.686 (5.577)	−0.168 (0.047)	0.968	0.745	0.615	0.993	0.939
TH	0.318 (0.178)	0.079 (0.021)	0.910 (0.023)	6.834 (1.492)	0.075 (0.044)	0.588	0.912	0.302	0.995	0.836

around 0.97 for nearly all currencies with the exception of the Swiss franc, while for the Asian equity indices this coefficient is lower and in most cases around 0.90. This difference implies that the dynamics of the conditional variance are more persistent for the currencies than for the Asian equity indices.[e] The sign of the estimated asymmetry coefficient of the conditional distribution, λ, is in agreement with the descriptive statistics of the unconditional distributions of our series of returns. The mostly negative skewness we find captures the fact that the tails of some of the marginal distributions are typically longer on the left side. The degrees of freedom parameters for the currencies range from 5.769 for the Canadian dollar to 10.497 for the US dollar. For the case of the five Asian countries, the degrees of freedom are in general slightly larger, ranging from 6.834 for Thailand to 13.686 for Taiwan.

We check that the marginal models are well-specified and subject them to a series of goodness-of-fit tests. We include three versions of the Kolmogorov–Smirnov test, as well as the Anderson–Darling and Kuiper tests of uniformity of the Probability Integral Transformation (PIT) of the marginal models. The p-values of the tests are reported in columns seven to twelve of Table 16.2. The models passed all the tests. It is very important that the marginal models be well-specified, since otherwise, the copula estimation that is conditional on the marginal models would be affected.

Figure 16.2 plots the time series of the conditional volatility for the five Asian countries.[f] This shows the dramatic increase in volatility due to the Asian crisis, which started at the beginning of July 1999 in Thailand when the Thai baht collapsed. The Asian crisis produced devaluations in the currencies and subsequent crashes in the stock markets of the region. Among the countries under study, South Korea and Thailand were the most affected, followed by Hong Kong, Singapore and Taiwan.

---

[e]This difference is due partly to the fact that we use a daily frequency for the currencies, while the stock indices have a weekly frequency.

[f]The time series of the conditional volatility of the six currencies are available upon request. We did not find any important economic event that produces a dramatic change in the conditional volatility. However, generally speaking, it seems that the conditional volatility increases whenever the euro is depreciating.

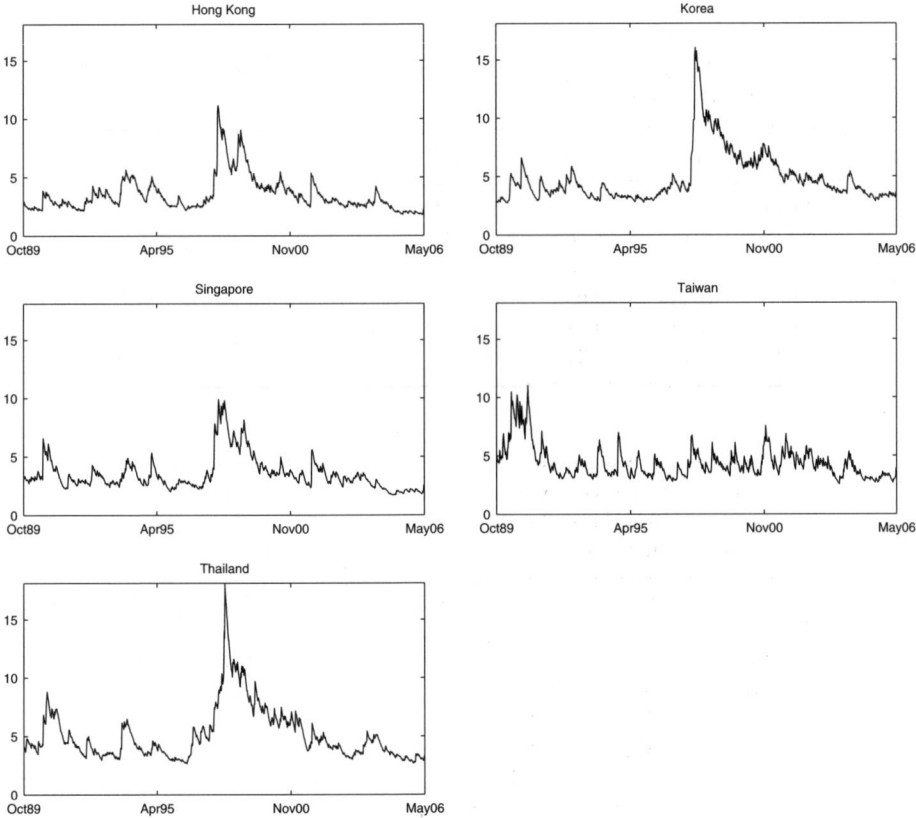

Figure 16.2. Time series of the conditional volatility for the five Asian equity indices.

16.4.3 Copula structure

We select the structure of the D-vine with the empirical Spearman rank correlations of the PIT of the marginal models shown in Table 16.3 for the exchange rates and Table 16.4 for the Asian equity indices. The idea is to rank pairs of series from the highest to the lowest Spearman rank correlation. Once a series has been selected twice, it cannot be used to form new pairs.

First of all, we focus on the case of the six currencies, whose correlations appear in Table 16.3. The highest Spearman rank correlation is between CAD and USD, with a value of 0.636. The second largest is between USD and JPY, with 0.508. This implies that we cannot use the USD to form a new pair. The next largest Spearman rank correlation is between AUD and CAD, with 0.466. Again, in that case the CAD series cannot be used anymore. If we continue with this method, the next pair is AUD–GBP and

Table 16.3. Spearman rank correlation matrix for the six exchange rates. This table contains the Spearman rank correlation matrix of the Probability Integral Transformation of the marginal models of the six exchange rates.

	AUD	CAD	JPY	USD	CHF	GBP
AUD	1.000					
CAD	0.466	1.000				
JPY	0.283	0.356	1.000			
USD	0.369	0.636	0.508	1.000		
CHF	−0.104	−0.070	0.151	−0.061	1.000	
GBP	0.343	0.374	0.320	0.499	0.067	1.000

Table 16.4. Spearman rank correlation matrix for the five Asian equity indices. This table contains the Spearman rank correlation matrix of the Probability Integral Transformation of the marginal models of the five Asian equity indices.

	HK	KR	SG	TW	TH
HK	1.000				
KR	0.375	1.000			
SG	0.531	0.382	1.000		
TW	0.358	0.351	0.347	1.000	
TH	0.447	0.337	0.486	0.301	1.000

the last one is JPY–CHF. The D-vine structure for the exchange rates for the first level is the following:

$$GBP - AUD - CAD - USD - JPY - CHF.$$

Following the same methodology with the Asian index returns, we obtain the following order:

$$TH - SG - HK - KR - TW.$$

We do not think that there is any economic reason to assume a factor structure for the data. However, in order to have an idea of the amount of dependence that is captured in the first tree of the vine, we compute the sum of the Spearman rank correlations of the pairs of the first tree. We do this for the D-vine structure we select, as well as for all possible canonical vine structures (see Table 16.5). For the exchange rates, the level of dependence captured in the first tree is higher for the D-vine than for any canonical vine. For the Asian equity indices, the sum of the Spearman rank correlations for

Table 16.5. Sum of the Spearman rank correlations for the first tree. This table shows the sum of the Spearman rank correlations for the first tree of the different canonical vine structures and the selected D-vine decomposition. The name of the columns indicates the pivot of the canonical vine. The sum of the Spearman rank correlations indicates the amount of dependence captured in the first tree by the different decompositions.

		Exchange Rates					
		AUD	CAD	JPY	USD	CHF	GBP
Canonical vine		1.356	1.763	1.618	1.951	−0.017	1.602
D-vine	2.105						

		Asian Equity				
		HK	KR	SG	TW	TH
Canonical vine		1.711	1.445	1.746	1.357	1.571
D-vine	1.743					

the canonical vine with Singapore as the pivot is 1.746, which is slightly higher than for the D-vine, 1.743. Moreover, when Hong Kong is the pivot, the sum of the Spearman rank correlations of the pairs of the first tree is also close to the one where Singapore is the pivot. This is an indication that the data does not lend itself to an obvious factor decomposition.

Tables 16.6 and 16.7 show estimates of the bivariate copulae that were selected according to the BIC criterion for the exchange rates and the Asian equity indices, respectively. The estimates have been calculated using a sequential estimation procedure.[g] With this procedure the parameters are consistent but not fully efficient. Fully efficient estimates of the marginals and D-vine copula parameters could be obtained by performing one step of the Newton–Raphson algorithm. For the currencies, we only select symmetric copulae, mainly Student t with degrees of freedom between 6.783 and 15.707. This suggests that a multivariate Student t-copula would not be a good approximation, since it implies that the degrees of freedom are the same for all copulae in the same tree, and get incremented by one with each tree. Time-varying copulae are selected mainly in the first tree. The dependence between Asian equity indices is quite different, as they reveal important asymmetries, which are reflected in the selection of asymmetric

---

[g]At each step, estimation is carried out conditionally on the parameters estimated in earlier steps, starting with the marginals and then the trees of the vine copula.

Table 16.6. Structure and estimation results of the D-vine copula for the six exchange rates. This table shows the estimates of the D-vine copula for the six exchange rates, obtained from a sequential estimation procedure. Standard deviations of the parameters are in brackets. The second column indicates the copula that has been selected according to the BIC criterion. When the name of the copula is followed by "tv", it means that the copula has a time-varying parameter. The column labeled with θ shows the parameters of the constant copula. When the copula parameter is time-varying we use ω^C, α^C and β^C that appear in $Q_t = \Omega(1 - \alpha^C - \beta^C) + \alpha^C \epsilon_{t-1}\epsilon'_{t-1} + \beta^C Q_{t-1}$, where Ω is a 2×2 matrix with ones on the diagonal and ω^C off-diagonal. Finally ν is the degrees of freedom parameter of the Student t-copula.

	Model	θ	ω^C	α^C	β^C	ν
		Tree 1				
GBP–AUD	Normal	0.362				
		(0.017)				
AUD–CAD	Student t tv		0.580	0.013	0.983	11.673
			(0.078)	(0.003)	(0.003)	(3.067)
CAD–USD	Student t tv		0.759	0.025	0.972	15.707
			(0.053)	(0.004)	(0.004)	(5.086)
USD–JPY	Student t tv		0.538	0.011	0.981	6.783
			(0.043)	(0.004)	(0.006)	(1.077)
JPY–CHF	Student t tv		0.139	0.027	0.964	15.045
			(0.089)	(0.007)	(0.010)	(5.012)
		Tree 2				
GBP–CAD|AUD	Independent					
AUD–USD|CAD	Student t	0.069				10.005
		(0.022)				(2.281)
CAD–JPY|USD	Independent					
USD–CHF|JPY	Student t tv		−0.162	0.025	0.968	10.723
			(0.094)	(0.007)	(0.011)	(2.595)
		Tree 3				
GBP–USD|AUD,CAD	Independent					
AUD–JPY|CAD,USD	Student t tv		0.153	0.022	0.968	10.608
			(0.074)	(0.006)	(0.011)	(2.549)
CAD–CHF|USD,JPY	Frank	−0.443				
		(0.125)				
		Tree 4				
GBP–JPY|AUD,CAD,USD	Independent					
AUD–CHF|CAD,USD,JPY	Student t	−0.088				13.948
		(0.022)				(4.095)
		Tree 5				
GBP–CHF|AUD,CAD,USD,JPY	Independent					

copulae like the Gumbel and rotated Gumbel. Like before, time-varying copulae are only selected in the first tree.

Figure 16.3 shows the time series of the Kendall's tau for the first level of the D-vine structure when time-varying copulae are selected for the exchange rates. Given the estimate of a specific time-varying copula and using the equations in Section 16.2.3, we can compute the Kendall's tau at each period. Figure 16.3 shows that the dependence between the Australian

Table 16.7. Structure and estimation results of the D-vine copula for the five Asian equity indices. This table shows the estimates of the D-vine copula for the five Asian equity indices, obtained from a sequential estimation procedure. Standard deviations of the parameters are in brackets. The second column indicates the copula that has been selected according to the BIC criterion. When the name of the copula is followed by "tv", it means that the copula has a time-varying parameter. The column labeled with θ shows the parameters of the constant copula. When the copula parameter is time-varying we use ω^C, α^C and β^C that appear in $Q_t = \Omega(1 - \alpha^C - \beta^C) + \alpha^C \epsilon_{t-1} \epsilon'_{t-1} + \beta^C Q_{t-1}$, where Ω is a 2×2 matrix with ones on the diagonal and ω^C off-diagonal. Finally ν is the degrees of freedom parameter of the Student t-copula.

	Model	θ	ω^C	α^C	β^C	ν
			Tree 1			
TH–SG	Student t tv		0.527	0.036	0.930	7.802
			(0.056)	(0.011)	(0.020)	(2.145)
SG–HK	Rgumbel tv		0.581	0.059	0.882	
			(0.039)	(0.020)	(0.052)	
HK–KR	Normal tv		0.173	0.014	0.986	
			(0.347)	(0.004)	(0.004)	
KR–TW	Rgumbel tv		0.240	0.015	0.985	
			(0.186)	(0.004)	(0.005)	
			Tree 2			
TH–HK|SG	Gumbel	1.151				
		(0.028)				
SG–KR|HK	Gumbel	1.172				
		(0.028)				
HK–TW|KR	Normal	0.257				
		(0.031)				
			Tree 3			
TH–KR|SG,HK	Gumbel	1.105				
		(0.025)				
SIF–TW|HK,KR	Frank	0.814				
		(0.205)				
			Tree 4			
TH–TW|SG,HK,KR	Independent					

and Canadian dollar experiences a steady decline over the sample period, from around 0.45 to 0.2. The dependence between the Canadian and US dollar drops from 0.6 to nearly zero. The dramatic drop in the dependence at the end of the sample is due to the depreciation of the US dollar against most currencies after June 2006. The dependence between the US dollar and the Japanese yen fluctuates around 0.37, with a maximum at a little

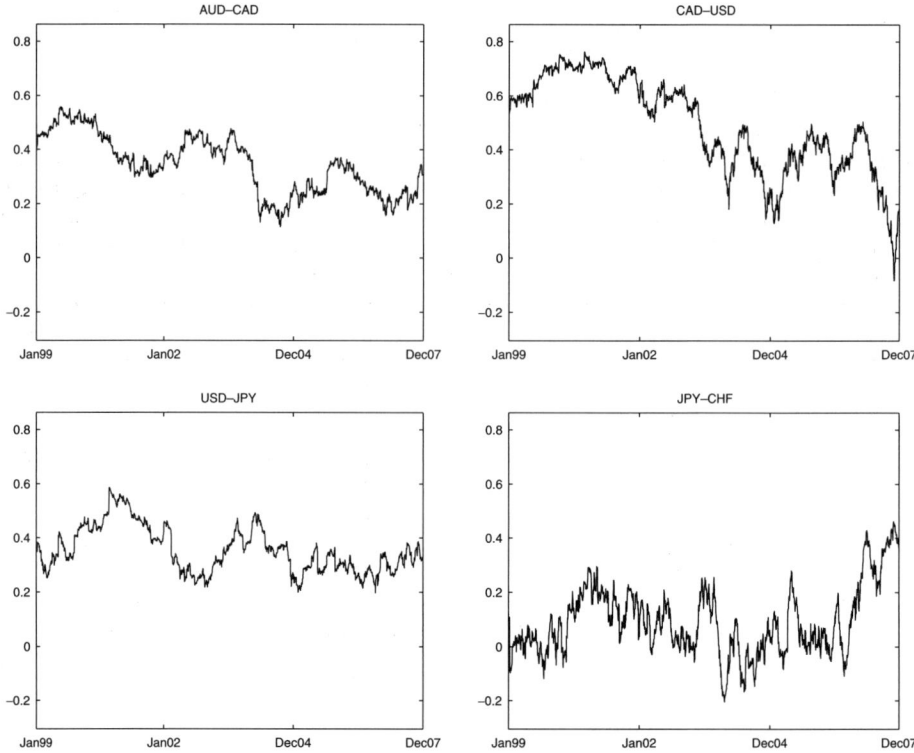

Figure 16.3. Time series of the Kendall's tau for the exchange rates. This figure shows the implied dynamics of the Kendall's tau for the first level of the D-vine structure, when we have selected a time-varying copula.

above 0.6 and a minimum of 0.2. The Kendall's tau of the Japanese yen and Swiss franc against the euro is around 0.1 most of the time but it increases significantly at the end of the sample to a value slightly above 0.4, due to the strengthening of the euro against all currencies during the course of 2007. Figure 16.4 shows the results for the Asian countries, with basically two types of evolution of Kendall's tau. For Thailand and Singapore, the dependence fluctuates around 0.3, with some periods of high dependence up to 0.5 and some periods of near independence. A similar pattern emerges for Singapore and Hong Kong, where the dependence fluctuates around 0.35. The two other pairs of exchange rates experience very clear increasing trends. The dependence between Hong Kong and Korea and between Korea and Taiwan increase significantly over the sample period from values around 0.1 to values around 0.4.

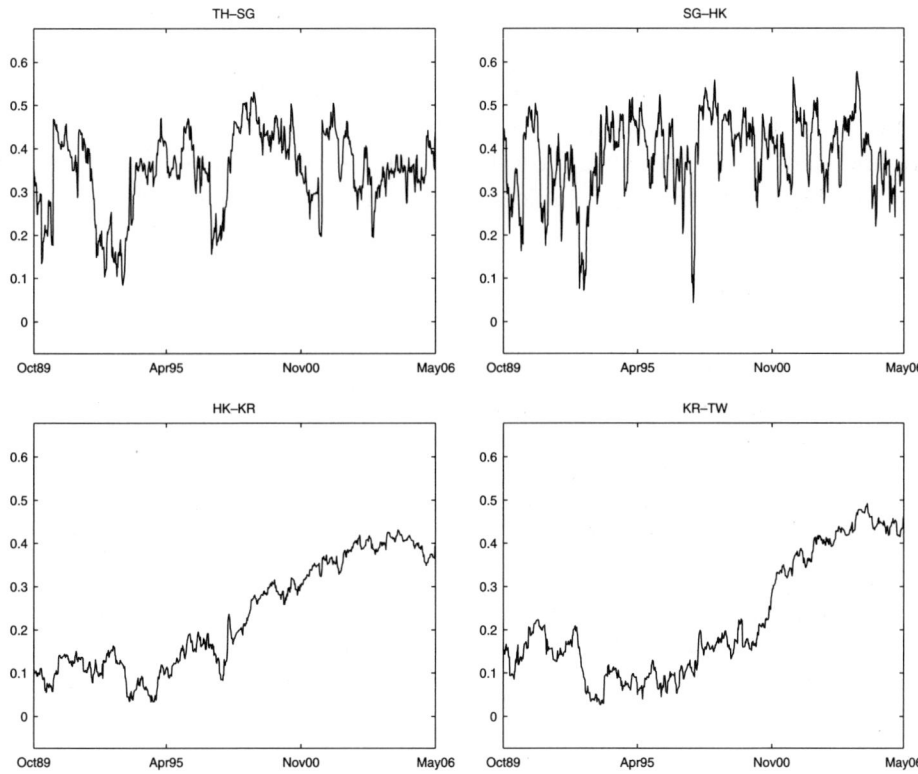

Figure 16.4. Time series of the Kendall's tau for the Asian equity indices. This figure shows the implied dynamics of the Kendall's tau for the first level of the D-vine structure, when we have selected a time-varying copula.

16.5 Conclusion

In this chapter we develop a new dynamic model of dependence for multivariate densities. The approach is based on introducing time variation in a D-vine copula, which is a very flexible multivariate copula, specified by a series of iteratively conditioned bivariate copulae. The advantage of such a model is that it can capture departures from symmetry and time variation in the dependence structure of a multivariate set of financial returns. We apply this model to two different data sets. The first sample comprises daily returns of six exchange rates (all against the euro), whereas the second data set consists of weekly returns on five Asian equity indices. We ask two questions: Is the dependence structure constant over time? Is the dependence symmetric? Our findings are that in both data sets, the D-vine structure is composed of some time-varying bivariate copulae. This corroborates previous findings in the financial literature that the dependence structure is not constant over time, e.g., Ref. 14. Moreover the time-varying copulae are mainly found in the first tree. Regarding the second question, we find that the dependence structure is symmetric for the exchange rates, whereas for the Asian equity indices the dependence structure presents clear signs of asymmetry.

References

1. Aas K., Czado C., Frigessi A. and Bakken H. (2009) Pair-copula constructions of multiple dependence. *Insurance: Mathematics and Economics*, 44:182–198.
2. Ang A. and Chen J. (2002). Asymmetric correlations of equity portfolios. *Journal of Financial Economics*, 63(3):443–94.
3. Bedford T.J. and Cooke R.M. (2002). Vines: A new graphical model for dependent random variables. *Annals of Statistics*, 30(4):1031–1068.
4. Cherubini U., Luciano E. and Vecchiato W. (2004). *Copula Methods in Finance*. Wiley, West Sussex, England.
5. Dias A. and Embrechts P. (2004). Dynamic copula models for multivariate high-frequency data in finance. Technical report.
6. Embrechts P., Klüppelberg C. and Mikosch T. (1997). *Modelling Extremal Events for Insurance and Finance*. Springer, New York.
7. Embrechts P., McNeil A.J. and Straumann D. (2002). Correlation and dependence in risk management: Properties and pitfalls, In *Risk Management: Value at Risk and Beyond*, M. Dempster, editor, pp. 176–223, Cambridge University Press.
8. Engle R. (2002). Dynamic conditional correlation: A simple class of multivariate generalized autoregressive conditional heteroskedasticity models. *Journal of Business and Economic Statistics*, 20(3):339–350.
9. Hansen B. (1994). Autoregressive conditional density estimation. *International Economic Review*, 35:705–730.
10. Heinen A. and Valdesogo A. (2009). Asymmetric CAPM dependence for large dimensions: The canonical vine autoregressive copula model. Technical report, available at SSRN.
11. Joe H. (1997). *Multivariate Models and Dependence Concepts*. Chapman & Hall/CRC. London; New York.
12. Jondeau E. and Rockinger M. (2006). The copula-GARCH model of conditional dependencies: An international stock market application. *Journal of International Money and Finance*, 25:827–853.
13. Kurowicka D. (2008). Choices of vines. Technical report, Delft University of Technology.
14. Longin F. and Solnik B. (1995). Is the correlation in international equity returns constant: 1960–1990? *Journal of International Money and Finance*, 14(1):3–26.
15. Longin F. and Solnik B. (2001). Extreme correlation of international equity markets. *Journal of Finance*, 56(2):649–76.
16. Nelsen R.B. (2006). *An Introduction to Copulas*. 2nd ed., Springer Series in Statistics. Springer, New York.

17. Patton A. (2004). On the out-of-sample importance of skewness and asymmetric dependence for asset allocation. *Journal of Financial Econometrics*, 2(1):130–168.
18. Patton A. (2006). Estimation of multivariate models for time series of possibly different lengths. *Journal of Applied Econometrics*, 21(2):147–173.
19. Patton A. (2006). Modelling asymmetric exchange rate dependence. *International Economic Review*, 47(2):527–556.
20. Sklar A. (1959). Fonctions de répartition à n dimensions et leurs marges. *Publications de l'Institut de Statistique de l'Université de Paris*, 8:229–231.
21. Tse T.K. and Tsui A.K.C. (2002). A multivariate generalized autoregressive conditional heteroscedasticity model with time-varying correlations. *Journal of Business and Economic Statistics*, 20(3):351–362.

CHAPTER 17

Summary and Future Directions

Dorota Kurowicka
Delft University of Technology
Mekelweg 4, 2628CD Delft, the Netherlands
D.Kurowicka@tudelft.nl

17.1 Summary 355
17.2 Future Research Directions 356

17.1 Summary

This book promotes vine copulae as a model of joint associations. The focus is on multivariate dependence modeling, as the transition from the well-developed theory for bivariate associations to multivariate dependence is a challenging step. Bivariate copulae have become a very popular way of specifying distributions of two random variables. Different bivariate copulae, capable of modeling various features of a joint distribution (correlation, tail dependence in the upper and lower corner), are available. They cover a wide range of bivariate dependence, are easy to sample and their parameters can be inferred from data. The rich theory of bivariate copulae will surely expand and develop as new features of bivariate distributions capture researchers' attention. Moving from the bivariate to the multivariate case is challenging for several reasons:

(1) Different types of unconditional bivariate copulae are often observed in multivariate data sets. Models that allow various types of bivariate copulae to be combined into a multivariate copula are of great interest. Graphical or hierarchical models are best suited to this setting.

(2) Combining unconditional copulae into one multivariate model cannot be done in an unconstrained way. For this reason, methods based on connecting unconditional copulae are quite limited (for example, n-dimensional model Markov trees allow the specification of only $n-1$ unconditional bivariate copulae while hierarchical Archimedean construction allows the stipulation of only $n-1$ different generators).

(3) For high-dimensional copulae, the possibility of partial specification is essential. Models that can handle full as well as partial specification would be preferred.

The vine-copula method allows a joint distribution to be built from bivariate and conditional bivariate copulae arranged together according to the graphical structure of a regular vine. This avoids problems of compatibility and leverages bivariate copulae to enable extensions to arbitrary dimensions. For n-dimensional models, we can independently specify $n(n-1)/2$ copulae, of which $n-1$ are unconditional. Vines can be sampled and inferred from data. The minimal information completion of any partially specified regular vine is trivially found by making the unspecified conditional copulae conditionally independent.

In this book, we study properties of regular vines and show that vine copulae realize a wide range of dependence structures with flexible asymmetry in the joint upper and lower tails. This flexibility can be achieved by the appropriate choice of bivariate copulae. We also showed that vines often perform better than other competing models in realizing dependence structures of multivariate data. Vines do require a bit of getting used to, but offer ample rewards in terms of flexibility, elegance and simplicity to those who persevere.

17.2 Future Research Directions

Vines are still young and there is much still to be learned. We conclude this book with a few research topics of interest. Further research on vines will be both application- as well as theory-driven.

The most interesting theoretical questions include:

(1) *Search for the "best" regular vine.* First of all, the question of what the "best" vine means is still open. So far, our experience extends only to D- and C-vines. As shown in this book, other vines provide interesting symmetries and properties. How can we utilize these properties in finding the "best" vine structure?

(2) *Model simplification.* Vine models with specified conditional independence statements will be of interest. Which conditional independence statements can be incorporated into a given vine? How to choose a regular vine which has all the specified conditional independence statements? For a given vine, conditional copulae at specified levels can be set to independent copulae and only the remaining copulae would have to be fitted to data. One line of research could go into efficient conditional independence tests that would allow us to find conditional copulae on a vine that could be set to independence. This probably would not be possible for high-dimensional vines; the fall-back position would be to use some heuristic search (e.g., as shown in this book based on finding small partial correlations) for the "best" vine structure.

(3) *Vines versus other graphical models.* In this book the relationship between vines and Bayesian belief nets was investigated. It would be interesting to compare vines to other graphical models, e.g., chain graphs.

(4) *Vine models with time-varying copula parameters.* Models with time-varying parameters are of great interest especially in the area of finance. Vines could offer their flexibility to modeling dependence in time. Some ideas in this direction have been presented in this book.

Vines so far have been applied mostly in the area of finance. We believe that many other applications will appear in the future. Clear expositions of the method, for a variety of users, is required as well as fast computer implementations. We hope that this book will standardize notation and provide a baseline for further theoretical work. More effort should be directed to developing professional vine software. This software must include:

(1) *Graphics.* Good graphical representation of vines is essential. A software package that will guide a user in representing and explaining a structure of a model is needed. Such software should also offer the possibility of performing calculations such as: inference, sampling, conditionalization.

(2) *Inference algorithms.* In Chapter 3 we showed inference algorithms for D- and C-vines. There is, however, a need for a general inference algorithm for regular vines.

(3) *Searching for the best vine structure.* "Greedy" algorithms to search for the "best" vine structure will have to be developed, implemented and thoroughly tested.

The rapid growth of the vine community in the last few years bodes well for the pursuit of this research agenda.

Index

Akaike information criterion (AIC), 115
algorithms
    construction of regular vines, 197, 202
    construction of vines, 195, 197, 206
    enumeration of vine arrays, 220
    generalized Toeplitz via partial correlations, 159
    generation of regular vine, 240
    inversion from vine array to binary vectors, 229
    likelihood for C-vine, 60
    likelihood for D-vine, 61
    optimal truncation of vine, 240, 242
    Prufer codes, 194
    simulation for 5-dimension vines, 150, 153
    simulation for regular vines, 147
application
    electricity load, 257, 270
    financial, 323, 332
    multivariate data analysis, 67
    risk analysis, 299
    weather data, 244

Bayesian belief net, 68, 282, 287
Blomqvist's beta, 160, 182

Cayley's theorem, 194
comonotonicity and countermonoticity, 53, 154
copula families
    1-factor, 171
    Archimedean, 27, 106, 330
    BB1, 181
    BB4, 182
    BB7, 183
    Clayton/MTCJ, 341
    elliptical, 22
    extreme value, 107
    Frank, 10, 106, 159, 340
    Galambos, 107
    Gaussian, 22, 331
    GT, 23
    Gumbel, 107, 340
    hierarchical or nested Archimedean, 24, 25, 308
    Koehler–Symanowski, 31
    mixture of max-id, 181
    MTCJ, 102, 159
    normal, 46
    Plackett, 106, 159
    student-t, 23, 307, 330
    t, 169
copula information criterion (CIC), 115

determinant of correlation matrix, 48, 156
directed acyclic graphs (DAG), 282

Fréchet class, 50, 52

generalized Toeplitz matrices, 156
glossary and notation, 14, 15

Kendall's tau, 159, 334

Laplace transform family
    Mittag-Leffler, 181
    Sibuya, 181
line graphs, 197

majorization, 66
Markov tree, 239

maximum pseudo-likelihood estimation, 118
microcorrelation, 90
model selection, 126, 250
mutual information or Kullback–Leibler divergence, 65, 119

Pareto distribution, 101, 102
partial correlation, 47
Prufer code, 194

random correlation matrices, 49
rank correlation, 49, 290
reflection symmetry, 179
relative correlation, 64

skewed t, 339
Sklar's theorem, 51
stochastic increasing positive dependence, 95

tail dependence, 92, 97, 168
tail dependence functions, 173

upper-lower tail dependence, 168

value-at-risk VaR, 332
vine
    array, 9, 196, 220
        B0–B3 5-dimensional vines, 144, 145, 150

Bayesian inference, 255, 266
C-vine or canonical vine, 39, 42
$C.i_5.i_6\ldots$ notation, 221
classification, 203
conditioned set, 40
conditioning set, 40
constraint set, 39
cumulative distribution function, 44, 45, 51
D-vine, 39, 42
$D.i_5.i_6\ldots$ notation, 221
definition, 39
density, 9, 44, 45, 238
dynamic, 337
enumeration of vine array, 193, 220
equivalence classes, 141, 222, 227
Gaussian or normal, 46, 156
graphs, 7, 9, 40, 41, 43
likelihood estimation, 59, 318
m-child, 41
m-descendant, 41
natural order, 199, 202, 220
number of regular vines, 42, 203
pair-copula construction PCC, 43, 253
partial correlation, 47
properties, 41, 46, 237
sampling algorithm, 55, 56, 147
truncated vine, 238
vines
    regular, 44

Institute for Statistics and Mathematics
Vienna University of Economics and Business
Augasse 2-6, 1090 Vienna